IMMUNOLOGY: A COMPARATIVE APPROACH

Edited by

R. J. TURNER
University of Wales, Aberystwyth, UK

JOHN WILEY & SONS
Chichester · New York · Brisbane · Toronto · Singapore

Telephone National Chichester (0243) 779777
International + 44 243 779777

Other Wiley Editorial Offices

John Wiley & Sons, Inc., 605 Third Avenue,
New York, NY 10158-0012, USA

Jacaranda Wiley Ltd, 33 Park Road, Milton,
Queensland 4064, Australia

John Wiley & Sons (Canada) Ltd, 22 Worcester Road,
Rexdale, Ontario M9W 1L1, Canada

John Wiley & Sons (SEA) Pte Ltd, 37 Jalan Pemimpin #05-04,
Block B, Union Industrial Building, Singapore 2057

Library of Congress Cataloging-in-Publication Data
Immunology : a comparative approach / edited by R.J. Turner.
p. cm.
Includes bibliographical references and indes.
ISBN 0 471 94400 9
1. Immunology, Comparative. I. Turner, R. J.
[DNLM: 1. Immune System—physiology. 2. Immunity—physiology.
3. Physiology, Comparative. QW 504 I363615 1994]
QR182. I473 1994
591.2'9—dc20
DNLM/DLC
for Library of Congress 93-47091
CIP

British Library Cataloguing in Publication Data

A catalogue record for this book is available from the British Library

ISBN 0 471 94400 9

Typeset in 10/12pt Times by Mackreth Media Services, Hemel Hempstead
Printed and bound in Great Britain by Bookcraft (Bath)

IMMUNOLOGY: A COMPARATIVE APPROACH

CONTENTS

CONTRIBUTORS

J.D. HORTON	Department of Biological Sciences, University of Durham, South Road, Durham DH1 3LE, UK
R.D. JURD	Department of Biology, University of Essex, Colchester CO4 3SQ, UK
M.J. MANNING	Department of Biological Sciences, University of Plymouth, Drake Circus, Plymouth PL4 8AA, UK
J.P. MÉTRAUX	Institut de Biologie Végétale, Université de Fribourg, 3 Rue Albert Gockel, CH-1700 Fribourg, Switzerland
D.A. MILLAR	Biomedical and Physiological Research Group, School of Biological Sciences, University College of Swansea, Singleton Park, Swansea SA2 8PP, UK
N.A. RATCLIFFE	Biomedical and Physiological Research Group, School of Biological Sciences, University College of Swansea, Singleton Park, Swansea SA2 8PP, UK
R.J. TURNER	Institute of Biological Sciences, University of Wales, Aberystwyth, Dyfed SY23 3DA, UK

PREFACE

This book is written for students and research workers with some knowledge of 'mainstream' immunology who are curious about the advances, opportunities and challenges in comparative work. We know a great deal about the immune system of the laboratory mouse and draw heavily upon it, but just how good a representative is it? What immunological similarities can we detect across different orders, classes, phyla and kingdoms? How do immunological characteristics change with overall structure, physiology and way of life? How might significant and perhaps insurmountable technical obstacles in one immunological model be bypassed by using another? How much progress has been made in the study of defence mechanisms in animals and plants of commercial importance?

In attempting to answer such questions, the book has been divided into six chapters, each focusing on a different group of organisms, and each written by people who are conducting research on species within that group. The inclusion of a chapter on plants is unorthodox for an immunology textbook; it has been done here not simply to tempt immunologists who happen to be fond of gardening, but to illustrate the extent to which plant and animal defences differ, and to point out where parallels between the two may be drawn. Invertebrates too (Chapter 2) are often disregarded in this context, a pity in view of their economic significance (e.g. as vectors, pests, sources of food and medical products) and the questions they raise about the origins and evolution of the immune system. The remaining four chapters consider the immunology of vertebrates (fishes, amphibians, reptiles and birds, mammals), including animals of veterinary interest and those currently favoured as immunological models.

Since there are wide variations in the amount and type of research information available for these different groups of organism, no attempt has been made to hold the chapters to a rigid format, rather to allow major themes for each group to be developed, and some of the enthusiasms and prejudices of the respective authors to emerge. However, an effort has been made to minimize jargon and to make each chapter accessible to the non-specialist.

R.J. Turner

ACKNOWLEDGEMENTS

I would like to acknowledge the efforts not only of the authors themselves and those who donated illustrations to the book, but also research immunologists whose original work (for reasons of space) has been cited only briefly or indirectly. Finally, I am indebted to Richard Baggaley, Lynda Gulvin, Lucy Jepson and Nicola Reavley at the publishers for their support and endless patience.

1

PLANTS

J.-P. Métraux

Institut de Biologie Végétale, Université de Fribourg, Switzerland

1.1 INTRODUCTION

Although plants do not have an immune system similar to animals, they are clearly able to develop defence reactions against invading pathogens. The scope of this chapter is to familiarize the reader with the different disease resistance mechanisms of plants as well as the basis of recognition and specificity. Whenever possible, comparisons will be made with animal defences.

Plant disease can be defined as all malfunctions of plants which result in loss in performance or reduced ability to survive and maintain their ecological niche[1]. In plants, disease is the exception rather than the rule. In most cases the interaction between plants and a potential pathogen does not lead to disease nor to symptoms, in which case the plant is immune. The penetration and the development of the pathogen can also be prevented to various extents by resistance mechanisms of the host plant. This is called an incompatible interaction. If the pathogen can penetrate its host and fulfil its life cycle, the plant is susceptible and the interaction is compatible. In this case the disease resistance mechanisms of the host plant have been inactivated or circumvented with success by the pathogen, and disease occurs. The amount of damage to a plant will be determined by two forces: the natural resistance of the host plant and the virulence of the pathogen.

1.2 THE DISEASE RESISTANCE MECHANISMS OF PLANTS

In order to survive, plants must be resistant to many hundreds of different microorganisms which they are likely to encounter in their lifetime. Plant pathogens include diverse microorganisms: among these are fungi, bacteria, viruses and viroids, mycoplasmas and nematodes. Resistance is not based upon specific mechanisms against each individual microorganism, but rather on an arsenal of defence reactions pre-existing in the host plant or induced at the time of contact with an invading organism. In fact it is often very

Immunology: A Comparative Approach. Edited by R.J. Turner

difficult to determine which individual defence reaction is responsible for the resistance against a particular pathogen.

In many cases it is likely that defence mechanisms we interpret to be solely against microorganisms might in fact be part of an ensemble of responses against a wide variety of biotic or abiotic stresses. The mechanisms existing in the plant before infection are referred to as pre-existing or constitutive, as opposed to resistance reactions which are induced in the plant by the infection.

1.2.1 Preformed Structures

Structural Barriers

Structural features of the plant might function as a physical barrier to invading microorganisms. All plant cells are surrounded by a cell wall, a composite polymeric material made of a matrix phase of non-cellulosic polysaccharides interwoven with crystalline cellulose, and in dicotyledonous species with a glycoprotein network[2] (Figure 1.1). In addition, plant cell walls contain varying

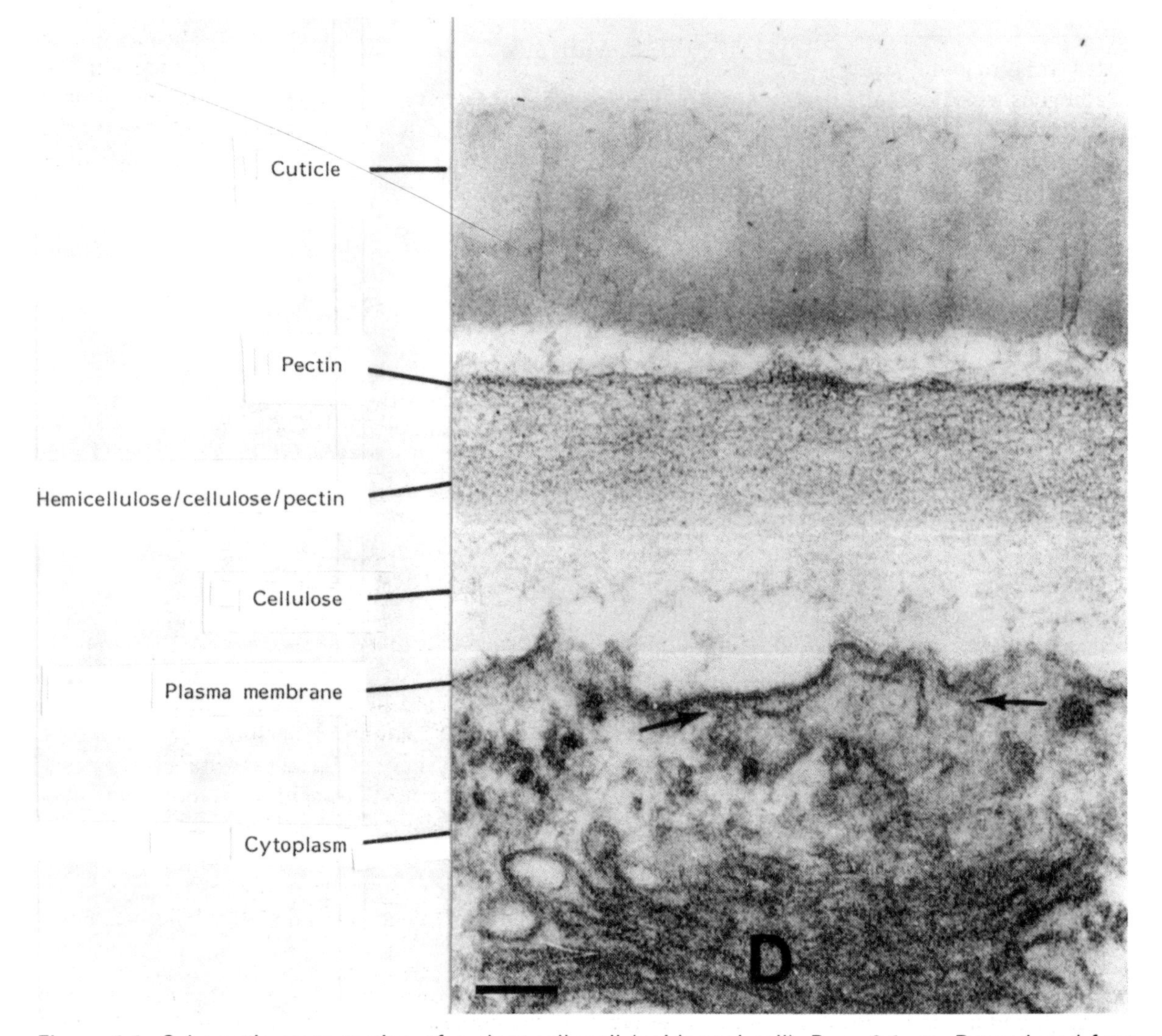

Figure 1.1. Schematic cross-section of a plant cell wall (epidermal cell). Bar = 0.1 μm. Reproduced from reference 3 with permission.

amounts of the hydrophobic phenolic polymer lignin. In aerial parts of plants, cell walls of epidermal cells are coated with a cuticle, consisting of a water repellent biopolyester of Ω-hydroxylated fatty acids. By analogy, this outer layer could be compared to waterproofing polymers such as proteins or chitin lining the outer surface of animals. The cell wall provides a structural barrier against organisms which gain access by penetrating unwounded tissue. Breakage of wall polymers weakens the cell wall and might facilitate the entry of microorganisms in the tissue. In fact, cell wall components are often substrates for hydrolytic enzymes of pathogens.

Pathogens which can successfully penetrate the host through the outer surface secrete cutinase, an enzyme capable of digesting the cuticle (Figure 1.2). Although only demonstrated for a few pathogens *in vitro*, cutinase might well be produced in many more cases at the site of penetration and only during the limited time-frame of penetration, thus making its *in vivo* detection difficult[5]. The importance of cutinase for pathogen invasion can be demonstrated by pretreatments of plants with specific inhibitors of cutinase, such as diisopropylfluorophosphate or antibodies against cutinase[5]. Microorganisms such as fruit rot-causing *Mycosphaerella* spp. can only cause disease by penetrating through wounds in the fruit skin. A cutinase gene was introduced into *Mycosphaerella* which yielded transformants able to penetrate through the cuticle of intact fruits, and the infection could be prevented by cutinase antibodies[6]. Such experiments illustrate the importance of the cuticle as one of the first lines of plant defence.

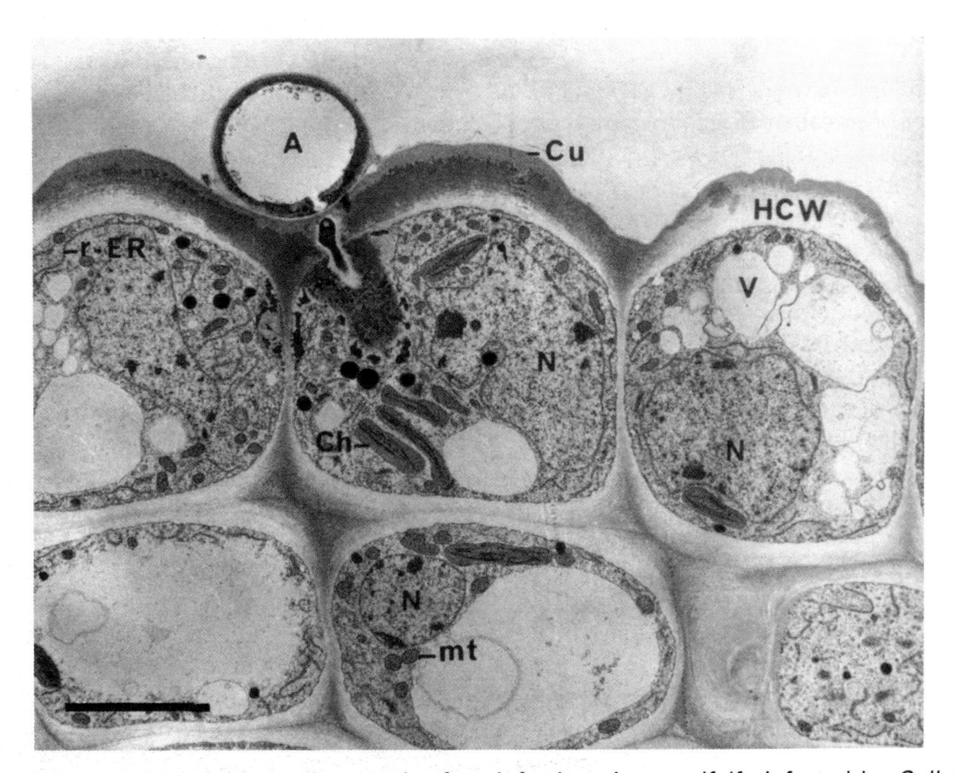

Figure 1.2. Transmission electron micrograph of an infection site on alfalfa infected by *Colletotrichum trifolii*. The infected cells show a large nucleus (N), mitochondria (mt), chloroplasts (Ch), small vacuoles (V), rough endoplasmic reticulum (r-ER), cisternae and lipid bodies. The host cell wall (HCW) and cuticle (Cu) are electron-dense around the penetration point. The fungal appressorium (A) (infection structure) is visible on the surface of the plant. Bar = 5 μm. Reproduced from reference 4 by permission.

Pathogens penetrating through cell walls also produce hydrolytic enzymes such as pectinase, which are capable of macerating the cell wall. Deletion of genes for pectate lyase in the soft rotting bacteria *Erwinia chrysanthemi* resulted in a significant loss of virulence[7,8]. The importance of the pectic component of the cell wall as a barrier is further substantiated by the presence of pectinase inhibitors in the cell wall[9–11]. Such inhibitors operate only against microbial enzymes and are part of a further chemical barrier against invading pathogens.

Other anatomical features which can be limiting for the pathogen are natural openings in the plant such as stomata, lenticels or hydathodes: morphological features and size and number of such structures might contribute to resistance.

Preformed Antimicrobial Compounds

Besides structural barriers, plants accumulate a large number of so-called secondary metabolites, which often have antimicrobial properties. The term 'secondary' would indicate that such compounds are only of minor importance for the plant, and it was a standing controversy whether such compounds were in fact waste products. However, over recent years this opinion has been gradually replaced by the view that secondary metabolites are beneficial and even essential for the survival of the plant[12]. Thus, secondary metabolites might function as a chemical defence against microbes, insects or herbivores or as allelopathic compounds influencing germination and growth of neighbouring plants. The list of such compounds is ever increasing but the evidence that they are major determinants of differences in susceptibility is often rather circumstantial. To be relevant to the resistance against pathogens, antimicrobials must be shown to occur in those cells or tissues which are in contact with the invading microorganism, their concentration must be high enough to affect the pathogen and they must exist in a form available to the pathogen[13]. Other lines of evidence would be cases where the amounts of antimicrobial metabolites are experimentally modified which should lead to corresponding changes in the resistance of the host plant. The chemical nature of such antimicrobial compounds covers a wide chemical spectrum; some examples, referred to below, are given in Figure 1.3.

Cyanogenic Glucosides These are encountered in about 2000 species, one typical example being the herbage legume white clover[14]. Hydrocyanic acid, a potent inhibitor of cytochrome oxidase, is released by specific glucosidases which come into contact with the cyanogenic glucoside upon loss of cell integrity caused by infection or mechanical damage (Figure 1.3). In a number of plant pathogens low levels of hydrocyanic acid can induce the enzyme formamide hydrolyase, which converts HCN to formamide thus providing a means of detoxifying HCN. The occurrence of formamide hydrolyase was analysed in 31 fungal species; this showed that the enzyme was produced by all 11 pathogens of cyanogenic plants, by nine of 14 pathogens of non-cyanogenic plants and by one out of six non-pathogens of plants[15]. The presence of a tolerance mechanism in pathogens of cyanogenic plants supports the hypothesis that cyanogenic glucosides might be preformed chemicals providing resistance to the plant against microorganisms unable to cope with HCN. Clearly, not all fungi producing formamide hydrolyase are able to colonize cyanogenic plants, indicating that other defence mechanisms must exist in such plants. Interestingly, the rate of detoxification of HCN by the plant itself might also be critical: beyond certain concentrations HCN is not detoxified rapidly enough by the plant and can inhibit its own defence mechanisms[16]. Hydrocyanic acid is also a known defence compound in arthropods[17].

Terpenoids Saponins are perhaps the most important and widely distributed class of terpenoids in plants. Their name stems from their surfactant activity: they can decrease the surface

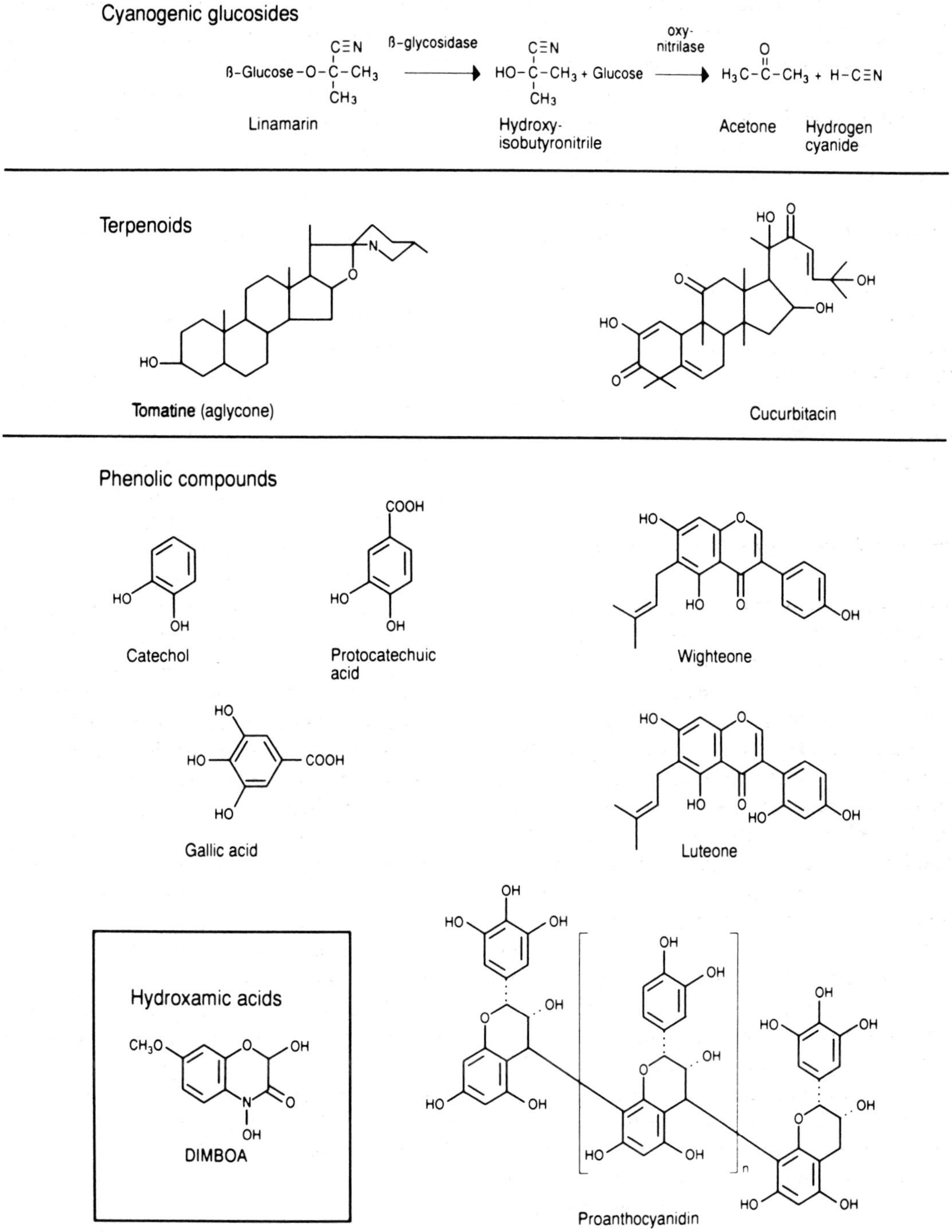

Cyanogenic glucosides

ß-Glucose–O–C(C≡N)(CH$_3$)–CH$_3$ —ß-glycosidase→ HO–C(C≡N)(CH$_3$)–CH$_3$ + Glucose —oxy-nitrilase→ H_3C–C(=O)–CH_3 + H–C≡N

Linamarin — Hydroxy-isobutyronitrile — Acetone — Hydrogen cyanide

Terpenoids

Tomatine (aglycone)

Cucurbitacin

Phenolic compounds

Catechol

Protocatechuic acid

Wighteone

Gallic acid

Luteone

Hydroxamic acids

DIMBOA

Proanthocyanidin

Figure 1.3. Some examples of constitutive secondary metabolites in plants. DIMBOA: 2,4-dihydroxy-7-methoxy-1,4-benzoxazin-3-one.

tension of water to make it foam like soap (Latin: *sapos*). Many tropical and subtropical plants contain saponins which are used for pharmaceutical products such as hormones. A number of diverse agricultural and horticultural crops also contain saponins. Saponins often occur as non-toxic glycosides and are only biologically active upon the action of specific β-glycosidases. Their action is based upon the binding of their aglycones to sterols in biological membranes. With sterols the saponin aglycones form insoluble complexes thus changing membrane properties leading to cell death. Studies on *Fusarium solani*, the causal agent of root and stem rot in tomato, revealed an isolate unable to colonize green fruits rich in the saponin tomatine. This strain was mutagenized and mutants insensitive to high levels of tomatine were isolated. These tomatine-insensitive mutants were able to rot green tomatoes and had low sterol content in their membranes[18]. Crosses between tomatine-sensitive wild-type strains and tolerant mutants gave progeny where virulence on green fruits, tolerance to tomatine and low membrane sterol content are always inherited together[19]. These and other studies provide evidence that tomatine is involved in chemical defence against fungi.

Cucurbitaceae (squashes, etc.) contain tetracyclic triterpenes referred to as cucurbitacins. These compounds are extremely bitter and render plants unpalatable or toxic to invertebrate and vertebrate herbivores. Ironically, cucurbitacins are strong kairomones for phytophagous diabroticite beetles, in which they stimulate arrest and compulsive feeding, and are important determinants of host-plant selection[20]. Cucurbitacins sequestered by the beetles can repel predators such as the chinese manta. Eggs of beetles can also contain cucurbitacins, which are deterrents against egg predators in the soil[20]. Cucumbers treated with cucurbitacins are protected against the grey mould fungus *Botrytis cinerea*[21]. It would be interesting to know if the levels of cucurbitacin present naturally in the plant are high enough to constitute a defence mechanism against certain fungi.

Phenolic Compounds The occurrence of phenolic compounds is amazingly diverse and extremely widespread in all plants[22]. These compounds include simple phenolics, coumarins and flavonoids as well as complex phenols such as tannins or lignins. Besides antimicrobial and allelopathic effects, experimental evidence increasingly suggests that phenolics can function as signals in plant development and in plant–microbe interactions[23,24].

Fungistatic and fungitoxic compounds occur in the cuticle where they influence the first and crucial step in the propagation of fungal spores[25]. Examples of such phenolic inhibitors are luteone and wighteone in lupin leaves[26,27], or gallic acid in *Acer* leaves[28]. Catechol and protocatechuic acid occur in the outer scales of red and yellow onion varieties and are linked to resistance against the fungus *Colletotrichum circinans*, the causal agent of onion smudge disease. When the scales are exposed to water, these water-soluble phenolics diffuse to the surface of the scales where they inhibit germination of fungal spores[29].

Proanthocyanidins are enzyme-inhibitory tannins present in high concentrations in leaves and immature fruits of many wild and cultivated plants. They are made of subunits of polyhydroxy flavans and form complexes with proteins or enzymes. These compounds have no direct antimicrobial effect but restrict invading microorganisms by denaturing their cell wall-degrading enzymes[30]. In strawberry fruits, *Botrytis cinerea* invades the flower at an early stage but remains confined to the receptacle, which contains high levels of proanthocyanidins. At maturation the protein denaturing potential of the proanthocyanidins decreases due to a higher degree of polymerization of proanthocyanidins and the entire fruit becomes rapidly invaded[31]. Breeders have developed cultivars with higher resistance to *B. cinerea*. One typical characteristic exhibited by the better cultivars is the presence in mature fruits of a white immature ring of tissue around the stem. In this collar proanthocyanidins are polymerized to a lesser extent and maintain their inhibitory potential against the grey mould fungus[31].

Hydroxamic Acids These compounds occur mainly in the shoots and roots of the Gramineae (grasses and cereals) with the exception of *Avena*, *Hordeum* and *Oryza*[32]. Hydroxamic acids occur as glycosides which are hydrolysed by a β-glucosidase upon loss in compartmentation after infection or mechanical injury. Correlations have been observed between hydroxamic acid levels and resistance against fungal diseases in wheat, corn and rye[32]. Cereals are also plagued by aphids, which are important vectors of viral diseases. Hydroxamic acid contents in corn and wheat leaves were inversely related to aphid infestations of these plants. When incorporated into diets of cereal aphids, hydroxamic acid had both antibiotic and antifeedant effects[33]. Other biological effects of hydroxamic acids include triggering of the reproduction of grass-feeding mammals and allelopathic effects on cereals[32]. Hydroxamic acids have been proposed as an agronomically useful trait to be considered in breeding for aphid resistance in wheat[34]. Such compounds exemplify the broad ecological role which can be played by plant secondary metabolites, extending far beyond the defence against microbes.

1.2.2 Induced Resistance Mechanisms

Local Versus Systemic Resistance

Like animals, plants produce local structural and biochemical barriers in response to infections by pathogens. An invading microorganism is blocked or killed at the site of attempted invasion and only a few plant cells are involved in the production of defence structures (Figure 1.4).

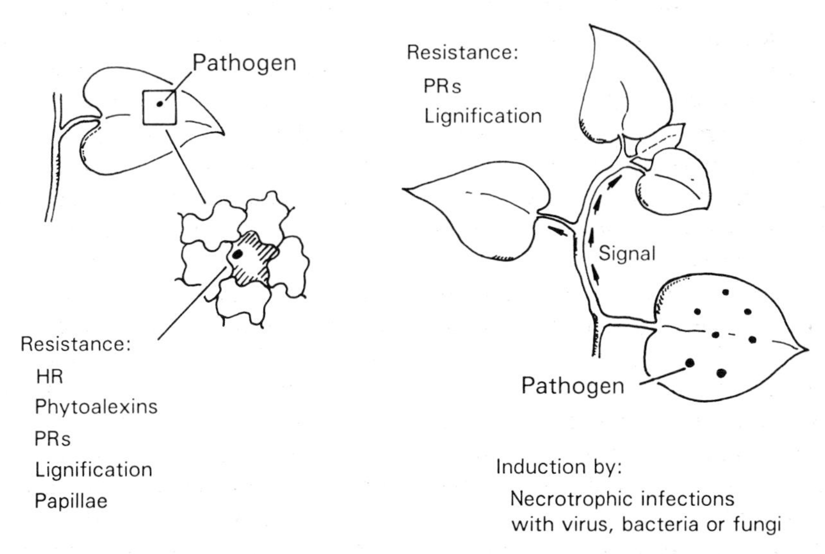

Figure 1.4. Local versus systemic disease responses in plants. In the local response, defence mechanisms are induced at/around the infected cell. In systemic induced resistance, localized infections with pathogens causing necroses result in subsequent resistance which can extend throughout the plant. A hypothetical signal is likely to be involved in the systemic activation of resistance. HR = hypersensitive response; PRs = pathogenesis-related proteins.

Moreover, the resistance response of plants can extend to regions untouched by the pathogen infection (Figure 1.4). This induced or acquired systemic resistance is also known as immunization. A first inoculation with pathogens capable of causing necrotic lesions results in resistance towards subsequent infections, even in tissue unaffected by the pathogen, but the resistance acts against infections by a wide variety of pathogens that need not be related to the inducing organism. For example, in cucumber or tobacco, a first infection with either fungi, bacteria or viruses protects the plant against subsequent infections by fungal, bacterial and viral pathogens both in the infected and uninfected parts of the plant[35]. The level of protection is related to the size and number of lesions produced during the first infection. Protection is not an all or none phenomenon as observed in genetically-determined resistance to a specific race of a pathogen. Despite this variation in expression, induced resistance can be extremely effective and long-lasting. For instance, in cucumber plants infected on the first leaf with tobacco necrosis virus or the fungus *Colletotrichum lagenarium*, induced resistance lasts for several weeks. A second 'booster' inoculation following the first inoculation gives protection up to the time of fruiting[36]. The effectiveness of systemic induced resistance has been confirmed in field trials for bean, cucumber and tobacco[37–39]. No deleterious effects were reported on the yield. Indeed, tobacco plants immunized by subepidermal stem injections with spores of the blue mould fungus *Peronospora tabacina*, the growth of the plants can be enhanced, even in the absence of the challenge pathogen[40]. To a certain extent, then, immunization in plants is analogous to immunization in animals and humans, bearing in mind that the resistance mechanisms of plants are quite different.

Induced Structural Barriers

A number of structural barriers can play an important role in containing or excluding invading pathogens.

Cell Wall Lignification Besides polysaccharides and proteins, cell walls contain lignin, a complex phenolic polymer. Lignin is a tough water-repellent network of substituted cinnamyl alcohols[41]. The encrustation of lignin strengthens the cell wall physically and allows plants an erect growth habit. Lignin is of special importance in resistance against various fungal pathogens because of their inability to degrade this complex polymer. Exceptions are white rot fungi and litter-decomposing basidiomycetes[42]. Qualitative or quantitative changes as well as changes in the speed of lignin formation were observed in incompatible plant–pathogen interactions[23,43–47].

The composition of lignin induced in response to infection can be markedly different from that in healthy plants. In roots of Japanese radish, infection by *Peronospora parasitica* results in a shift from guaiacyl-syringyl lignin to lignin enriched in *p*-hydroxyphenyl and guaiacyl propane units[48]. Changes in the composition of lignin induced after infection were also observed in wheat[49], muskmelon[50] and potato[51], indicating that different pathways for lignin biosynthesis are activated after infection. Such compositional changes could affect the cell wall properties at the site of infection. Whether this is sufficient to prevent or slow down the penetration is not fully documented.

The timing and localization of induced lignification is critical. In wheat lines carrying a resistance gene against leaf or stem rust, lignin and lignin-like materials are deposited at the sites of infection of the rust pathogen as well as at sites of infection of non-pathogens[52–56]. Infection of potato tubers by non-pathogens results in a faster accumulation of lignin[51]. Lignification was also shown to increase faster and to a larger amount in systemically protected leaves of cucumber[57,58]. Using radiolabelled phenylalanine or cinnamic acid as precursors for lignin, lignification was detected at an early stage at the sites of attempted invasion in resistant wheat varieties[52,55,56]. Labelled lignin precursors are incorporated more rapidly into lignin in response to wounding or challenge infection in systemically protected cucumber leaves[59].

Another approach to assess the importance of lignification is by measurement of enzymes involved in the synthesis and polymerization of lignin precursors such as phenylalanine ammonia lyase (PAL), 4-coumarate CoA ligase (4CL), cinnamyl alcohol dehydrogenase (CAD), *o*-methyltransferase (OMT) and peroxidase (PO) (Figure 1.5). These enzymes are associated with the incompatible reaction in wheat[60–64]. In tobacco leaves reacting hypersensitively to tobacco mosaic virus, the activities of PAL, cinnamic acid-4-hydroxylase, 4CL and OMT increase in the cells around the necroses[65,66].

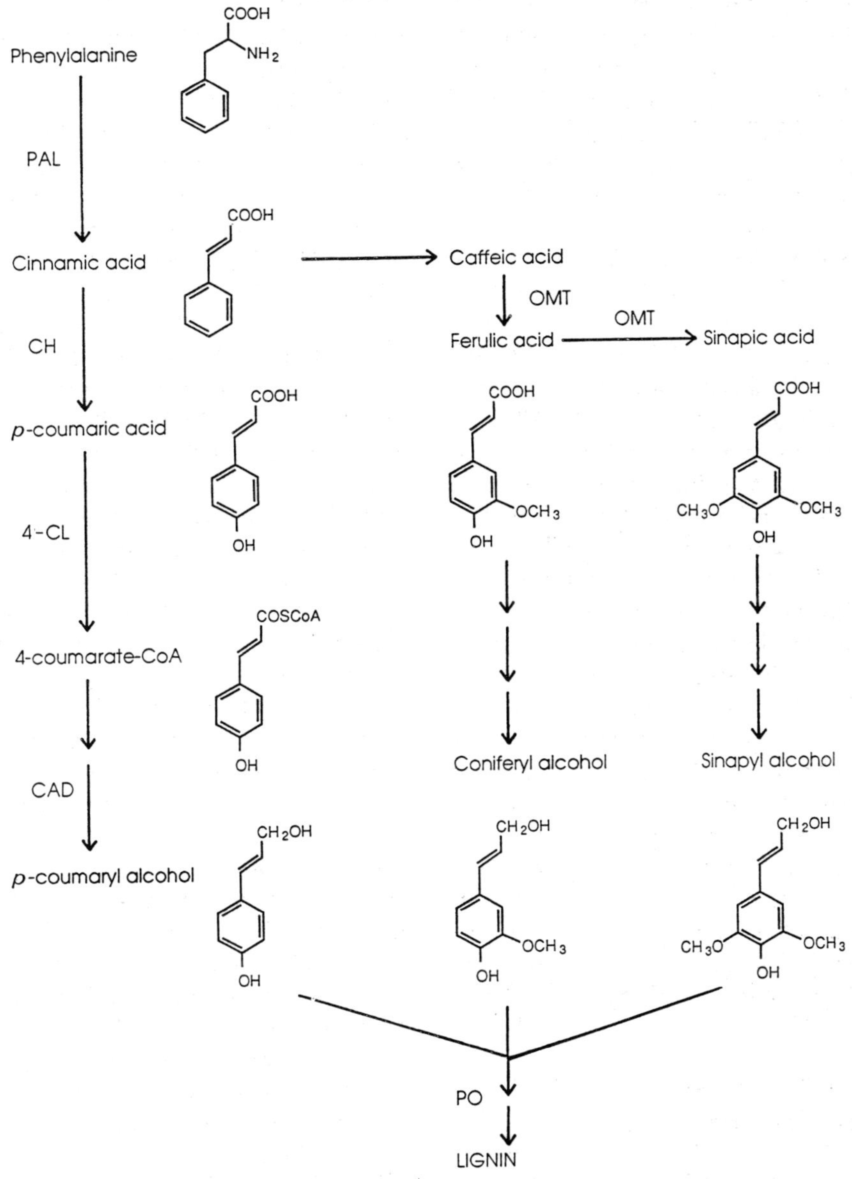

Figure 1.5. The pathway for lignin biosynthesis in plants. PAL = phenylalanine ammonia lyase; CH = cinnamate hydroxylase; 4CL = 4-coumarate CoA ligase; CAD = cinnamyl alcohol dehydrogenase; OMT = *o*-methyl transferase; PO = peroxidase.

Treatment of tobacco tissue with inhibitors of PAL such as α-aminooxyacetic acid (AOA) or α-aminooxy-β-phenylpropionic acid (AOPP) increases the size of the necrotic lesions in the case of several resistant plant-virus combinations[67]. However, these experiments are only of indicative value since AOA and AOPP are not specific inhibitors of PAL and could affect other biochemical pathways branching off from cinnamic acid. The causality of lignification was examined using specific suicide inhibitors of CAD, *N*(*o*-aminophenyl)sulphinamoyltertiobutyl acetate (NH_2-PAS) and *N*(*o*-hydroxyphenyl)sulphinamoyltertiobutyl acetate (OH-PAS)[68,69]. Resistant wheat pretreated with these inhibitors showed a decrease in the frequency of lignified host cells and an increase in cell penetration by stem rust pathogens[53].

It is still unclear how pathogen invasion is really prevented by lignification, but it strengthens the cell wall mechanically and provides a hydrophobic barrier which might block pathogen penetration. For example, cell wall-degrading enzymes or toxins produced by the pathogen might not reach their targets. Also unpolymerized lignin precursors such as coniferyl alcohol might be fungitoxic and could inhibit a pathogen even before a lignin barrier is established[57].

Hypersensitive Response (HR) This is characterized by a rapid death of host cells during an incompatible interaction with a pathogen. Pathogens which can induce it remain confined to the hypersensitive necrosis, which is expressed as localized browning and cellular collapse. The hypersensitive response appears to be an elegant restriction mechanism involving an energy-dependent sequence of biochemical and physiological events leading to the collapse of host cells[70].

A number of studies indicate that membrane damage is among the first events taking place during the hypersensitive response. In cultured tobacco cells, infection with the incompatible bacterium *Pseudomonas syringae* pv. *pisi* results in an increase in the H^+ influx/K^+ efflux[71]. The importance of the H^+/K^+ response was demonstrated with Tn5 mutants of *Pseudomonas syringae* pv. *syringae*: mutants which had lost the ability to induce HR could not induce the H^+/K^+ and vice versa[72]. Electrolyte leakage is linked with lipid peroxidation in cucumber cotyledons infected with the incompatible pathogen *Pseudomonas syringae* pv. *pisi*[73]. Lipoxygenase and lipolytic acyl hydrolase, which are involved in the membrane breakdown, are both induced during the first hours of the response of bean leaves infected with an incompatible race of *Pseudomonas syringae* pv. *phaseolica*[74]. In potato tuber tissue and protoplasts, NADPH-dependent production of superoxide anions was observed within minutes after contact with *Phytophthora infestans* or hyphal wall components derived from it. It was proposed that O_2^- radicals cause cell death and/or activate resistance mechanisms[75,76]. Similarly, O_2^- radicals initiate lipid peroxidation in cucumber cotyledons undergoing the hypersensitive response leading to electrolyte leakage[73]. Lipid peroxidation and lipoxygenase induction were both observed in tomato leaves treated with elicitor or with a superoxide-generating system[77]. Interestingly, O_2^- radicals are also produced in immunocompetent cells of animals after recognition of foreign cells or materials. The active oxygen species can cause membrane damage such as lipid peroxidation and result in local inflammation, or in some cases carcinogenesis[78].

At later stages, other defence reactions can take place such as browning of the cells and production of lignin, phytoalexins or pathogenesis-related proteins[79]. Thus the hypersensitive response is a pleiotropic response which results in a complex array of metabolic changes during an incompatible interaction. Its early timing is a key element for a successful resistance response. This implies that processes such as recognition and the control of specificity are of foremost importance for the survival of the plant.

Papillae Formation Papillae are thickenings of the plant cell walls occurring directly underneath the penetration peg of infecting organisms.

They are also referred to as callosities or lignitubers[80]. Papillae contain callose (a polymer of β-1,3-glucan), as well as lignin. In addition proteins, gums, suberin and silicon were also reported[80]. Formation of papillae was often associated with the resistance of various plants to fungi, particularly in cereals[81]. In the upper leaves of cucumber systemically protected by a first inoculation, papilla formation was observed at sites of unsuccessful penetration[82]. However, unsuccessful penetration sites were also observed without papilla deposition, indicating that in this situation papillae were not the exclusive mechanism of defence. Powdery mildew-resistant barley lines induce papillae at the infection site at an earlier time than susceptible lines. In addition, papillae in resistant interactions contained a light-absorbing component rich in phenylpropanoids which is absent in compatible interactions[81]. Ca^{++} ions were also associated in larger amounts with papillae in resistant lines[81,83]. Since callose synthesis is strictly Ca^{++}-dependent, it is tempting to speculate that localized callose synthesis may at least be one component contributing to papilla formation and resistance.

Hydroxyproline-rich Cell Wall Glycoproteins (HPRGs) These proteins (also termed extensins) have a conspicuous repeat of ser-hyp-hyp-hyp-hyp sequences where the ser or the hyp residue can be linked to glycosyl residues[84]. The amount of HPRG in the cell wall was observed to increase after wounding or infection with various pathogens[85,86]. The function of these proteins in defence is still unclear. Crosslinking of the cell wall by isodityrosine links either within HPRGs or to other cell wall components is hypothesized to provide additional sites for lignification[85,87]. The evidence for intermolecular links is, however, missing so far. Using antibodies against HPRG it could be shown that these proteins are deposited in increased amounts at sites of fungal infection[88]. HPRGs are coded by a family of genes which were found to be differentially regulated depending on whether the interaction with the pathogen was compatible or resistant. Differential regulation of HPRG genes was also found after wounding or infection[85]. The intriguing association between HPRGs and induced resistance is a topic for further research.

Vascular Occlusions Infections of plant vessels caused by bacteria or fungi cause the vessels to be occluded by gels, gums or tyloses. Tyloses are balloon-like protrusions of vascular parenchyma cell walls which penetrate through wall pits into the lumen of vessel elements. Blockage of vessels by occlusion hinders the transpiration stream and results in wilting and water stress. It also limits the spread of a pathogen or its spores. Resistance in elms and oaks was associated with a more rapid build up of tyloses[89,90]. Vascular occlusions might also limit the bacterial spread in Pierce's disease of grapes, the bacteria becoming entrapped in gums and tyloses[91].

Induced Biochemical Barriers

Phytoalexins Antimicrobials that are produced or accumulate at or around the site of pathogen infection are called phytoalexins[92] and include over 300 natural products distributed among phenolics, terpenoids, polyacetylenes, fatty acid derivatives and other classes[93–99] (Figures 1.6 and 1.7).

Two lines of investigation support the notion that phytoalexins are factors involved in the resistance of plants against pathogens. One line concerns the timing and location of their production in relation to the position of the pathogen. Microscopic observations combined with biochemical determinations of glyceollin, the main phytoalexin in soybean, were carried out in microtome sections of soybean hypocotyls resistant or susceptible to race 1 of *Phytophthora megasperma*[100]. Resistant plants accumulate inhibitor levels of glyceollin around invading hyphae 8h after inoculation. At the same time glyceollin is hardly detectable in susceptible plants. In these experiments, hypocotyls were inoculated using mycelial plugs placed over small

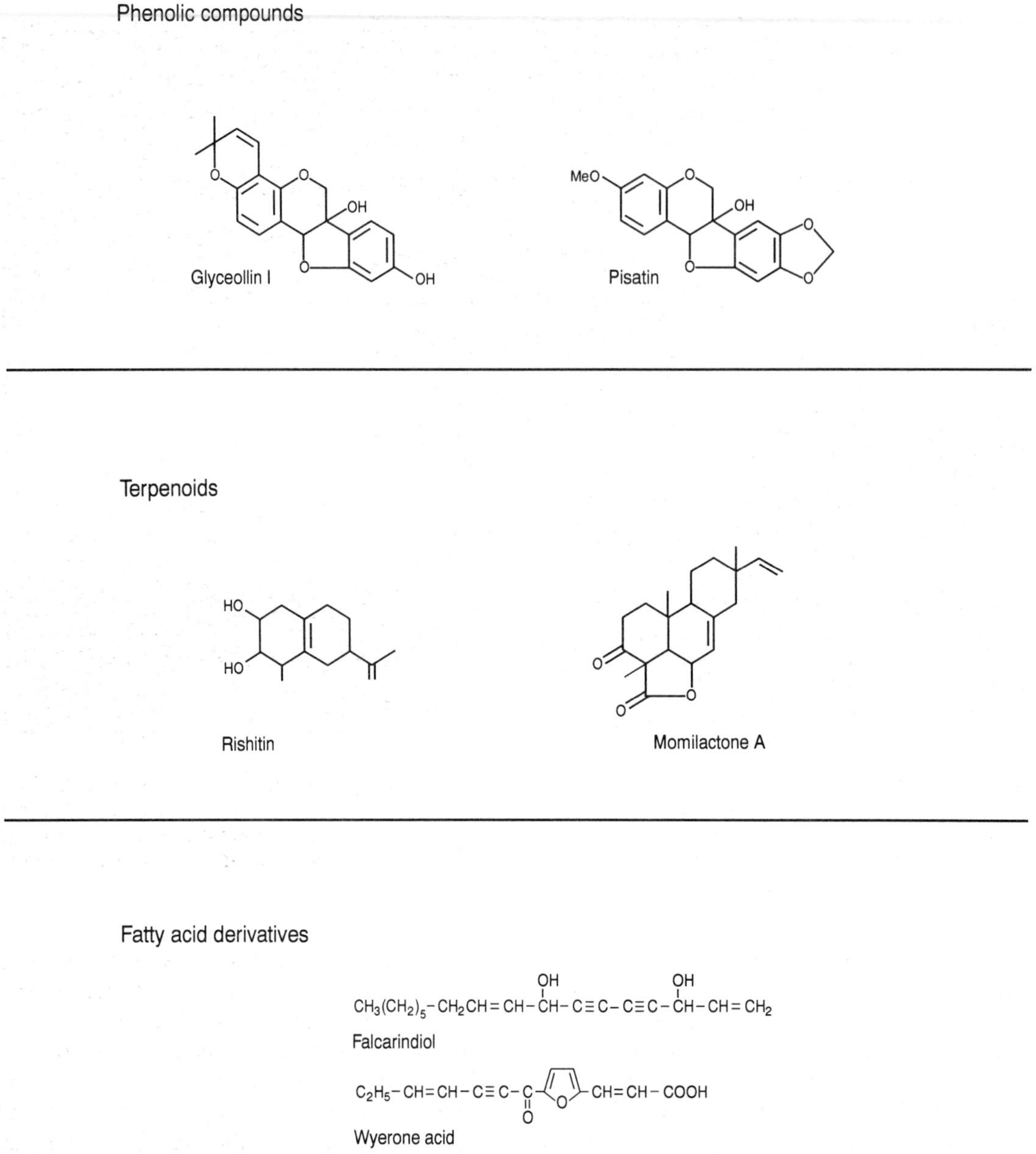

Phenolic compounds

Glyceollin I

Pisatin

Terpenoids

Rishitin

Momilactone A

Fatty acid derivatives

$CH_3(CH_2)_5-CH_2CH=CH-\underset{|}{\overset{OH}{C}}H-C\equiv C-C\equiv C-\underset{|}{\overset{OH}{C}}H-CH=CH_2$

Falcarindiol

$C_2H_5-CH=CH-C\equiv C-\underset{\parallel \atop O}{C}-(\text{furan})-CH=CH-COOH$

Wyerone acid

Figure 1.6. Some phytoalexins in plants.

slash wounds in the epidermis, which is not the natural infection path for a soil-borne pathogen. The very early steps in the host–pathogen interaction might escape observation. Another study used small unwounded soybean seedlings inoculated on the roots with virulent or avirulent races of *P. megasperma*[101]. Glyceollin was determined by radioimmunoassay in cryotome sections of the roots. At the same time, the fungal progression was monitored using immunofluorescent stains for *P. megasperma* hyphae. In the resistant interaction glyceollin accumulate significantly in the

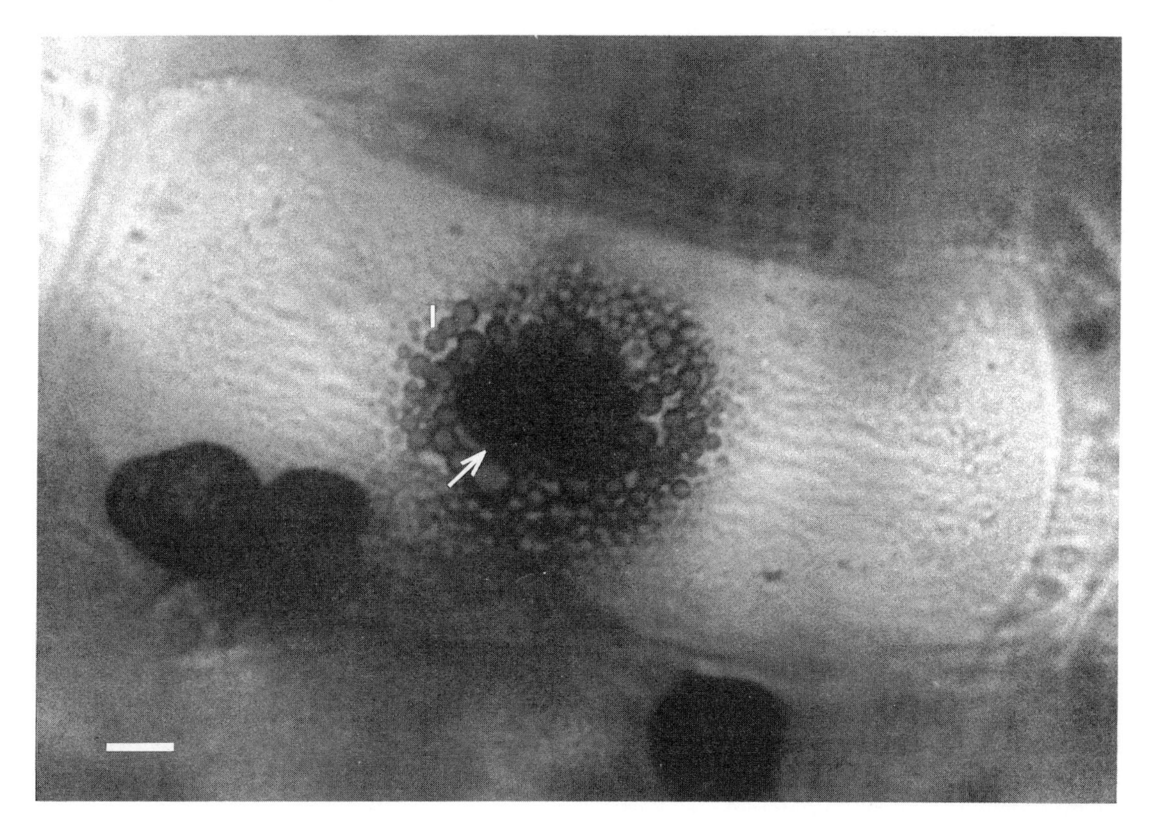

Figure 1.7. Visualization of localized phytoalexin synthesis in a sorghum leaf inoculated with the fungus *Colletotrichum graminicola*. The surface view shows an epidermal cell responding to fungal invasion (arrow) by the production of pigmented inclusions (I) which consist of apigeninidin, a flavonoid phytoalexin. The picture was taken 26 h after inoculation. At later stages the inclusions coalesce and the cell dies. Bar = 5 μm. Photograph kindly provided by R. Nicholson, Purdue University.

epidermis and in the cortex around the site of infection. Hyphae were not detected in advance of measurable levels of glyceollin. In the compatible interaction extensive colonization extends rapidly to the root stele. Little or no glyceollin was detected in the cortex but glyceollin can reach inhibitory levels in the epidermis. These careful observations support the notion that glyceollin accumulation is an important factor in the defence of the plant. They do not exclude the possibility that other resistance mechanisms might be triggered by the incompatible fungal race, preventing spread of the fungus in the root stele.

The detoxification of phytoalexins by pathogenic fungi provides another line of evidence which suggests that phytoalexins are important components in the defence of the plant[102]. In pea the detoxifying activity is due to an inducible cytochrome P-450 monooxygenase referred to as pisatin demethylase. Analyses of crosses between pathogenic and non-pathogenic strains of *Nectria haematococca* indicate that pisatin demethylation is necessary for pathogenicity[102]. Expression of pisatin demethylase in the corn pathogen *Cochliobolus heterostrophus* results both in the ability to demethylate pisatin and to infect pea[103]. This provides experimental evidence indicating that phytoalexins are a plant defence factor and that their detoxification is required for pathogenicity.

The production of phytoalexins after infection

results from an increased synthesis from precursors. This increase in synthesis is often related to an increase in the activities of the biosynthetic enzymes which are synthesized *de novo* soon after the initial steps of the plant–pathogen interaction[97,99]. A well-documented example is the synthesis of isoflavonoid phytoalexins, where a number of enzymes from the early steps such as phenylalanine ammonia lyase (PAL), chalcone synthase and chalcone isomerase are transcriptionally regulated[97–99] (Figure 1.8). Phytoalexins can also be produced by hydrolysis of direct precursors. In soybean cotyledons, the precursor of glyceollin, daidzein, is constitutively present in large quantities as a conjugate, which is rapidly hydrolysed during an incompatible interaction[104] (Figure 1.8). Glyceollin accumulation in soybean cotyledons can result both from *de novo* synthesis and conjugate hydrolysis. The relative contribution of each pathway is not yet elucidated.

The availability of antibodies and cDNA for several enzymes of the phenylpropanoid phytoalexin pathway has yielded information on the spatial distribution of phytoalexin biosynthesis. In bean inoculated with *Colletotrichum lindemuthianum*, transcripts of PAL were found in cells adjacent to the site of infection as well as in healthy tissue several millimetres away from the point of application of the fungal spores[105], suggesting the presence of some mechanism of intercellular signalling. *In situ* hybridization in leaf tissue of young potato plants shows the accumulation of PAL mRNA in healthy cells around the site of an incompatible interaction with *Phytophthora infestans*[106]. Similar observations were carried out in parsley leaves infected with an incompatible pathogen *Phytophthora megasperma* f.sp. *glycinea*[107]. These studies support the notion that enzymes for phytoalexin biosynthesis are transcribed at the site of infection. They do not rule out the possibility that other processes such as lignification, which require similar biochemical routes, are also involved.

Little is known about the molecular mode of action of phytoalexins, despite abundant work demonstrating their toxic activity against fungal, plant and even mammalian cells. One possible mechanism could be the uncoupling of oxidative phosphorylation[108]. The transport of H^+ ions at the tonoplast membrane around the vacuole is also strongly affected by phytoalexins[109].

Pathogenesis-related Proteins (PRs) Induction of these proteins is one of the best documented biochemical changes associated with induced resistance[110–112]. They were first observed in tobacco and defined initially as acid extractable low-molecular weight and proteinase-resistant proteins which accumulate in the intercellular fluid of infected leaves[113] (Table 1.1). Recently basic homologues of acidic pathogenesis-related proteins were found and their

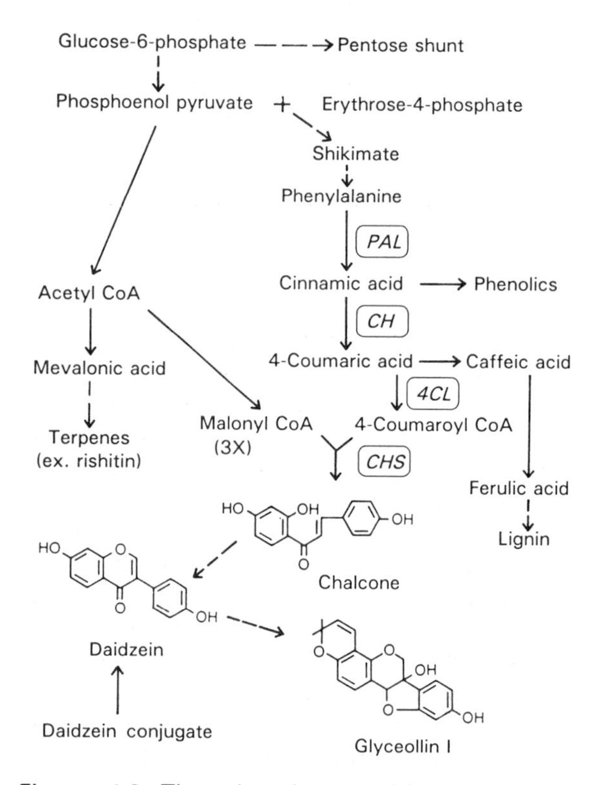

Figure 1.8. The phenylpropanoid pathway for glyceollin biosynthesis in soybean tissue. Phenylalanine ammonia lyase (PAL), cinnamic acid-4-hydroxylase (CH), 4-coumarate-CoA-ligase (4CL) and chalcone synthase (CHS) are enzymes induced during infection or by treatments with various elicitors.

Table 1.1 Pathogenesis-related proteins induced in tobacco (*Nicotiana tabacum* cv. Samsun NN) by infection with tobacco mosaic virus (TMV)[110].

	Acidic proteins		Basic proteins		
Group	Class	MW (kDa)	Name	MW (kDa)	Function
1	1a	15.8	16	16.0	Unknown
	1b	15.5			Unknown
	1c	15.6			Unknown
2a	2a	39.7	glucanase b	33.0	β-1,3-glucanase
	N	40.0			β-1,3-glucanase
	O	40.6			β-1,3-glucanase
	Q′	36.0			β-1,3-glucanase
2b	O′	25.0			β-1,3-glucanase
3	P	27.5	Ch 32	32.0	Chitinase
	Q	28.5	Ch 35	34.0	Chitinase
4	s1	14.5			Unknown
	r1	14.5			Unknown
	s2	13.0			Unknown
	r2	13.0			Unknown
5a	R	24.0	Osmotin	32.0	Unknown
	S	24.0			Thaumatin-like protein
5b			45 kDa	45.0	Unknown

compartmentalization is typically intracellular, in the vacuole[110–112] (Table 1.1). A correlation between the appearance of pathogenesis-related proteins, induced resistance and the hypersensitive response has been observed in a multitude of dicotyledonous as well as monocotyledonous plant species[110,111,114]. The function of most of the proteins is still unknown. Two classes, however, have been shown to have chitinase and β-1,3-glucanase activity *in vitro*[115–117]. Interestingly, the substrates for these enzymes, chitin and β-1,3-glucan, occur in cell walls of fungi; indeed, mixtures of chitinase and glucanase are fungicidal *in vitro*[173]. Another class are structurally similar to proteinase inhibitors, and are more likely to function in resistance to insects[118]. The timing and accumulation of the mRNA coding for pathogenesis-related proteins correlates well with the magnitude and timing of the induced resistance response, which suggests a role in the disease[114]. The proteins are synthesized in response to various necrotizing pathogens but also after treatment with chemicals such as polyacrylic acid, salicylate, amino acid derivatives or salts of heavy metals[110,111,119]. In all cases described, their induction is transcriptionally regulated[110,111]. Thus the data available up to now support the hypothesis that the accumulation of pathogenesis-related proteins is a mechanism of disease resistance. The different classes are perhaps directed against different pathogens. However, proving that these proteins play a causal role in resistance will await further experiments. Chimaeric genes for PR proteins including the coding region of PR-1, GRP and PR-S coupled to the strong constitutive cauliflower mosaic virus-promoter were used to transform tobacco plants. Although the proteins were targeted properly into the intercellular fluids in transformed plants, no resistance was found against viral infection[120,121]. Further tests using other proteins or combinations of proteins, other expression patterns and other diseases are envisaged. Difficulties in interpreting results from constitutively expressing transformants comes from the fact that the complicated regulatory systems controlling the expression of pathogenesis-related proteins *in vivo* are not readily accessible by such experiments.

1.3 RECOGNITION AND SPECIFICITY

The interaction between plants and microorganisms is a process which involves highly specific recognition events. On one hand, plants can readily discriminate amongst a very broad spectrum of potential pathogens, of which only very few fail to be recognized and cause disease. On the other hand, pathogens induce genes for pathogenicity mostly after recognition of specific signals from their host plants. These initial recognition events which occur after host-pathogen contact are decisive for the outcome of the interaction. Our working models are based on cell-to-cell interactions in animals where signals are recognized by specific membrane-bound receptors and subsequently transduced into a cellular response.

1.3.1 Recognition of Plants by Pathogens

Available evidence shows that the genes necessary for the early steps in infection are induced in a pathogen once it has detected a suitable host plant. This early recognition event is mediated by detecting chemical or physical signals produced by the plant. *Agrobacterium tumefaciens*, the causal agent of crown gall disease, induces extensive tumours in dicotyledonous plants[122]. This soil-borne bacterium responds to products secreted from the roots of its host plant. Seven phenolic compounds, including catechol and *p*-hydroxybenzoic acid, are able to induce bacterial genes for virulence, the products of which are essential for the early steps of infection[123].

Another elegant example is provided by the symbiotic interaction between nitrogen-fixing bacteria and leguminous plants[124]. *Rhizobium meliloti* enters its specific host alfalfa through the tip of root hairs. Once entered, the bacterium grows through the infection thread towards cortical cells. The latter are induced to divide actively and form a meristematic structure, the nodule, in which bacteria become surrounded by a host membrane to form nitrogen-fixing bacteroids. The host plant possesses the entire genetic information for nodule formation but nodulation is only induced by a nodule-inducing principle produced by the bacteria. This active principle is the product of a cluster of bacterial genes for nodulation (*nod* genes). The expression of *nod* genes requires the presence of luteolin, a flavonoid exuded from alfalfa roots. The product formed upon induction of *nod* genes by luteolin was shown to be a sulphated and acylated oligosaccharide, which is a host-specific nodulation factor[125].

A number of fungi such as the pea pathogen *Fusarium solani* pv. *pisi* infect their hosts by direct penetration through the leaf surface[126]. Recognition of the plant surface by the fungus leads to the induction of the cutinase necessary for penetration. Recognition is mediated by cutin monomers released from the cuticle by basal levels of cutinase produced by the germinating spores[127]. Cutin monomers are potent chemical signals which activate the fungal cutinase gene[128] (Figure 1.9).

Topographical features of the leaf surface can also have a signalling function for the formation of fungal infection structures. Uredospore germlings of the bean rust fungus *Uromyces appendiculatus* have evolved sensitive mechanisms to recognize the location of the stomata over which they differentiate into an appressorium. The signal for differentiation is the presence of a sharp ridge of an average height of 0.5 μm such as the lip of guard cells[129]. The signal is a purely physical stimulus since germlings will differentiate even when growing over artificial silicon wafers containing ridges of similar sizes.

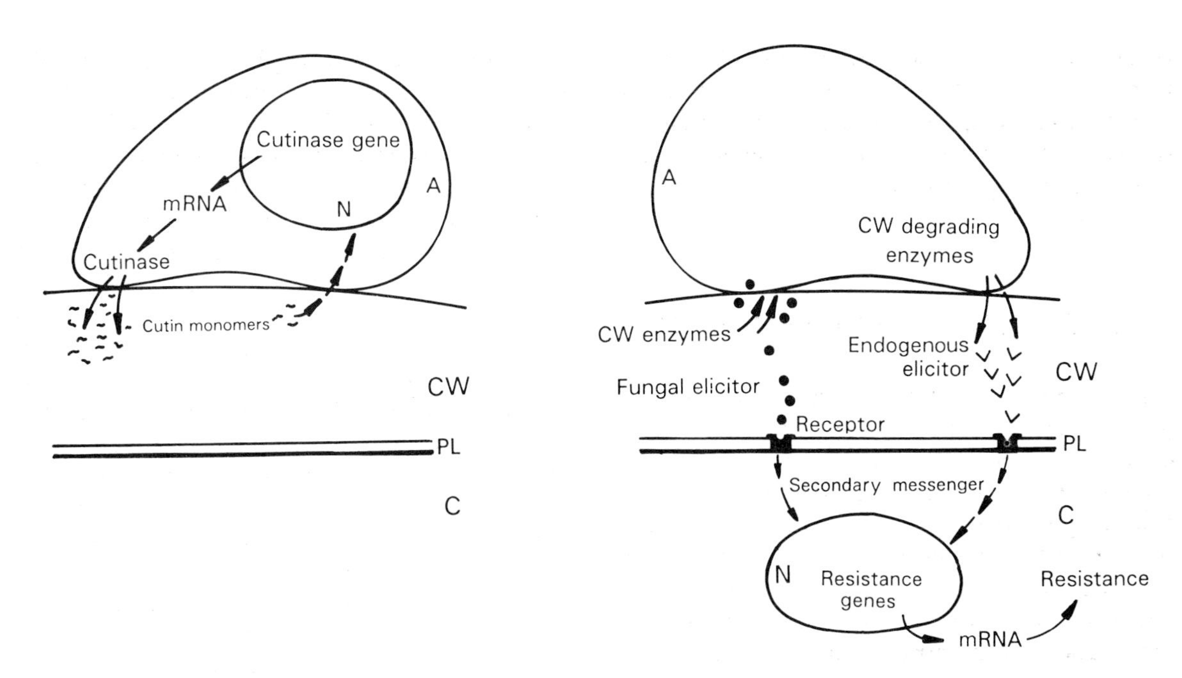

Figure 1.9. Schematic representation of recognition events in plant-pathogen interactions. Left side: recognition of a plant by a fungus. Cutin monomers are released from the cuticle by the basal cutinase activity produced by a germinating spore. These monomers are recognized by the fungus presumably at the nuclear envelope, and induce the activation of a cutinase gene which provides for increased cutinase production necessary for fungal penetration. Right side: Plants recognize fungi by elicitors interacting with receptors, for example at the plasma membrane. Secondary messengers relay the information to the nucleus where genes for resistance are activated. Note that elicitors may be released either from the fungal cell wall by plant cell wall enzymes or from the plant cell wall by fungal enzymes. Additional elicitors (not represented) might be compounds produced directly by the fungus. Key: A = fungal appressorium; C = cytoplasm; PL = plasma membrane; CW = cell wall; N = nucleus.

1.3.2 Recognition of Pathogens by Plants

A current model based on numerous observations holds that components liberated during the plant–pathogen interaction—called elicitors—are recognized by plants where they induce defence reactions (Figure 1.9). Originally elicitors were defined as inducers of phytoalexins only; the term is now more widely used and applies to any inducible defence mechanism. Although a large number of molecules are known to have elicitor activity, little information is available on their interaction with plant receptors and the subsequent transduction into defence reactions[95,97,98]. Biotic elicitors include complex carbohydrates from plant and fungal cell walls, microbial enzymes, polypeptides, proteins and lipids. Abiotic elicitors are purely physical or chemical stimuli such as salts of heavy metals, cold, detergents, or UV light.

Soluble elicitors such as polysaccharides or glycoproteins were discovered in the culture filtrates of pathogens grown in liquid culture[95,130]. The treatment of plant cell walls with cell wall-degrading enzymes such as those produced by microbial pathogens (endopolygalacturonase, endopolygalacturonic acid lyase) during infection releases fragments which have elicitor activity[95]. Such oligouronides are called endogenous elicitors. A variety of elicitors can be released from fungal cell walls by chemical or enzymatic treatments[95]. A classic example is the release of elicitors by mild hydrolysis of the cell wall of *Phytophthora megasperma*. This elicitor preparation consists of a complex mixture of oligo- and polysaccharides which can induce phytoalexins in host as well as in non-host plants of *P. megasperma*. From this mixture the minimum structure needed for elicitor activity is a hepta-β-glucoside which can elicit the synthesis of glyceollin in soybean cotyledons in the nanomolar range[131,132]. The structure of this extremely active heptaglucoside was confirmed using chemical synthesis[133]. The availability of a pure elicitor preparation made the search for specific plant receptors possible. Radiolabelling of a partially purified glucan elicitor was used to demonstrate a specific binding site localized at the plasma membrane of soybean cotyledons[134,135]. A binding site for a pure preparation of the highly active hepta-β-glucoside was found on soybean microsomal membranes[136]. The affinity of this receptor for the heptasaccharide was comparable with the activity of the elicitor in the bioassay, and closely related isomers showed neither elicitor activity nor binding affinity for the receptor. The receptor could be inactivated by heat and proteolysis and may either be a protein or a glycoprotein.

Formal proof that either one of the described elicitors is indeed produced in sufficient amounts during the plant–pathogen interaction is missing. The high specific activity of the hepta-β-glucoside and the presence of selective high-affinity membrane receptors would suggest a role for such molecules in the recognition of the pathogen by the plant and offer a tool for further investigations.

1.3.3 Specificity

All plants are resistant against most potential pathogens. This type of resistance is called unspecific, non-host or basal resistance and involves some of the constitutive and inducible mechanisms presented above[137]. Thus non-host resistance is a complex multigenic resistance. For a successful infection, a pathogen must overcome constitutive and inducible barriers. A pathogen is unlikely to breach these barriers for a new host by one or a few mutations only, since there are many genes involved in non-host resistance which is generally considered a stable resistance.

The resistance of a plant cultivar towards a pathogen biotype can often vary. Such pathogen biotypes are referred to as races or isolates and are characterized by the outcome of the interaction with individual plant cultivars. In the early 1930s, Flor started a series of fundamental observations on the interactions between different flax cultivars infected with various races of the flax rust fungus *Melampsora lini*. He explained his results by postulating that resistance only occurs when single genes for resistance in plants are matched by single genes for avirulence in the pathogen[138]. When a plant harbours many genes for resistance specific for a given pathogen, disease only occurs if all the corresponding avirulence genes in the pathogen race are recessive. The analysis of a number of host-pathogen systems where different pathogen races are known yields results which can be interpreted by this gene-for-gene concept (Table 1.2).

In molecular terms, one possible way to interpret gene-for-gene relationships is a receptor-ligand model where the receptor is the product of the gene for resistance in the host and the ligand a race-specific elicitor produced by the genes for avirulence in the pathogen. Efforts are currently

Table 1.2 Possible interactions between alleles of a plant disease resistance gene (*R* or *r*) and complementary genes for avirulence in the pathogen (*A* or *a*). Incompatibility (I) or resistance occurs only if both dominant genes for resistance and avirulence are matched, indicating that the specificity is controlled by the interaction of the product of the *R* and *A* alleles. The other combinations result in compatibility (C) or susceptibility. This arrangement is termed the quadratic check.

	Genotype of the host cultivar	
Genotype of the pathogen race	*R*	*r*
A	I	C
a	C	C

aimed at cloning plant resistance genes. The complication is that the products of resistance genes have not been identified and the only method of detection is based on rating the response of an infection with a pathogen containing the corresponding genes for avirulence. The approaches comprise shotgun cloning of genes by screening of the acquired resistance or avirulence, gene tagging with transposable elements or linkage analysis with known markers on the chromosome[139].

Cloning of avirulence genes has met with more success. The first gene for avirulence was cloned from the race 6 of *Pseudomonas syringae* pv. *glycinea*, a bacterium causing soybean blight. This gene encodes a 100kDa protein of unknown function. Expression of this avirulence gene in virulent strains resulted in transconjugants exhibiting the same host range incompatibility as the wild-type race 6[140]. A fungal gene product for avirulence was isolated from the intercellular fluid of tomato infected with the biotrophic fungus *Cladosporium fulvum*, the causal agent of tomato leaf mould[141]. The product of avirulence gene *Avr9* was found to be a peptide of 27 amino acids. This peptide could reproduce the same hypersensitive response symptoms as those expressed during the incompatible interaction in a host harbouring the corresponding resistance genes[142]. A cDNA clone for *Avr9* was isolated using oligonucleotide probes derived from the peptide elicitor[143]. Virulent races of *C. fulvum* transformed with *Avr9* clones showed an incompatible reaction on tomato cultivars carrying the resistance gene complementary to *Avr9*[174]. This confirms that *Avr9* is a race-specific gene for avirulence. The nature of the receptor for this ligand is not yet known but current work is aimed at identifying the receptor using radiolabelled elicitor. This approach will eventually lead to a characterization of a gene for resistance.

Race-specific elicitor molecules were also isolated from Gram-negative bacteria expressing the avirulence gene *AvrD* from *Pseudomonas syringae* pv. *tomato*[144]. Leaf injection of small amounts (10ng per injection site for a visible lesion) of this elicitor causes a hypersensitive response in soybean leaves carrying the corresponding gene (*Rpg4* gene) for resistance[145]. The structure of this elicitor is not yet fully characterized, but once known it will be instrumental in the search for the gene product of the resistance gene *Rpg4*.

1.3.4 Signal Transduction Pathways in Plants

The induction of resistance mechanisms involves transcription of new sets of genes in response to stimuli generated during the plant–pathogen interaction[146]. Most hydrophilic signals cannot readily cross membranes but interact with surface receptors. The binding relays intracellular signals across the cytoplasm, and the activation of gene expression is modulated by secondary messengers[147]. Receptors for extracellular signals with a lipophilic character may be located intracellularly, for example at the surface of the nucleus. The signal-receptor complex interacts directly with specific genes to regulate their expression (typical examples in animals are steroid hormones[148]). In the interaction between *Fusarium solani* pv. *pisi* and pea the synthesis of cutinase in the fungus is triggered by cutin oligomers at the plant surface[127]. These oligomers, together with a soluble protein fraction, activate the transcription of the cutinase gene in isolated nuclei[128]. The protein factor is likely to be of nuclear origin and is presumably lost during preparation. At high ionic strengths it behaves like a 100kDa protein. Together with cutin monomers, this protein factor causes phosphorylation of a 55kDa protein fraction in the presence of the fungal nuclei[126]. It is tempting to speculate that the phosphorylated protein might be a transacting factor mediating gene transcription.

Evidence is available showing that signals such as elicitors interact with membrane-bound receptors (see discussion above). Attention has been focused on possible secondary messengers with known functions in animal systems such as phosphoinositides, Ca^{++} ions or cAMP. Many surface receptors in animal systems activate a phospholipase C which catalyses the formation of inositol-1,4,5-triphosphate (IP3) and diacylglycerol

(DAG) from phosphatidyl inositol-4,5-biphosphate (IP2). IP3 and DAG are potent secondary messengers triggering various metabolic events[149]. One of the responses invoked by IP3 is the release of sequestered Ca^{++} ions[150]. The increase in cytoplasmic calcium levels can affect calcium and calcium-calmodulin-dependent kinases, lipases or ATPases[151,152]. Phosphoinositides and the related enzymes are present in plants and can affect calcium efflux from microsome or tonoplast vesicles *in vitro*[153]. The role of phosphoinositides as a possible link between the receptor of elicitors and phytoalexin biosynthesis was tested in cultured parsley and soybean cells[154]. The levels of IP3 did not vary after treatment of cultured parsley cells with elicitors, which leads to the conclusion that, at least in this system, inositides may not be involved in signal transduction.

In soybean cells, elicitation of glyceollin by glucan elicitors requires calcium[155,156] These experiments suggest that an early event after the elicitor–receptor interaction is an effect on ion fluxes, which could include calcium. The induction of callose is a typical response to fungal host penetration and depends on an enzyme, β-1,3-glucanase which is strictly calcium-dependent. Elicitors which induce callose synthesis also induce Ca^{++} uptake and results obtained with calcium ionophores indicate that calcium is important but not essential for callose synthesis as well as for induction of phytoalexin synthesis[156,157]. Cyclic AMP occurs at very low levels in uninfected plants and is usually detected by radioimmunoassay. The level of cAMP was found to increase after a hypersensitive response caused by viral infections on *Nicotiana glutinosa*[158]. In cultured cells, or in hypocotyls of soybean, infection or elicitor treatment raised the level of cAMP several fold over that of controls[159]. Larger increases were detected in cultured carrot cells treated with elicitor[160]. Although cAMP applied exogenously can induce phytoalexins, it remains questionable whether the endogenous levels are sufficient to act as secondary messengers.

Induction of resistance distal from the infection sites indicates the existence of intercellular signals. Stem girdling and grafting experiments indicate that the putative signal for induced systemic resistance moves through the phloem tissue of the vascular system of the plant[161,162]. Analyses of cucumber phloem collected at various times after an inducing infection showed an increase in salicylic acid which preceded the development of resistance against infection[163,164]. Similar results were also obtained in tobacco where the level of salicylic acid increased in the uninfected leaves of tobacco mosaic virus-infected plants[165]. Exogenous application of salicylic acid can induce resistance in cucumber and tobacco, which is not due to a direct fungicidal effect of this compound nor of its metabolites[119,164,166]. Salicylic acid induces the same set of PR genes as those expressed in response to a resistance-inducing pathogen infection[114,165]. The level of salicylic acid in induced tobacco leaves is sufficient to account for the induction of PR-1 accumulation, a protein which is strongly associated with induced resistance[167]. Salicylic acid might thus be a promising candidate for the signal that activates induced systemic resistance, but little is known about its synthesis, metabolism, transport and mode of action in plants.

Wounding of tomato or potato by predator feeding or by other mechanical stress leads to the production of proteinase inhibitors both locally and systemically[168]. The putative signal for the systemic transmission was given the name PIIF (proteinase inhibitor inducing factor) and was thought for some time to be a pectic fragment with a degree of polymerization of 20 units[169]. This pectic fragment was however found to be non-mobile and the enzymes necessary to release the cell wall fragments were not detected in tomato[170]. A new effort was undertaken to find the systemic signal using tomato leaf extracts and a small peptide was found which induced proteinase inhibitor genes when supplied to plants[171]. The peptide, called systemin, is phloem-mobile and has a surprisingly strong activity at the level of fmol/plant. This important observation can now be followed up by studies on the function of systemin in the signal transduction pathway.

1.4 CONCLUSION

The structural and anatomical differences between plants and animals are reflected by the different ways they have evolved to cope with pathogens. Elements such as the cell wall or the presence of well compartmentalized secondary metabolites and the various physical and chemical barriers induced at the time of infection seem to be important to plants.

In recent years attention has focused increasingly on the general process of plant–pathogen recognition, the perception of signals and the subsequent signal transduction cascade which leads eventually to the activation of new sets of genes for resistance. The basic principles underlying these processes such as ligand–receptor interactions, secondary messengers or the activation of transcription might be broadly similar in plants and animals, though from the discussion above it might be appreciated by the reader that complex networks of signals operating in plant–pathogen interactions are not likely to be explained uniquely by Ca^{++}, phosphoinositides or by cAMP. The central role played by the cell wall or the vacuole and the intricate relationship of the plant with a constantly changing environment reinforces the view that additional molecules such as oligosaccharides or phenolics are likely to be involved in signalling.

While plant–pathogen interactions is a marvellous topic for the scientific investigation of the biological reality of life, these interactions can also be a plague for the farmer trying to grow a good crop. Traditional knowledge in plant pathology has led to the production of plant protectants which are all based on the eradication of the microorganisms. With increasing knowledge of plant–microbe interactions, conventional methods for plant protection might be complemented by new approaches. Systemic induced resistance characterized by a broad resistance against a large number of pathogens has become an attractive model for the development of chemicals which trigger the defence mechanisms of the plant in a way similar to the endogenous signal[172]. Treatment of crop plants to increase their resistance is based on the natural potential of plants to defend themselves and does not directly impact on the microbial flora.

Disease can also be avoided by using resistant varieties obtained by traditional breeding. Improvement of crop plants by genetic engineering might complement this approach in the future[112]. More needs to be known about genes and gene products for resistance as well as the underlying mechanisms governing recognition and the responses thereof. Substantial progress has been made in this direction and it will be exciting to follow future developments in this interesting area.

Acknowledgements

The financial support of the Swiss National Science Foundation (Grant 31-34098) to defray expenses for the publication of this article is gratefully acknowledged.

1.5 REFERENCES

1. Wheeler, M. (1975). *Plant Pathogenesis.* Springer Verlag, New York.
2. Cassab, G.I. and Varner, J.E. (1988). Cell wall proteins. *Ann. Rev. Plant Physiol. Plant Molec. Biol.* **39,** 321–353.
3. Gunning, B.E.S. and Steer, M.W. (1977). *Biologie der Pflanzenzelle.* G. Fischer Verlag, Stuttgart.
4. Mould, M.J.R., Boland, G.J. and Robb, J. (1991). Ultrastructure of the *Colletotrichum trifolii-Medicago sativa* pathosystem. II. Post penetration events. *Physiol. Molec. Plant Pathol.* **38,** 195–210.
5. Kolatukudy, P.E. (1985). Enzymatic penetration of the plant cuticle by fungal pathogens. *Ann. Rev. Phytopathol.* **23,** 223–250.
6. Dickman, M.B., Podila, G.K. and Kolatukudy, P.E. (1989). Insertion of cutinase gene into a wound pathogen enables it to infect intact host. *Nature* **342,** 446–448.
7. Boccara, M., Diolez, A., Rouve, M. and Koutoujansky, A. (1988). The role of individual pectate lyases of *Erwinia chrysanthemi* strain 3937 in pathogenicity on saintpaulia plants. *Physiol. Molec. Plant Pathol.* **33,** 95–104.

8. Reid, J.L. and Collmer, A. (1988). Construction and characterization of an *Erwinia chrysanthemi* mutant with directed deletions in all of the pectate-lyase structural genes. *Molec. Plant-Microbe Interact.* **1,** 32–38.
9. Lafitte, C., Barthe, J.P., Montillet, J.L. and Touze, A. (1984). Glycoprotein inhibitors of *Colletotrichum lindemuthianum* endopolygalacturonase in near isogenic lines of *Phaseolus vulgaris* resistant and susceptible to anthracnose. *Physiol. Plant Pathol.* **25,** 39–53.
10. Abou-Goukh, A.A. and Labavitch, J.M. (1983). The *in vivo* role of Bartlett pear fruit polygalacturonase inhibitors. *Physiol. Plant Pathol.* **23,** 123–135.
11. De Lorenzo, G., Ito, Y., d'Ovidio, R. *et al.* (1990). Host–pathogen interactions XXXVII. Abilities of the polygalacturonase-inhibiting proteins from four cultivars of *Phaseolus vulgaris* to inhibit the endopolygalacturonases from three races of *Colletotrichum lindemuthianum. Physiol. Molec. Plant Pathol.* **36,** 421–435.
12. Harborne, J.B. (1990). Role of secondary metabolites in the chemical defence mechanisms in plants. In *Bioactive Compounds from Plants* (Ciba Foundation Symp. 154). Wiley, Chichester, pp. 126–139.
13. Wood, R.K.S. (1967). *Physiological Plant Pathology*. Blackwell, Oxford.
14. Conn, E.E. (1981). Cyanogenic glucosides. In *Biochemistry of Plants, Vol. 7*, Stumpf, P.K. and Conn, E.E. (eds). Academic Press, New York, pp. 479–500.
15. Fry, W.E. and Evans, P.H. (1977). Association of formamide hydrolyase with fungal pathogenicity to cyanogenic plants. *Phytopathology* **67,** 1001–1006.
16. Lieberei, R., Böhle, B., Giesemann, A. and Junqueira, N.T.V. (1989). Cyanogenesis inhibits active defence reactions in plants. *Plant Physiol.* **90,** 33–36.
17. Schildknecht, H., Maschwitz, U. and Krauss, D. (1968). Blausäure im Wehrsekret des Erdläufers *Pachymerium ferrugineum. Naturwissensch.* **55,** 230.
18. Défago, G. and Kern, H. (1983). Induction of *Fusarium solani* mutants insensitive to tomatine, their pathogenicity and aggressiveness to tomato fruits and pea plants. *Physiol. Plant Pathol.* **22,** 29–37.
19. Défago, G., Kern, H. and Sedlar, L. (1983). Genetic analyses of tomatine insensitivity, sterols content and pathogenicity for green tomato fruits in mutants of *Fusarium solani. Physiol. Plant Pathol.* **22,** 39–43.
20. Ferguson, J.E. and Metcalf, R.L. (1985). Cucurbitacins. Plant-derived defence compounds for diabroticities. *J. Chem. Ecol.* **11,** 311–318.
21. Bar-Nun, N. and Mayer, A.M. (1990). Cucurbitacins protect cucumber tissue against infection by *Botrytis cinerea. Phytochemistry* **29,** 787–791.
22. Harborne, J.B. (1980). Plant phenolics. In *Secondary Plant Products*, Bell, E.A. and Charlwood, B.V. (eds). Springer Verlag, Berlin, pp. 329–402.
23. Friend, J. (1985). Phenolic substances and plant disease. In *Ann. Proc. Phytochem. Soc. Eur. 25*, van Sumere, C.F. and Lea, P.J. (eds). Clarendon, Oxford, pp. 367–392.
24. Lynn, D.G. and Chang, M. (1990). Phenolic signals in cohabitation: implications for plant development. *Ann. Rev. Plant Physiol. Plant Molec. Biol.* **41,** 497–526.
25. Blakeman, J.P. and Atkinson, P. (1976). Antimicrobial substances associated with the aerial surfaces of plants. In *Microbial Ecology of the Phylloplane*, Blakeman, J.P. (ed.). Academic Press, New York, pp. 245–263.
26. Harborne, J.B., Ingham, J.L., King, L. and Payne, M. (1976). The isopentenyl isoflavone luteone as a pre-infectional antifungal agent in the genus *Lupinus. Phytochemistry* **15,** 1485–1487.
27. Ingham, J.L., Keen, N.T. and Hymowitz, T. (1977). A new isoflavone phytoalexin from fungus inoculated stems of *Glycine wightii. Phytochemistry* **16,** 1943–1946.
28. Dix, N.J. (1979). Inhibition of fungi by gallic acid in relation to growth on leaves and litter. *Trans. Brit. Mycol. Soc.* **73,** 329–336.
29. Walker, J.C. and Stahmann, M.A. (1955). Chemical nature of disease resistance in plants. *Ann. Rev. Plant Physiol.* **6,** 351–366.
30. Schlösser, E. (1988). Preformed chemical barriers in host parasite incompatibility. In *Experimental and Conceptual Plant Pathology*, Singh, R.S., Singh, U.S., Hess, W.M. and Weber, D.J. (eds). Gordon & Breach, New York, pp. 465–476.
31. Jersch, S., Scherrer, C., Huth, G. and Schlösser, E. (1989). Proanthocyanidins as basis for quiescence of *Botrytis cinerea* in immature strawberry fruits. *Zeitschrift Pflanzenkrankheiten Pflanzenschutz* **96,** 365–378.
32. Niemeyer, H.M. (1988). Hydroxamic acids (4-hydroxy-1,4-benzoxazin-3-ones), defence chemicals in the graminae. *Phytochemistry* **27,** 3349–3358.
33. Argandona, V.H., Corcuera, L.J., Niemeyer, H.M. and Campbell, B.C. (1983). Toxicity and

feeding deterrency of hydroxamic acids from graminae in synthetic diets against the greenbug, *Schizaphis graminum*. *Ent. Exp. Appl.* **34,** 134–138.
34. Copaja, S.V., Niemeyer, H.M. and Wratten, S.D. (1991). Hydroxamic acid levels in Chilean and British wheat seedlings. *Ann. Appl. Biol.* **118,** 223–227.
35. Madamanchi, N.R. and Kuc, J. (1991). Induced systemic resistance in plants. In *The Fungal Spore and Disease Initiation in Plants and Animals*, Cole, G.T. and Hoch, H.C. (eds). Plenum Press, New York, pp. 347–362.
36. Kuc, J. and Richmond, S. (1977). Aspects of the protection of cucumber against *Colletotrichum lagenarium* by *Colletotrichum lagenarium. Phytopathology* **67,** 533–536.
37. Caruso, F. and Kuc, J. (1977). Field protection of cucumber, watermelon and muskmelon against *Colletotrichum lagenarium* by *Colletotrichum lagenarium. Phytopathology* **67,** 1290–1292.
38. Tuzun, S., Nesmith, W., Ferriss, R.S. and Kuc, J. (1986). Effects of stem injections with *Peronospora tabacina* on growth of tobacco and protection against blue mould in the field. *Phytopathology* **76,** 938–941.
39. Sutton, D.C. (1982). Field protection of bean against *Colletotrichum lindemuthianum* by *Colletotrichum lindemuthianum. Aust. Plant Pathol.* **11,** 50–51.
40. Tuzun, S. and Kuc, J. (1985). A modified technique for inducing systemic resistance to blue mould and increasing growth in tobacco. *Phytopathology* **75,** 1127–1129.
41. Lewis, N.G. and Yamamoto, E. (1990). Lignin: occurrence, biogenesis and biodegradation. *Ann. Rev. Plant Physiol. Plant Molec. Biol.* **41,** 455–496.
42. Kirk, T.K. (1971). Effects of microorganisms on lignin. *Ann. Rev. Phytopathol.* **9,** 185–210.
43. Bird, P. (1988). The role of lignification in plant disease. In *Experimental and Conceptual Plant Pathology*, Singh, R.S., Singh, U.S., Hess, W.M. and Weber, D.J. (eds). Gordon & Breach, New York, pp. 523–535.
44. Legrand, M. (1983). Phenylpropanoid metabolism and its regulation in disease. In *Biochemical Plant Pathology*, Callow, J.A. (ed.). Wiley, Chichester, pp. 367–384.
45. Matern, U. and Kneusel, R.E. (1988). Phenolic compounds in plant disease resistance. *Phytoparasitica* **16,** 153–170.
46. Ride, J.P. (1983). Cell walls and other structural barriers in defence. In *Biochemical Plant Pathology*, Callow, J.A. (ed.). Wiley, Chichester, pp. 215–236.
47. Vance, C.P., Kirk, T.K. and Sherwood, R.T. (1980). Lignification as a mechanism of disease resistance. *Ann. Rev. Phytopathol.* **18,** 259–288.
48. Asada, Y. and Matsumoto, I. (1972). The nature of lignin obtained from downy mildew-infected Japanese radish roots. *Phytopathol. Z.* **73,** 208–214.
49. Ride, J.P. (1975). Lignification in wounded wheat leaves in response to fungi and its possible role in resistance. *Physiol. Plant Pathol.* **5,** 125–134.
50. Touzé, A. and Rossignol, M. (1977). Lignification and the onset of premunition in muskmelon plants. In *Cell Wall Biochemistry Related to Specificity in Host-Plant Pathogen Interactions*, Solheim, B. and Raa, J. (eds). Universitetsforlaget, Oslo, pp. 289–293.
51. Hammerschmidt, R. (1984). Rapid deposition of lignin in potato tuber tissue as a response to fungi non-pathogenic on potato. *Physiol. Plant Pathol.* **24,** 33–42.
52. Beardmore, J., Ride, J.P. and Granger, J.W. (1983). Cellular lignification as a factor in the hypersensitive resistance of wheat to stem rust. *Physiol. Plant Pathol.* **22,** 209–220.
53. Moersbacher, B.M., Noll, U., Gorrichon, L. and Reisener, H.J. (1990). Specific inhibition of lignification breaks hypersensitive resistance of wheat to stem rust. *Plant Physiol.* **93,** 465–470.
54. Southerton, S.G. and Deverall, B.J. (1990). Histochemical and chemical evidence for lignin accumulation during the expression of resistance to leaf rust fungi in wheat. *Physiol. Molec. Plant Pathol.* **36,** 483–494.
55. Ride, J.P. and Pearce, R.B. (1979). Lignification and papilla formation at sites of attempted penetration of wheat leaves by non-pathogenic fungi. *Physiol. Plant Pathol.* **15,** 79–92.
56. Tiburzy, R. and Reisener, H.J. (1990). Resistance of wheat to *Puccinia graminis* f. sp. *tritici*: Association of the hypersensitive reaction with the cellular accumulation of lignin-like material and callose. *Physiol. Molec. Plant Pathol.* **36,** 109–120.
57. Hammerschmidt, R. and Kuc, J. (1982). Lignification as a mechanism for induced systemic resistance in cucumber. *Physiol. Plant Pathol.* **20,** 61–71.
58. Kovats, K., Binder, A. and Hohl, H.R. (1991). Cytology of induced systemic resistance of cucumber is *Colletotrichum lagenarium. Planta* **183,** 484–490.
59. Dean, R.A. and Kuc, J. (1987). Rapid lignification in response to wounding and infection as a mechanism for induced systemic protection in cucumber. *Physiol. Molec. Plant Pathol.* **31,** 69–81.

60. Maule, A.J. and Ride, J. (1976). Ammonia-lyase and *o*-methyltransferase activities related to lignification in wheat leaves infected with *Botrytis. Phytochemistry* **15,** 1661–1664.
61. Maule, A.J. and Ride, J. (1983). Cinnamate 4-hydroxylase and hydroxy-cinnamate:CoA ligase in wheat leaves infected with *Botrytis cinerea. Phytochemistry* **22,** 1113–1116.
62. Moersbacher, B.M., Noll, U.M., Flott, B.E. and Reisener, H.J. (1988). Lignin biosynthetic enzymes in stem rust infected, resistant and susceptible near-isogenic wheat lines. *Physiol. Molec. Plant Pathol.* **33,** 33–46.
63. Moersbacher, B.M., Witte, U., Königs, D. and Reisener, H.J. (1989). Changes in the level of enzyme activities involved in lignin biosynthesis during the temperature sensitive resistant response of wheat (Sr6) to stem rust (P6). *Plant Sci.* **65,** 183–190.
64. Thorpe, J.R. and Hall, J.L. (1984). Chronology and elicitation of changes in peroxidase and phenylalanine ammonia-lyase activities in wounded wheat leaves in response to inoculation by *Botrytis cinerea. Physiol. Plant Pathol.* **25,** 363–379.
65. Legrand, M., Fritig, B. and Hirth, L. (1976). Enzymes of the phenylpropanoid pathway and the necrotic reaction of hypersensitive tobacco to tobacco mosaic virus. *Phytochemistry* **15,** 1353–1359.
66. Legrand, M., Fritig, B. and Hirth, L. (1978). *o*-diphenyl *o*-methyltransferase of healthy and tobacco mosaic virus-infected hypersensitive tobacco. *Planta* **144,** 101–108.
67. Massala, R., Legrand, M. and Fritig, B. (1987). Comparative effects of two competitive inhibitors of phenylalanine ammonia-lyase on the hypersensitive resistance of tobacco to tobacco mosaic virus. *Plant Physiol. Biochem.* **25,** 217–225.
68. Baltas, M., Cazaux, L., Gorrichon, L. *et al.* (1988). Sulphinamoylacetates as sulphine precursors. Mechanism of basic hydrolysis and scheme of irreversible inactivation of cinnamoyl alcohol dehydrogenase, an enzyme of the lignification process. *J. Chem. Soc. Perkin Trans. II*, 1473–1478.
69. Grand, C., Sarni, F. and Boudet, A.M. (1985). Inhibition of cinnamyl-alcohol-dehydrogenase activity and lignin synthesis in poplar (*Populus* × *euramericana* Dode) tissues by two organic compounds. *Planta* **163,** 232–237.
70. Slusarenko, A.J., Croft, K.P. and Voisey, C.R. (1991). Biochemical and molecular events in the hypersensitive response of bean to *Pseudomonas syringae* pv. *phaseolica*. In *Proc. Phytochem. Soc. Europe, Biochemistry and Molecular Biology of Plant Pathogen Interactions*, Smith, C.J. (ed), pp. 127–143.
71. Atkinson, M.M., Huang, J.S. and Knopp, J.A. (1985). The hypersensitive reaction of tobacco to *Pseudomonas syringae* pv. *pisi. Plant Physiol.* **79,** 843–847.
72. Atkinson, M.M. and Baker, C.J. (1987). Association of host plasma membrane K^+/H^+ exchange with multiplication of *Pseudomonas syringae* pv. *syringae* in *Phaseolus vulgaris. Phytopathology* **77,** 1273–1279.
73. Keppler, L.D. and Novacky, A. (1987). The initiation of membrane lipid peroxidation during bacteria-induced hypersensitive reaction. *Physiol. Molec. Plant Pathol.* **30,** 233–245.
74. Croft, K.P.C., Voisey, C.R. and Slusarenko, A.J. (1990). Mechanisms of hypersensitive cell collapse: correlation of increased lipoxygenase activity with membrane damage in leaves of *Phaseolus vulgaris* L. cv. Red Mexican inoculated with an avirulent race I of *Pseudomonas syringae* pv. *phaseolica. Physiol. Molec. Plant Pathol.* **36,** 49–62.
75. Doke, N. (1983). Involvement of superoxide anion generation in the hypersensitive response of potato tuber tissue to infection with an incompatible race of *Phytophthora infestans* and to the hyphal wall components. *Physiol. Plant Pathol.* **23,** 345–357.
76. Doke, N. (1983). Generation of superoxide anion by potato tuber protoplasts during the hypersensitive response to hyphal wall components of *Phytophthora infestans* and specific inhibition of the reaction by suppressors of hypersensitivity. *Physiol. Plant Pathol.* **23,** 359–367.
77. Peevers, T.L. and Higgins, V.J. (1989). Electrolyte leakage, lipoxygenase, and lipid peroxidation induced in tomato leaf tissue by specific and non-specific elicitors from *Cladosporium fulvum. Plant Physiol.* **90,** 867–875.
78. Halliwell, B. and Gutteridge, J.M.C. (1985). *Free Radicals in Biology and Medicine*. Clarendon Press, Oxford.
79. Fritig, B., Kaufmann, S., Dumas, B. *et al.* (1987). Mechanisms of the hypersensitivity reaction of plants. In *Plant Resistance to Viruses* (CIBA Foundation Symp. 133). Wiley, Chichester, pp. 92–108.
80. Aist, J.R. (1976). Papillae and related wound plugs of plant cells. *Ann. Rev. Phytopathol.* **14,** 145–163.
81. Aist, J.R., Gold, R.E., Bayles, C.J. *et al.* (1988). Evidence that molecular components of papillae

may be involved in ml-o resistance to barley powdery mildew. *Physiol. Molec. Plant Pathol.* **33,** 17–32.
82. Stumm, D. and Gessler, C. (1986). Role of papillae in the induced systemic resistance of cucumbers against *Colletotrichum lagenarium. Physiol. Molec. Plant Pathol.* **29,** 405–410.
83. Kunoh, H., Aist, J.R. and Israel, H.W. (1986). Elemental composition of barley coleoptile papillae in relation to their ability to prevent penetration by *Erysiphe graminis. Physiol. Molec. Plant Pathol.* **29,** 69–78.
84. Lamport, D. (1977). Structure, biosynthesis and significance of cell wall glycoproteins. In *Recent Advances in Phytochemistry 11*, Loewus, F.A. and Runeckles, V.C. (eds). Plenum, New York, pp. 79–111.
85. Corbin, D.R., Sauer, N. and Lamb, C.J. (1987). Differential regulation of a hydroxyproline-rich glycoprotein gene family in wounded and in infected plants. *Molec. Cell Biol.* **7,** 4337–4344.
86. Mazau, D. and Esquerré-Tugayé, M.T. (1986). Hydroxyproline-rich glycoprotein accumulation in the cell walls of plants infected by various pathogens. *Physiol. Mole. Plant Pathol.* **29,** 147–157.
87. Whitmore, F.W. (1978). Lignin-protein complex catalyzed by peroxidase. *Plant Sci. Letters* **13,** 241–245.
88. O'Connel, R.J., Brown, I.R., Mansfield, J.W. *et al.* (1990). Immunocytochemical localization of hydroxyproline-rich glycoproteins accumulating in melon and bean at sites of resistance to bacteria and fungi. *Molec. Plant-Microbe Interact.* **3,** 33–40.
89. Elgersma, D.M. (1973). Tylose formation in elms after inoculation with *Ceratocystis ulmi*, a possible resistance mechanism. *Netherlands J. Plant Pathol.* **79,** 218–220.
90. Jacobi, W.R. and MacDonald, W.L. (1980). Colonization of resistant and susceptible oaks by *Ceratocystis fagacearum. Phytopathology* **70,** 618–623.
91. Mollenhauer, H.H. and Hopkins, D.L. (1976). Xylem morphology of Pierce's disease-infected grapevines with different levels of tolerance. *Physiol. Plant Pathol.* **9,** 95–100.
92. Paxton, J.D. (1981). Phytoalexins—a working redefinition. *Phytopathol. Z.* **101,** 106–109.
93. Bailey, J.A. and Mansfield, J.W. (1982). *Phytoalexins*. Wiley, New York.
94. Barz, W., Daniel, S., Hinderer, W. *et al.* (1988). Elicitation and metabolism of phytoalexins in plant cell cultures. In *Applications of Plant Cell and Tissue Culture* (CIBA Foundation Symp. 137). Wiley, Chichester, pp. 178–198.
95. Darvill, A.G. and Albersheim, P. (1984). Phytoalexins and their elicitors a defence against microbial infection in plants. *Ann. Rev. Plant Physiol.* **35,** 243–275.
96. Dixon, R.A. (1986). The phytoalexin response: elicitation, signalling and control of host gene response. *Biol. Rev.* **61,** 239–291.
97. Dixon, R.A. and Harrison, M.J. (1990). Activation, structure and organization of genes involved in microbial defence in plants. *Adv. Genetics* **28,** 165–234.
98. Ebel, J. (1986). Phytoalexin synthesis: the biochemical analysis of the induction process. *Ann. Rev. Phytopathol.* **24,** 235–264.
99. Hahlbrock, K. and Scheel, D. (1989). Physiology and molecular biology of phenylpropanoid metabolism. *Ann. Rev. Plant Physiol. Plant Molec. Biol.* **40,** 347–369.
100. Yoshikawa, M., Yamauchi, K. and Masago, H. (1978). Glyceollin: its role in restricting fungal growth in resistant soybean hypocotyls infected with *Phytophthora megasperma* var. *sojae. Physiol. Plant Pathol.* **12,** 73–82.
101. Hahn, M.G., Bonhoff, A. and Griesebach, H. (1985). Quantitative localization of the phytoalexin glyceollin in relation to fungal hyphae in soybean roots infected with *Phytophthora megasperma* f.sp. *glycinea. Plant Physiol.* **77,** 591–601.
102. VanEtten, H.D., Matthews, D.E. and Matthews, P.S. (1989). Phytoalexin detoxification: importance for pathogenicity and practical implications. *Ann. Rev. Phytopathol.* **27,** 143–164.
103. Schäfer, W., Straney, D., Ciuffetti, L. *et al.* (1989). One enzyme makes a fungal pathogen, but not a saprophyte, virulent on a new host plant. *Science* **246,** 247–249.
104. Graham, T.L., Kim, J.E. and Graham, M.Y. (1990). Role of constitutive isoflavone conjugates in the accumulation of glyceollin in soybean infected with *Phytophthora megasperma. Molec. Plant-Microbe Interact.* **3,** 157–166.
105. Bell, J., Ryder, T.B., Wingate, V.P.M. *et al.* (1986). Differential accumulation of plant defence gene transcripts in a compatible and an incompatible plant-pathogen interaction. *Molec. Cell Biol.* **6,** 1615–1623.
106. Cuypers, B., Schmelzer, E. and Hahlbrock, K. (1988). *In situ* localization of rapidly accumulated phenylalanine ammonia-lyase mRNA around penetration sites of *Phytophthora infestans* in potato leaves. *Molec. Plant-Microbe Interact.* **1,** 157–160.
107. Schmelzer, E., Krüger-Lebus, S. and Hahlbrock, K. (1989). Temporal and spatial patterns of gene expression around sites of attempted fungal

infection in parsley leaves. *The Plant Cell* **1,** 993–1001.
108. Laks, P.E. and Pruner, M.S. (1989). Flavonoid biocides: structure-activity relations of flavonoid phytoalexin analogues. *Phytochemistry* **28,** 87–91.
109. Giannini, J.L., Holt, J.S. and Briskin, D.P. (1991). The effect of glyceollin on soybean (*Glycine max* L.) tonoplast and plasma membrane vesicles. *Plant Sci.* **74,** 203–211.
110. Bol, J.F., Linthorst, H.J.M. and Cornelissen, B.J.C. (1990). Plant pathogenesis-related proteins induced by virus infection. *Ann. Rev. Phytopathol.* **28,** 113–138.
111. Linthorst, H.J.M. (1991). Pathogenesis-related proteins of plants. *Crit. Rev. Plant Sci.* **10,** 123–150.
112. Ryals, J., Métraux, J.P. and Ward, E. (1991). Systemic acquired resistance: an inducible defence mechanism in plants. In *The Biochemistry and Molecular Biology of Inducible Enzymes and Proteins in Higher Plants*, Wray, J.L. (ed.). University Press, Cambridge.
113. Van Loon, L.C. (1985). Pathogenesis-related proteins. *Plant Molec. Biol.* **4,** 111–116.
114. Ward, E.R., Ukness, S.J., Williams, S.C. *et al.* (1991). Coordinate gene activity in response to agents that induce systemic acquired resistance. *The Plant Cell* **3,** 1085–1094.
115. Kaufmann, S., Legrand, M., Geoffrey, P. and Fritig, B. (1987). Biological function of pathogenesis-related proteins: four PR proteins of tobacco have β-1,3-glucanase activity. *EMBO J.* **11,** 3209–3212.
116. Legrand, M., Kaufmann, S., Geoffroy, P. and Fritig, B. (1987). Biological function of pathogenesis-related proteins: four tobacco pathogenesis-related proteins are chitinases. *Proc. Nat. Acad. Sci. USA* **84,** 6750–6754.
117. Métraux, J.P., Streit, L. and Staub, T. (1988). A pathogenesis-related protein in cucumber is a chitinase. *Physiol. Plant Pathol.* **33,** 1–9.
118. Richardson, M., Valdes-Rodrigues, S. and Blanco-Labra, A. (1987). A possible function for thaumatin and a TMV-induced protein suggested by homology to a maize inhibitor. *Nature* **327,** 432–434.
119. Van Loon, L.C. and Antoniw, J.F. (1982). Comparison of the effects of salicylic acid and etephon with virus-induced hypersensitivity and acquired resistance in tobacco. *Netherlands J. Plant Pathol.* **88,** 237–256.
120. Cutt, J.R., Harpster, M.H., Dixon, D.C. *et al.* (1989). Disease response to tobacco mosaic virus in transgenic tobacco plants that constitutively express the pathogenesis-related *PR-lb* gene. *Virology* **173,** 89–97.
121. Linthorst, H.J.M., Meuwissen, R.L.J., Kaufmann, K.D. and Bol, J.F. (1989). Constitutive expression of pathogenesis-related proteins PR-l, GRP and PR-S in tobacco has no effect on virus infection. *The Plant Cell* **1,** 285–291.
122. Nester, E.W., Gordon, M.P., Amasino, R.M. and Yanofski, M.F. (1984). Crown-gall, a molecular and physiological analysis. *Ann. Rev. Plant Physiol.* **35,** 387–413.
123. Bolton, G.W., Nester, E.W. and Gordon, M.P. (1986). Plant phenolic compounds induce expression of the *Agrobacterium tumefaciens* loci needed for virulence. *Science* **232,** 983–985.
124. Long, S. (1989). *Rhizobium*-legume nodulation: life together in the underground. *Cell* **56,** 203–214.
125. Lerouge, P., Roche, P., Faucher, C. *et al.* (1990). Symbiotic host-specificity of *Rhizobium meliloti* is determined by a sulphated and acylated glucosamine oligosaccharide signal. *Nature* **344,** 781–784.
126. Kolatukudy, P.E., Podila, G.K., Sherf, B.A. *et al.* (1991). Mutual triggering of gene expression in plant-fungus interactions. In *Advanced Molecular Genetics of Plant-Microbe Interactions*, Hennecke, H. and Verma, D.P.S. (eds). Kluwer, Dordrecht, pp. 242–249.
127. Woloshuk, C.P. and Kolatukudy, P.E. (1986). Mechanism by which contact with plant cuticle triggers cutinase gene expression in the spores of *Fusarium solani* f. sp. *pisi. Proc. Nat. Acad. Sci. USA* **83,** 1704–1708.
128. Podila, G.K., Dickman, M.B. and Kolatukudy, P.E. (1988). Transcriptional activation of a cutinase gene in isolated fungal nuclei by plant cutin monomers. *Science* **242,** 922–925.
129. Hoch, H.C., Staples, R.C., Whitehead, B. *et al.* (1987). Signaling for growth orientation and cell differentiation by surface topography in *Uromyces. Science* **235,** 1659–1662.
130. West, C.A. (1981). Fungal elicitors of the phytoalexin response in higher plants. *Naturwissenschaften* **68,** 447–457.
131. Sharp, J.K., McNeil, M. and Albersheim, P. (1984). The primary structures of one elicitor-active and seven elicitor-inactive hexa(β-D-glucopyranosyl)-D-glucitols isolated from the mycelial walls of *Phytophthora megasperma* f.sp. *glycinea. J. Biol. Chem.* **259,** 11321–11336.
132. Sharp, J.K., Valent, B. and Albersheim, P. (1984). Purification and partial characterization of a β-glucan fragment that elicits phytoalexin accumulation in soybean. *J. Biol. Chem.* **259,** 11312–11320.

133. Sharp, J.K., Albersheim, P., Ossowski, P. *et al.* (1984). Comparison of the structures and elicitor activities of a synthetic and a mycelial wall-derived hexa(β-D-glucopyranosyl)-D-glucitol. *J. Biol. Chem.* **259,** 11341–11345.
134. Schmidt, W.E. and Ebel, J. (1987). Specific binding of a fungal glucan phytoalexin elicitor to membrane fractions from soybean *Glycine max. Proc. Nat. Acad. Sci. USA* **84,** 4117–4121.
135. Cosio, E.G., Popperl, H., Schmidt, W.E. and Ebel, J. (1988). High-affinity binding of fungal β-glucan fragments to soybean (*Glycine max* L.) microsomal fractions and protoplasts. *Eur. J. Biochem.* **175,** 309–315.
136. Cheong, J.J. and Hahn, M. (1991). A specific, high-affinity binding site for the hepta-β-glucoside elicitor exists in soybean membranes. *The Plant Cell* **3,** 137–147.
137. Heath, M.C. (1987). Evolution of plant resistance and susceptibility to fungal invaders. *Can. J. Plant Pathol.* **9,** 389–397.
138. Flor, H.H. (1942). Inheritance of pathogenicity in *Melampsora lini. Phytopathology* **32,** 653–669.
139. Garber, R.C. (1991). Molecular approaches to the analysis of pathogenicity genes from fungi causing plant disease. In *The Fungal Spore and Disease Initiation in Plants and Animals*, Cole, G.T. and Hoch, H.C. (eds). Plenum Press, New York, pp. 483–502.
140. Staskawicz, B.J., Dahlbeck, D. and Keen, N.T. (1984). Cloned avirulence gene of *Pseudomonas syringae* pv. *glycinea* determines race-specific incompatibility on *Glycine max* (L.) Merr. *Proc. Nat. Acad. Sci. USA* **81,** 6024–6028.
141. De Wit, P.J.G.M. and Spikman, G. (1982). Evidence for the occurrence of race and cultivar-specific elicitors of necrosis in intercellular fluids of compatible interactions of *Cladosporium fulvum* and tomato. *Physiol. Plant Pathol.* **21,** 1–11.
142. Scholtens-Toma, I.M.J. and De Wit, P.J.G.M. (1988). Purification and primary structure of a necrosis inducing peptide from apoplastic fluids of tomato infected with *Cladosporium fulvum* (syn. *Fulvia fulva*). *Physiol. Molec. Plant Pathol.* **33,** 59–67.
143. Van Kan, J.A.L., Van den Ackerveken, G.F.J.M. and De Wit, P.J.G.M. (1991). Cloning and characterization of the avirulence gene *Avr9* of the fungal pathogen *Cladosporium fulvum*, causal agent of tomato leaf mould. *Molec. Plant–Microbe Interact.* **4,** 52–59.
144. Keen, N.T., Tamaki, S., Kobayashi, D. *et al.* (1990). Bacteria expressing avirulence gene *D* produce a specific elicitor of the soybean hypersensitive reaction. *Molec. Plant-Microbe Interact.* **3,** 112–121.
145. Keen, N.T., Kobayashi, D., Tamaki, S. *et al.* (1991). Avirulence gene *D* from *Pseudomonas syringae* pv. *tomato* and its interaction with resistance gene *Rpg4* in soybean. In *Advanced Molecular Genetics of Plant–Microbe Interactions*, Hennecke, H. and Verma, D.P.S. (eds). Kluwer, Dordrecht, pp. 37–44.
146. Lamb, C.J., Lawton, M.A., Dron, M. and Dixon, R.A. (1989). Signals and transduction mechanisms for activation of plant defences against microbial attack. *Cell* **56,** 215–224.
147. Boller, T. (1989). Primary signals and secondary messengers in the reaction of plants to pathogens. In *Second Messengers in Plant Growth and Development*, Boss, W.F. and Morré, D.J. (eds). Liss, New York, pp. 227–255.
148. Evans, R.M. (1988). The steroid and thyroid hormone receptor superfamily. *Science* **240,** 889–895.
149. Rana, R.S. and Hokin, L.E. (1990). Role of phosphoinositides in transmembrane signalling. *Physiol. Rev.* **70,** 115–168.
150. Streb, H., Irvine, R.F., Berridge, M.J. and Schulz, I. (1983). Release of Ca^{++} from a non-mitochondrial intracellular store in pancreatic acinar cells by inositol-1,4,5-triphosphate. *Nature* **306,** 67–69.
151. Kauss, H. (1987). Some aspects of calcium-dependent regulation in plant metabolism. *Ann. Rev. Plant Physiol.* **38,** 47–72.
152. Marmé, D. (1989). The role of calcium and calmodulin in signal transduction. In *Second Messengers in Plant Growth and Development*, Boss, W.F. and Morré, D.J. (eds). Liss, New York, pp. 57–80.
153. Drobak, B.K. and Ferguson, I.B. (1985). Release of Ca^{++} from plant hypocotyl microsomes by inositol-1,4,5-triphosphate. *Biochem. Biophys. Res. Commun.* **130,** 1241–1246.
154. Strasser, H.C., Hofmann, C., Griesebach, H. and Matern, U. (1986). Are polyphosphoinositides involved in signal transduction of elicitor-induced phytoalexin synthesis in cultured plant cells? *Z. Naturforsch. (C)* **41,** 717–724.
155. Stäb, M.R. and Ebel, J. (1987). Effects of Ca^{++} on phytoalexin induction by a fungal elicitor in soybean cells. *Arch. Biochem. Biophys.* **257,** 416–423.
156. Scheel, D., Colling, C., Hedrich, R. *et al.* (1991). Signals in plant defence gene activation. In *Advanced Molecular Genetics of Plant-Microbe Interactions*, Hennecke, H. and Verma, D.P.S. (eds). Kluwer, Dordrecht, pp. 373–380.
157. Köhle, H., Jeblick, W., Poten, F. *et al.* (1985). Chitosan-elicited callose synthesis in soybean

cells as a Ca^{++} dependent process. *Plant Physiol.* **77,** 544–551.

158. Rosenberg, N., Pines, M and Sela, I. (1982). Adenosine 3′,5′-cyclic monophosphate: its release in a higher plant by an exogenous stimulus as detected by radioimmunoassay. *FEBS Letters* **137,** 105–107.
159. Hahn, M.G. and Griesebach, H. (1983). Cyclic AMP is not involved as a second messenger in the response of soybean to infection by *Phytophthora megasperma* f.sp. *glycinea. Z. Naturforsch.* **38C,** 578–582.
160. Kurosaki, F., Tsurusawa, Y. and Nishi, A. (1987). The elicitation of phytoalexins by Ca^{++} and cyclic AMP in carrot cells. *Phytochemistry* **26,** 1919–1923.
161. Gianninazzi, S. and Ahl, P. (1983). The genetic and molecular basis of b-proteins in the genus *Nicotiana. Netherlands J. Plant Pathol.* **89,** 275–281.
162. Guedes, M.E.M., Richmond, S. and Kuc, J. (1980). Induced systemic resistance to anthracnose in cucumber as influenced by the location of the inducer inoculation with *Colletotrichum lagenarium* and the onset of flowering. *Physiol. Plant Pathol.* **17,** 229–233.
163. Métraux, J.P., Signer, H., Ryals, J. *et al.* (1990). Increase in salicylic acid at the onset of systemic acquired resistance in cucumber. *Science* **250,** 1004–1006.
164. Rasmussen, J.B., Hammerschmidt, R. and Zook, M.N. (1991). Systemic induction of salicylic acid accumulation in cucumber after inoculation with *Pseudomonas syringae* pv. *syringae. Plant Physiol.* **97,** 1342–1347.
165. Malamy, J., Carr, J.P., Klessig, D.F. and Raskin, I. (1990). Salicylic acid: a likely endogenous signal in the resistance response of tobacco to viral infection. *Science* **250,** 1002–1004.
166. Mills, P.R. and Woods, R.K.S. (1984). The effects of polyacrylic acid, acetylsalicylic acid, and salicylic acid on resistance of cucumber to *Colletotrichum lagenarium. Phytopathol. Z.* **111,** 209–216.
167. Yalpani, N., Silverman, P., Wilson, T.M.A. *et al.* (1991). Salicylic acid is a systemic signal and an inducer of pathogenesis-related proteins in virus-infected tobacco. *The Plant Cell* **3,** 809–818.
168. Green, T.R. and Ryan, C.A. (1972). Wound-induced proteinase inhibitor in plant leaves: a possible defence mechanism against insects. *Science* **175,** 776–777.
169. Bishop, P.D., Pearce, G., Bryant, J.E. and Ryan, C.A. (1984). Isolation and characterization of the proteinase inhibitor-inducing factor from tomato leaves. *J. Biol. Chem.* **259,** 13172–13177.
170. Baydoun, E.A.H. and Fry, S.C. (1985). The immobility of pectic substances in injured tomato leaves and its bearing on the identity of the wound hormone. *Planta* **165,** 269–276.
171. Pearce, G., Strydom, D., Johnson, S. and Ryan, C.A. (1991). A polypeptide from tomato leaves induces wound-inducible proteinase inhibitor proteins. *Science* **253,** 895–898.
172. Métraux, J.P., Ahl-Goy, P., Staub, T. *et al.* (1991). Induced systemic resistance in cucumber in response to 2,6-dichloro-isonicotinic acid and pathogens. In *Advanced Molecular Genetics of Plant-Microbe Interactions*, Hennecke, H. and Verma, D.P.S. (eds). Kluwer, Dordrecht, pp. 432–439.
173. Mauch, F., Mauch-Mani, B. and Boller, T. (1988). Antifungal hydrolases in pea tissue. II. Inhibition of fungal growth by combinations of chitinase and β-1, 3-glucanase. *Plant Physiol.* **88,** 936–942.
174. Van den Ackervecken, G.F.J.M., Van Kan, J.A.L. and De Wit, P.J.G.M. (1992). Molecular analysis of the avirulence gene *Avr9* of the fungal tomato pathogen *Cladosporium fulvum* fully supports the gene-for-gene hypothesis. *Plant J.* **2,** 359–366.

2

INVERTEBRATES

D.A. Millar and N.A. Ratcliffe

Biomedical and Physiological Research Group, School of Biological Sciences, University College of Swansea, UK

2.1 INTRODUCTION

Considering that invertebrates make up approximately 96% of all animal species, it is surprising that only a very small proportion of immunologists have focused their attention on this group. Investigations into the defence strategies of invertebrates are important for several reasons. First, it is vital to understand the mechanisms by which parasites are able to survive and multiply in their invertebrate hosts without invoking an immune response. This is especially true in the case of vector species which carry organisms responsible for such terrible diseases as trypanosomiasis, malaria, onchocerciasis and schistosomiasis. Second, an understanding of the interaction between pathogens and invertebrate immune systems would be invaluable in assessing the value of microbes in the biological control of invertebrate pests which can cause widespread destruction of agricultural products. Third, a detailed knowledge of invertebrate immunity may assist in preventing the outcome of disease in those molluscs and crustaceans which are commercially farmed, often in overcrowded and stressful conditions. Fourth, it may be possible to monitor environmental pollution by carefully assessing changes in invertebrate immune function. Fifth, by examining the immune responses of invertebrates, we may gain greater insight into the functioning and evolution of the vertebrate immune system. Lastly, isolated and purified invertebrate immunoreactive molecules may provide novel chemotherapeutic agents for use in man[1].

The success of invertebrate animals and their ability to colonize every ecological niche must, in part, be due to the presence of an effective immune system capable of combating infection. Certainly, some environments such as the sea are teeming with potentially pathogenic microorganisms which, in the absence of an effective defence system, could threaten large numbers of invertebrate species.

As would be expected from a large assemblage of diverse species, the defence mechanisms

Immunology: A Comparative Approach. Edited by R.J. Turner

employed by invertebrates are varied and, for ease of discussion, will be divided into three main categories: the largely external barriers, and internal cellular and humoral components (Table 2.1). The external barriers, as described below, act as first lines of defence to prevent the entry of infective agents into the body. The division of internal components of immunity into cellular and humoral is somewhat arbitrary for, in many instances, the two components interact: e.g. humoral factors acting as recognition molecules to facilitate the phagocytosis of foreign material by blood cells. Furthermore, many humoral molecules such as agglutinins, lysins and antimicrobial factors, may originally have been synthesized and secreted by blood cells.

Table 2.1. Components of invertebrate immune systems (see text for details).

Defence reaction	Role in immunity
Physico-chemical barriers to invasion e.g.:	
Arthropod exoskeleton	Impenetrable barrier sometimes associated with antimicrobial factors (e.g. fungitoxic quinones), protease inhibitors and glycoprotein exudates
Test	Barrier with more flexibility than exoskeleton
Mucus	Entraps microorganisms which may be removed by muco-ciliary mechanism. Sometimes contains lectins, protease inhibitors and antibacterial factors
Mainly cellular:	
Clotting/coagulation	Prevent fluid loss at site of injury
Wound healing	Seals wounds and prevents fluid loss and intrusion of infectious agents. Cytokines may be involved
Phagocytosis	Ingests small numbers of invading microorganisms. Chemotactic and recognition factors (lectins, components of activated prophenol-oxidase?) involved
Nodule formation	Augments phagocytosis if high numbers of invasive microorganisms present
Encapsulation response	Deals with parasites, etc. too large to be phagocytosed. In a few insect species, capsules are acellular (humoral encapsulation) and blood cells may not participate
Cytotoxic reaction	Destroys encapsulated parasites and is involved in non-fusion reactions between incompatible strains of colonial animals (sponges, coelenterates, tunicates)
Transplant rejection	Prevention of fusion between incompatible colonial forms
Mainly humoral:	
Lysins	Naturally-occurring and antimicrobial in serum of many invertebrate groups. E.g. lysozyme
Agglutinins (lectins)	Self/non-self recognition, killing and development of insect vector parasites, protease inhibition, aids lysis of foreign cells, activator of insect haemocyte coagulation and prophenoloxidase
Antimicrobial factors	Kill or neutralize invading microorganisms. Naturally-occurring or induced
Other inducible factors	Antibody-like molecules in insects, earthworms and echinoderms. Function(s) uncertain
Complement-like factors	In annelids, horseshoe crabs, molluscs, insects and echinoderms. Opsonic (?) antimicrobial (?)
Components of activated prophenoloxidase system	Opsonization, killing, cell movement, and cell-degranulation

2.2 PHYSICAL AND CHEMICAL BARRIERS

The external barriers to infection have largely been taken for granted. Nevertheless, they are important in preventing damage to the underlying tissues which would inevitably lead to fluid loss and the entry of pathogenic microorganisms.

The main physical barriers are the tough exoskeleton of the arthropods (including, in insects, the cuticular lining of the foregut and hindgut and the peritrophic membrane of the midgut), the test or tunic of ascidians (sea squirts), the test of echinoderms, the shells of molluscs, and mucus which covers the soft bodies of coelenterates, annelids, molluscs and protochordates.

The arthropod exoskeleton offers a very hard, impenetrable barrier which protects the animal from both physical damage and attack by pathogens. The animals are, however, rather vulnerable during moulting prior to the hardening of the new cuticle. Although the exoskeleton itself constitutes a mechanical barrier, additional antimicrobial factors such as fungitoxic quinones produced during melanin synthesis may also be important[2]. In addition, protease inhibitors such as that found in the cuticle of the crayfish, *Astacus astacus*, may act by inhibiting parasitic invasion[3]. The horseshoe crab, *Limulus polyphemus*, apart from having a tough carapace, also secretes a glycoprotein through canals in the carapace in response to injection of bacterial endotoxin or exposure to fouled sea water[4]. Because of its viscosity the exudate acts as a mechanical barrier and it immobilizes bacteria by binding their flagella. In addition, the glycoprotein possesses erythrocyte- and algae-agglutinating properties.

The ascidian test, which is composed of cellulose, protein and organic compounds, and which varies in consistency from gelatinous to fibrous, offers more flexibility for the animal than the arthropod exoskeleton[5].

There are no problems of restricted movement for invertebrates enclosed within a mucus layer. Invertebrate mucus has been widely studied from many different aspects. Many invertebrates such as coelenterates, annelids, molluscs, echinoderms and protochordates secrete mucus which functions in diverse ways such as in navigation, defence against large predators, resistance to desiccation, structural support, feeding, locomotion[6] and as a possible aid in plasma clotting in earthworms[7]. Invertebrate mucus is also an effective first line of defence against infection. It is admirably suited to carry out this function for two reasons. First, because of its sticky, gel-like nature, it can entrap bacteria and other microorganisms and prevent them from penetrating into the underlying tissue. Furthermore, the coordinated beat of cilia on the skin, below the mucus layer, can effectively clear entrapped particles. Second, soluble molecules such as lectins, lysins, protease inhibitors and antibacterial factors secreted into the mucus may play an important role in defence by killing or otherwise rendering harmless the trapped microorganisms[8–12].

2.3 INTERNAL CELLULAR DEFENCES

Invertebrate host defence mechanisms involving blood cells include wound repair and coagulation, phagocytosis, nodule formation and encapsulation, cytotoxicity and transplant rejection[13] (Table 2.1). Apart from the latter, all these events occur naturally in response to injury or to infection by microorganisms or metazoan parasites. Tissue graft rejection, however, is an artificially induced situation used primarily to test for immunological specificity and memory. Related, naturally-occurring events are the non-fusion reactions observed following contact between incompatible colonial sponges, coelenterates or ascidians.

2.3.1 Invertebrate Blood Cells

A great deal of information has now accumulated on the structure and function of invertebrate blood cells[14], although the classification of the blood cell types in some groups, particularly the insects, remains controversial[15]. In many cases, attempts at blood cell classification have been frustrated by the variety of morphological types present throughout the different invertebrate phyla, and even within individual groups. Some cell types with different names may simply be a single cell type at different developmental stages. Attempts have been made to simplify the situation by classifying all invertebrate blood cells into five main groups, namely progenitor, phagocytic, haemostatic, nutritive and pigmented cells, based upon functional rather than morphological criteria[16] (Table 2.2). This classification scheme applies to both haemocytes and coelomocytes, which are the blood cells found, respectively, in animals whose body cavities are haemocoelic or coelomic in origin.

Progenitor cells are small in diameter (4–10 μm) and with their high nuclear:cytoplasmic ratio, superficially resemble vertebrate lymphocytes (Figures 2.1 and 2.2). They are the stem cells from which other blood cells are derived. Phagocytic cells occur in members from every

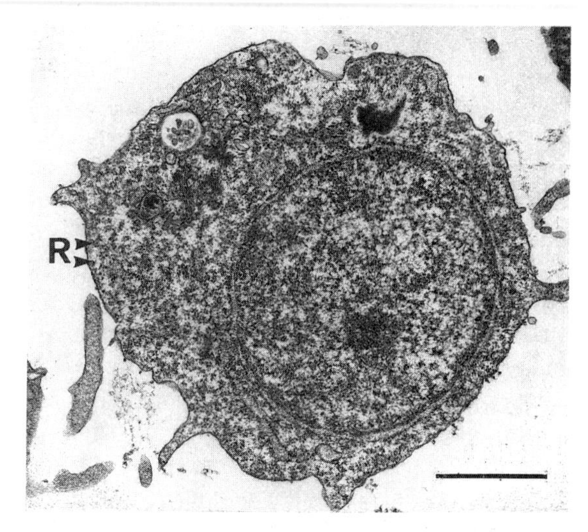

Figure 2.1. Electron micrograph of a progenitor cell from the tunicate, *Ciona intestinalis*, showing the undifferentiated cytoplasm containing numerous free ribosomes (R). Scale bar = 1 μm. From reference 13 with permission.

Table 2.2. Cells and tissues of the invertebrate immune system.

Cells/tissues	Role in immunity
Free and sessile haemocytes/coelomocytes	Cellular responses but may secrete humoral factors (e.g. agglutinins, antimicrobial factors, lysins, components of prophenoloxidase system)
Progenitor cells	Stem cells for other cell types
Phagocytic cells	Phagocytosis, nodule formation, encapsulation, wound closure, clotting and killing
Haemostatic cells	Blood coagulation, non-self recognition, lysozyme and agglutinin production
Nutritive cells	Role in defence uncertain. Encapsulation and wound healing?
Pigmented cells	Respiratory function? Antibacterial or anti-algal?
Permanently fixed cells. E.g. chloragogen cells of some annelids, reticular and pore cells of molluscs, reserve cells of crustaceans, nephrocytes of crustaceans, insects and tunicates, mid-gut and sinus-lining cells of gastropod molluscs and crustaceans, fat body of insects	Phagocytosis, pinocytosis of colloids and particulate material. Lysozyme synthesis (pericardial cells of insects) and antibacterial factors. Clearance of foreign particles (sinus-lining cells). Synthesis of immune proteins (fat body)
Phagocytic organs of insects, echinoderms and urochordates	Phagocytosis. Production of humoral antibacterial factors (?)
Haemopoietic organs.	Haemopoiesis, phagocytosis, synthesis of antimicrobial factors in some species

Based on Table I of Reference 17 with permission.

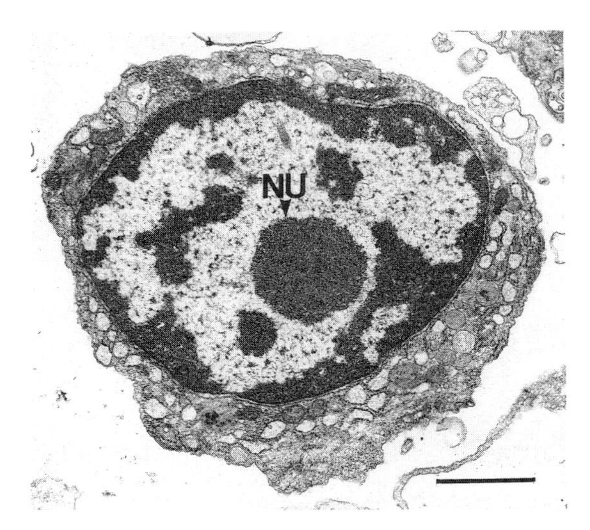

Figure 2.2. Electron micrograph of a progenitor cell from the echinoderm, *Cucumaria normani*, with characteristic prominent nucleolus (NU), high nuclear:cytoplasmic ratio and thin rim of cytoplasm. Scale bar = 1 μm. From reference 18 with permission.

invertebrate phylum and may be hyaline or granular in appearance (Figures 2.3 and 2.4). They have a very important role to play in internal defence including phagocytosis, wound healing and clotting reactions (see below).

Haemostatic cells (Figures 2.5 and 2.6) function in sealing wounds and preventing fluid loss. In most invertebrates, this process is performed simply by plugging the wound by amoebocytic, often phagocytic cells, although in arthropods this event is accompanied by plasma gelation (Figure 2.6), which occurs following degranulation of the haemostatic cells at the site of injury. Usually, only a small percentage of blood cells are nutritive although they may in some cases comprise up to 70% of the blood cell population. The role of blood cells in nutrition is doubtful and often based upon the presence of carbohydrate, protein and lipid within the various cell types[13]. Pigmented cells are associated with members of various invertebrate phyla such as sipunculids, echiuroids, chelicerates, crustaceans, lophophorates and echinoderms[14]. Pigmented cells, which

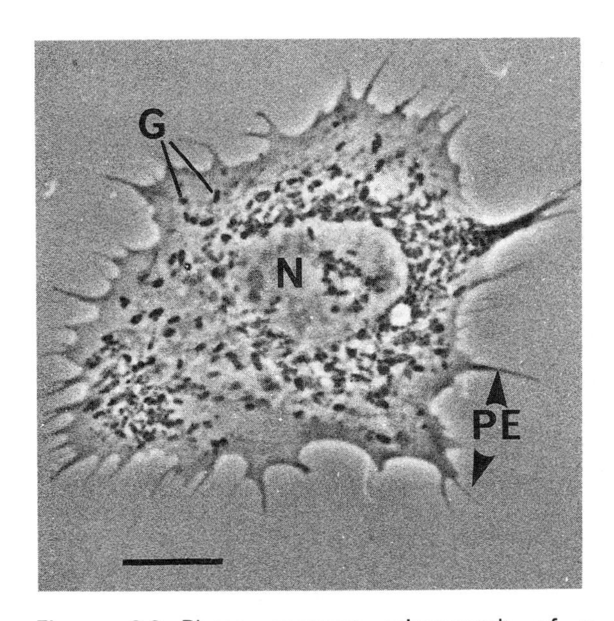

Figure 2.3. Phase contrast micrograph of a phagocytic cell from the stick insect, *Clitumnus extradentatus*, after 20 min *in vitro*. Cell contains numerous granules (G) and has formed spike-like protoplasmic extensions (PE). Nucleus (N). Scale bar = 10 μm. Reproduced by courtesy of Dr A.F. Rowley.

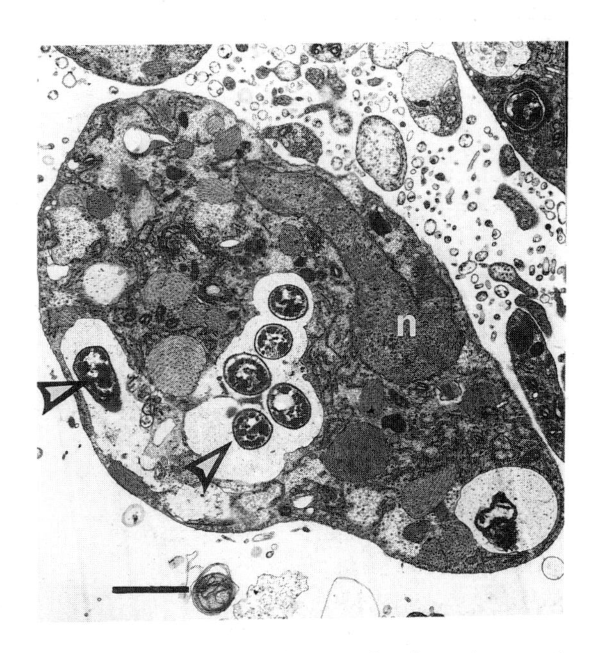

Figure 2.4. Electron micrograph of a phagocytic coelomocyte from the hemichordate, *Saccoglossus ruber*, showing phagocytosed bacteria (arrowheads). Nucleus (n). Scale bar = 2 μm.

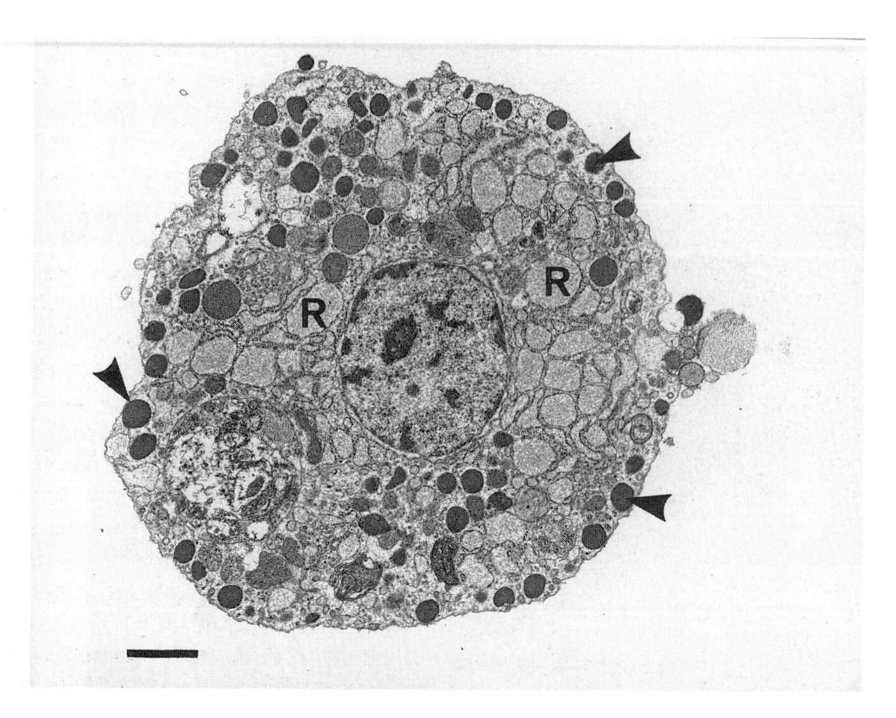

Figure 2.5. Electron micrograph of a haemostatic cell from the wax moth, *Galleria mellonella*, showing dense granules (arrowheads) and swollen rough endoplasmic reticulum (R). Scale bar = 1 μm. Reproduced by courtesy of Dr A.F. Rowley.

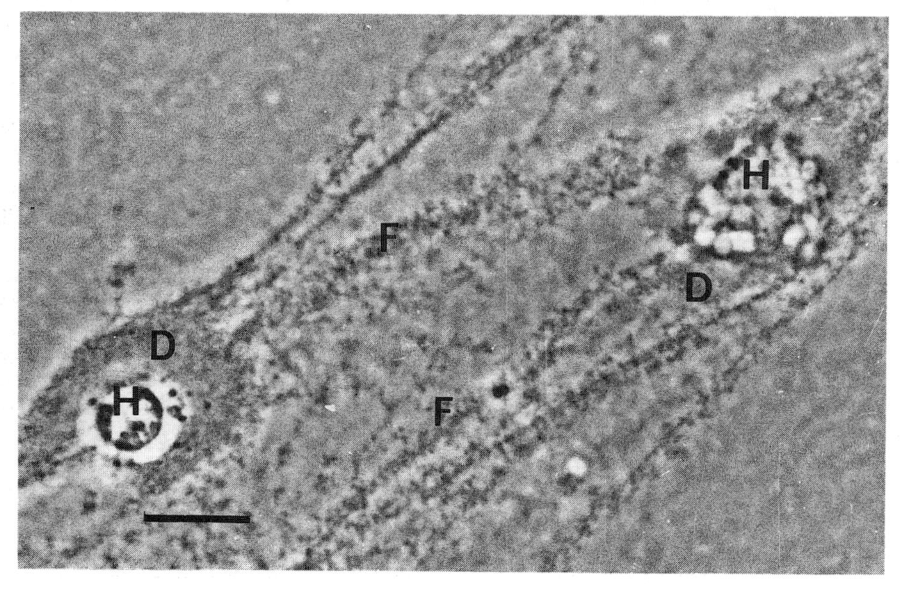

Figure 2.6. Phase contrast micrograph of two haemostatic cells (H) from the stick insect, *Clitumnus extradentatus*, after 11 min *in vitro*. Note densely coagulated haemolymph (D) and fibrous coagulum (F) interconnecting the two cells. Scale bar = 10 μm. From reference 198 by permission of Springer-Verlag.

sometimes contain haemoglobin or haemocyanin and may, thus, have a respiratory function, are not generally involved in defence and are similar functionally to vertebrate erythrocytes. In echinoderms, however, the pigment echinochrome may have antibacterial or anti-algal properties[18].

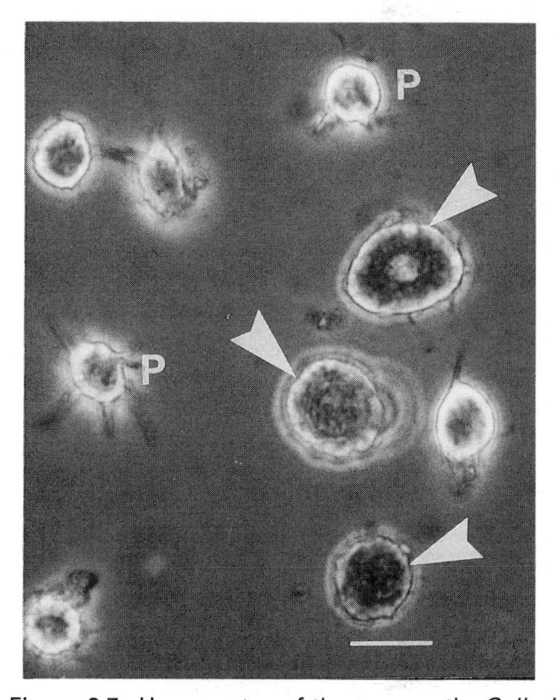

Figure 2.7. Haemocytes of the wax moth, *Galleria mellonella*, showing specific staining of the haemostatic cells (arrowheads) by a mouse anti-haemostatic cell monoclonal antibody. Note the unstained and spreading phagocytic cells (P). Phase contrast optics. Scale bar = 10 μm. From reference 202 with permission.

It is now possible to characterize invertebrate blood cells further in terms of their density, enzyme content, behaviour and affinity for lectins and monoclonal antibodies[19,20] (Figure 2.7). Thus, using a variety of techniques to distinguish physical and biochemical differences between blood cells and to identify subpopulations, it will now be possible to determine more accurately the role of the various cell types in immune reactions. For instance, blood cell separation by density gradient centrifugation provides an ideal method by which reasonable numbers of viable, relatively pure populations of blood cells can be obtained for study. Thus, using separated haemocytes in culture, it has been shown that, as in vertebrate immune responses, cell-cell co-operation is necessary in invertebrate cell-mediated immunity[21,22] (Table 2.3).

Great variation also exists in the origin of invertebrate blood cells (Table 2.4) including the progenitor cell type[13,14]. In some groups, such as the annelids, echinoderms, cephalochordates and probably the hemichordates (Millar and Ratcliffe, unpublished observations), the blood cells/coelomocytes arise from the coelomic lining (Figure 2.8). In contrast, distinct haemopoietic centres of varying complexity exist in insects, molluscs, crustaceans and urochordates, and in the latter group the centres are well developed as 'lymph nodules' in the pharyngeal wall and around the gut. In asteroid echinoderms, there is controversy as to whether the Tiedemann's

Table 2.3. Evidence for cell co-operation for *in vitro* phagocytosis by phagocytic cells (plasmatocytes) of the wax moth, *Galleria mellonella*.

Type of monolayer[a]	Cell type counted	No. of bacteria/100 cells after 1 h incubation
Phagocytic cells alone	Phagocytic cells	16.1 ± 1.6
Haemostatic cells alone	Haemostatic cells	0
Mixed	Phagocytic cells	103.9 ± 5.2[b]
	Haemostatic cells	0

From reference 22 with kind permission from Elsevier Science Ltd, The Boulevard, Langford Lane, Kidlington OX5 1GB, UK.
[a] The cell types were separated from whole haemolymph on a continuous Percoll gradient.
[b] Note the great increase in the number of bacteria taken up by the phagocytic cells when mixed with a population of non-phagocytic haemostatic cells.

Table 2.4. Haemopoietic/lymphoid tissues in selected coelomate invertebrates.

Group	Cell types	Haemopoietic site(s)
Annelids	Coelomocytes Chloragogen cells (eleocytes) Haemocytes	Coelomic lining cells Chloragogen tissues around intestine 'Blood glands' in haemocoelic system From coelomic lining (?)
Molluscs	Haemocytes	Varies, some gastropods have localized haemopoietic sites, although none reported for bivalves Specialized haemopoietic tissue, the white bodies, exists behind the eyes in cephalopods
Insects	Haemocytes	Varies, either in distinct haemopoietic areas or by division of freely circulating cells
Crustaceans	Haemocytes	Foci in various sites, including base of rostrum, dorsal and lateral walls of foregut and vicinity of ophthalmic artery
Echinoderms	Coelomocytes	Coelomic lining cells, axial organ, Tiedemann's bodies, polian vesicles
Hemichordates	Coelomocytes	Coelomic lining cells
Cephalochordates	Coelomocytes	Coelomic lining cells
Urochordates	Haemocytes	'Lymph nodules' in pharyngeal wall, around digestive tract and in the body wall

From reference 13 with permission.

bodies, polian vesicles or axial organs are sites of coelomocyte production or simply regions where coelomocytes can accumulate and/or mature[13] (Table 2.4). Generally, maturation of invertebrate haemocytes or coelomocytes, possibly from progenitor cells, takes place within the haemolymph or coelomic fluid, respectively.

Finally, as well as the freely circulating blood cells, there are also various fixed cells and organs which are important in clearing non-self material from the circulation[13] (Table 2.2). These include the sinus-lining cells of molluscs (Figure 2.9) and crustaceans, the reticular cells and pore cells of molluscs, reserve cells, branchial podocytes (nephrocytes) and other fixed cells of crustaceans, nephrocytes (ventral nephrocytes, garland cells, diaphragm cells, pericardial cells) of insects and nephrocytes of tunicates. In addition, phagocytic organs have been described in insects, echinoderms and urochordates.

2.3.2 Wound Repair and Coagulation

It is essential that following wounding or damage to the body wall or gut lining, effective measures occur to prevent excessive loss of fluid and the intrusion of possibly infectious microorganisms. The prevention of blood loss comes under the general heading of haemostasis, and the reactions at the site of injury, involving vasoconstriction, blood cell accumulation and haemolymph coagulation, constitute an inflammatory response. Because of the relatively low pressures of the haemolymph and coelomic fluid in most invertebrates, vasoconstriction is generally not of immediate importance, although in some soft-bodied invertebrates such as annelids, some molluscs and some echinoderms, haemostatic responses are accompanied by muscle contraction which reduces the size of the wound. In insects, wound closure is assisted by extrusion of the fat body,

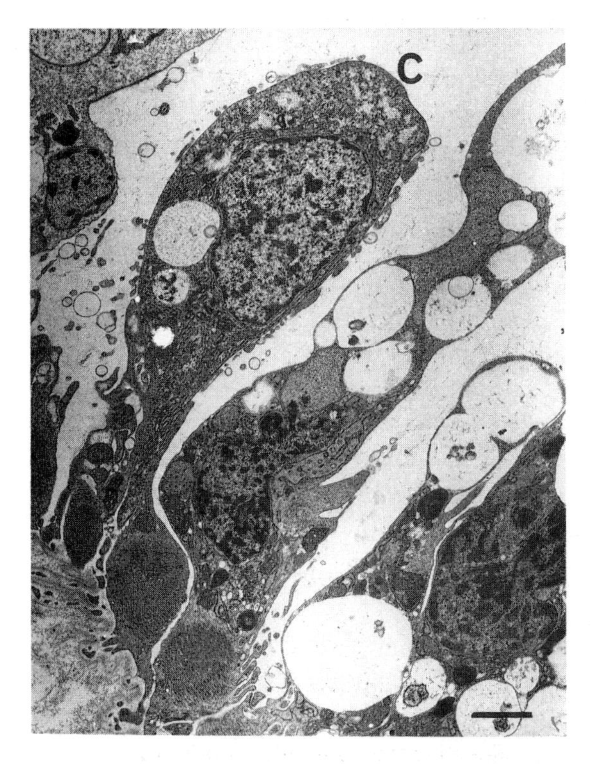

Figure 2.8. Electron micrograph of a coelomocyte (C) of the hemichordate, *Saccoglossus ruber*, appearing to bud from the coelomic lining of the trunk. Scale bar = 2 μm.

which together with the haemolymph clot are further strengthened by the deposition of melanin[23] (Figure 2.10). In molluscs, the cellular wound plug is stabilized with collagen[24].

In all invertebrates, including the lowly sponges, wound healing is initiated by the migration of haemocytes or coelomocytes to the site of injury where they aggregate to form a wound plug which prevents further blood loss and traps infecting microorganisms[13,14]. The migration of these blood cells is probably under the influence of chemotactic factors, released either from the blood cells themselves or from the damaged tissue, and which may include the 'wound factor' of insects[25,26] and the sea star factor of *Asterias forbesi*[27].

Although it has been suggested that plasma clotting may occur in earthworms[7], it is only in arthropods that cellular aggregation followed by true plasma coagulation (Figure 2.6) to reinforce the cellular plug has been observed[28]. Coagulation in arthropods is a complex series of events during which conversion of the clotting proteins (coagulogen) into an insoluble gel (coagulin) by an enzyme cascade occurs. It probably involves three or more serine proteases and is regulated by at least one protease inhibitor. For

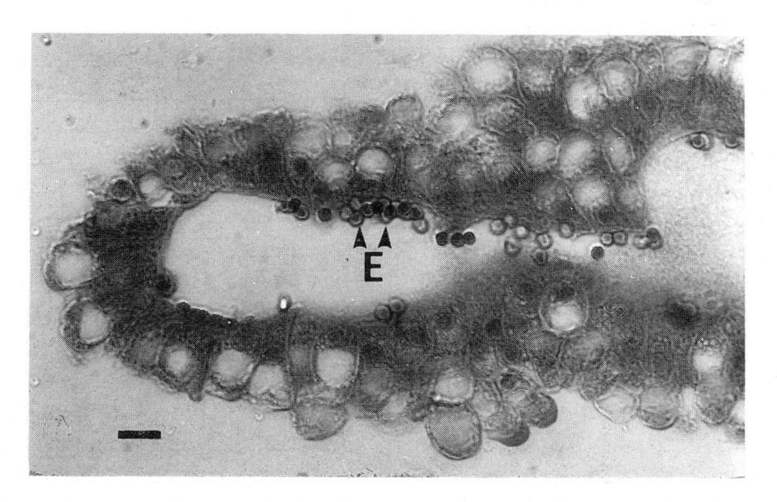

Figure 2.9. Light micrograph of sinus-lining cells of the snail, *Helix pomatia*, which have entrapped injected erythrocytes (E). Scale bar = 20 μm. Reproduced by courtesy of Professor L. Renwrantz from reference 13 with permission.

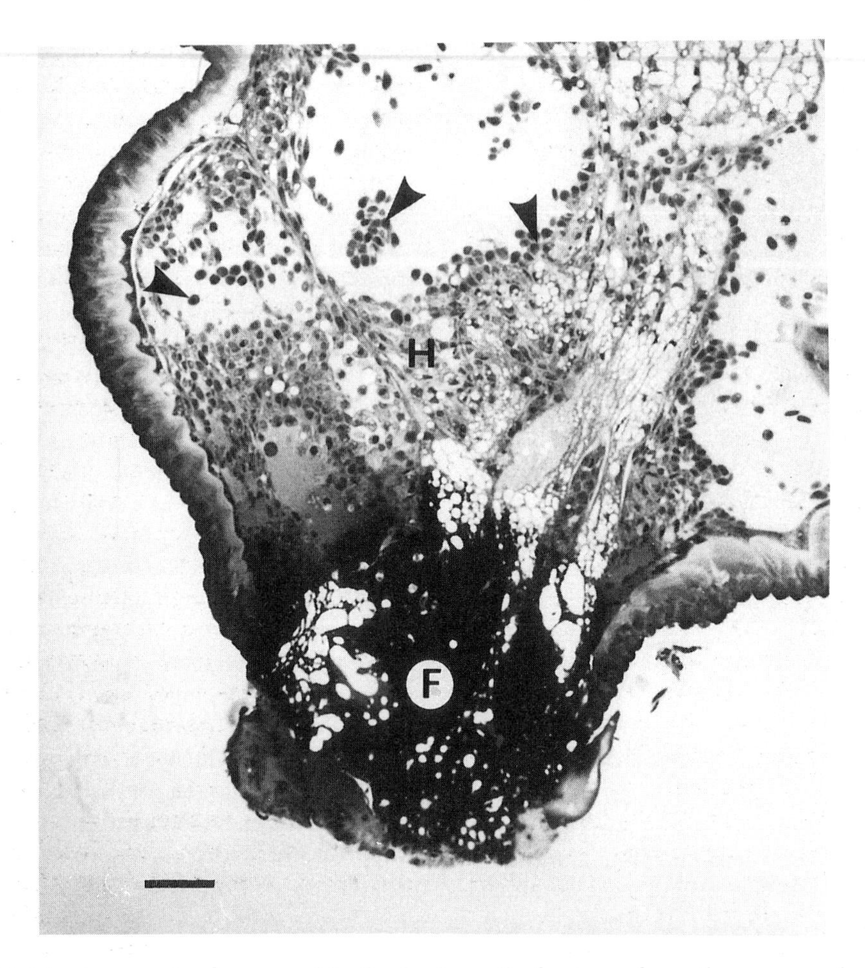

Figure 2.10. Light micrograph of a wound in the body wall of a larva from the wax moth, *Galleria mellonella*, 6 h after injury. Note extruded and melanized fat body (F), and haemolymph (H) sealing the wound and forming an amorphous scab. Numerous blood cells (arrowheads) are also present. Scale bar = 50 μm. From reference 199 by permission of the Wistar Institute Press.

details of the processes involved in coagulation by arthropods, the reader is directed to several reviews[13,28,29].

In both crustaceans and insects, coagulation is accompanied by melanization (Figure 2.10). Melanin and its precursors may be toxic and thus kill invading microorganisms[30]. Furthermore, the components of the prophenoloxidase-activating system which are involved in melanin synthesis, and are thought to be opsonic (see Section 2.5.2) possibly attract further blood cells to the injury site and enhance phagocytosis and wound repair.

2.3.3 Phagocytosis

Phagocytosis is ubiquitous within the animal kingdom, occurring in members of all invertebrate phyla and the vertebrates. In invertebrates, it may be regarded as the primary cellular response against invading microorganisms, with larger pathogens and parasites being enclosed within multicellular nodules or capsules. The main stages are recognition (attraction or chemotaxis and attachment), ingestion and killing. These processes have been variably studied in different invertebrate species, with most information coming from the molluscs and insects. Our present knowledge is thus far from complete.

Unfortunately, studies on chemotaxis in invertebrates are few. However, coelomocytes from the earthworm, *Lumbricus terrestris*, were shown to migrate towards foreign tissue in response to a chemoattractant which was heat-labile and of M_r below 10kDa[31]. Similarly, haemocytes from the oyster, *Crassostrea virginica*, were variably attracted towards live bacteria and the attractant was shown to be a chemical found in the bacterial cell walls and cell envelopes[32]. In insects and crustaceans, it has been suggested that components of the activated prophenoloxidase system may act as chemotactic foci[33,34], although this has yet to be substantiated. In insects, however, there is some experimental evidence that factors released during prophenoloxidase activation may have an effect on chemokinetic behaviour[35].

Following contact of the foreign particles with the phagocytic cell, the next step in the sequence of events is attachment. As with vertebrate phagocytes, attachment depends upon the surface properties of both the particle and phagocyte. Recognition and attachment may be mediated, in part, in a non-specific way by physicochemical properties of the foreign material such as surface charge and hydrophobicity[20]. Normally, however, recognition and attachment are specific and rely on the interaction of the foreign particle either directly with receptors on the phagocyte membrane, or indirectly via bridging molecules. Humoral factors which facilitate or enhance phagocytosis are generally referred to as opsonins, which, in vertebrates, are mainly immunoglobulins (usually IgG) and the third component of complement. There is no real proof that invertebrates possess antibody, although evidence is accumulating which strongly suggests that components of the complement system may exist in these animals[36,37] (see also Sections 2.4.4 and 2.5), and receptors for C3 have been described on coelomocytes of an echinoderm[38] and an annelid[39].

In vertebrates, specific recognition may also take place in the absence of antibody and complement through carbohydrate-binding proteins called lectins[40]. Lectins are also ubiquitous within body fluids and tissues of invertebrates and recent evidence has suggested that they, too, may act as recognition molecules[41–44]. Furthermore, in arthropods, a role for components of the activated prophenoloxidase (PpO) system has also been implicated in recognition and attachment[45,46]. The role of invertebrate lectins and the PpO system as self recognition molecules is discussed in more detail in Section 2.5.

From a phylogenetic point of view, it has been shown that substances serologically related to human fibronectin, a cell adhesion protein with opsonic properties, have been found in invertebrate tissues, including snail haemocytes[19]. It is possible that these invertebrate proteins may have similar adhesion functions and, as such, function as opsonins. Of relevance here is the recent isolation and purification of a 76kDa cell adhesion factor from the granular and semigranular haemocytes of the crayfish, *Pacifastacus leniusculus*[47]. This factor, associated with the PpO system, induces degranulation and cell adhesion of the granular haemocytes and contains an amino acid sequence, arg-gly-asp (RGD), which is apparently crucial for the activity of vertebrate adhesion factors, including fibronectin[48]. A more recent report showed that a 90kDa protein from the haemocytes of the insect, *Blaberus craniifer*, not only induces degranulation of the haemocytes and enhances adhesion to the substratum, but also exhibits immunological cross-reactivity with cell adhesion factor[49]. This finding suggests that very similar molecules may function in cell-cell communication in different arthropod groups.

Clearly, further investigations are required to determine the range and nature of factors acting as opsonins in invertebrates. Indeed, opsonic molecules may actually be quite diverse, especially since substances like the mammalian entero-neuropeptides, somatostatin and calcitonin, have been located in distinct haemocyte populations of the pond snail, *Physella heterostropha*, and synthetic somatostatin and calcitonin have been shown to stimulate the *in vitro* attachment of latex beads to the snail haemocytes[50].

Following attachment to the phagocyte surface, the foreign particle is internalized (Figure 2.4). In most cases, this internalization or endocytosis occurs either by invagination of the haemocyte surface or by the production of filopods or pseudopods of various shapes which surround the foreign body, enclosing it in a phagocytic vacuole or phagosome. By the use of microfilament inhibitors such as cytochalasin B, it has been demonstrated that invertebrate phagocytes, like their vertebrate counterparts, require intact microfilaments for endocytosis[51].

In contrast with attachment, ingestion of the foreign particle is an energy-requiring process. From studies with insectan[52] and molluscan[53] haemocytes, it appears that glycolysis provides the energy for phagocytosis. Similar events occur in mammalian leucocytes, but there is also a marked increase in oxygen utilization during phagocytosis arising from the metabolism of glucose via the hexose monophosphate shunt with the production of different reactive oxygen intermediates such as superoxide anion, hydrogen peroxide, hydroxyl radical and singlet oxygen[54]. Until recently, it was thought that phagocytosis by invertebrate cells was not accompanied by increased oxygen utilization, nor affected by potassium cyanide, an inhibitor of mitochondrial function, which suggested that stimulation of the hexose monophosphate shunt or Krebs cycle did not occur. Nevertheless, because of the paucity of information, such statements must now be treated with caution especially since it has been recently shown that superoxide can be produced by the haemocytes of *Biomphalaria glabrata* upon stimulation by the miracidia from *Schistosoma mansoni*[55], that phagocytosing haemocytes of another snail, *Lymnaea stagnalis*, generate reactive forms of oxygen and superoxide[56], and that amoebocytes of the scallop, *Patinopecten yessoensis*, produce hydrogen peroxide *in vitro*[57]. Furthermore, hydrogen peroxide and peroxidase activity have been found in the amoebocytes of several other invertebrate species[58] (Table 2.5, Figure 2.11).

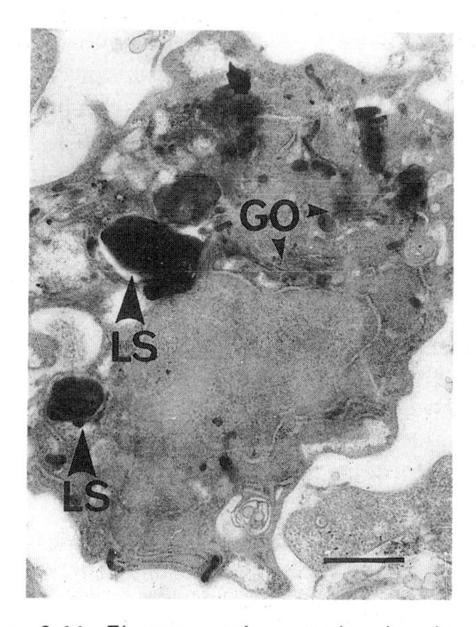

Figure 2.11. Electron micrograph showing the cytochemical localization of peroxidase in lysosomes (LS) and the Golgi complex (GO) of an amoebocyte from the snail, *Lymnaea stagnalis*. Scale bar = 1 μm. Reproduced by courtesy of Professor T. Sminia from reference 13 with permission.

Table 2.5. Possible killing factors present in invertebrate blood cells (see text for details).

Factor	Groups
Low pH	Molluscs, other invertebrate species(?)
Lysosomal enzymes including peroxidase	Most invertebrates
Lysozyme	Annelids, molluscs, insects, echinoderms
Immune proteins e.g. cecropins, attacins	Insects
Toxic oxygen metabolites/hydrogen peroxide	Molluscs, probably other species
Components of the activated prophenoloxidase system	Insects, crustaceans and other invertebrates(?)
Vanadium	Tunicates

Lysozyme, which is present in many invertebrate haemocytes, is also antibacterial. Toxic quinones and melanin, generated during activation of the prophenoloxidase system in crustaceans and insects, may also aid killing[2] (Table 2.5), as may other factors such as vanadium in tunicate blood cells[13] (Figure 2.12).

Generally, following internalization, the phagosome containing the foreign particle fuses with lysosomes to form a phagolysosome. Here, the phagocytosed particle is killed by lysosomal enzymes and other factors mentioned above (Table 2.5). Following killing, lysosomal enzymes such as acid and alkaline phosphatase, lipase, non-specific esterase, proteases, amylase and glucosidase, digest the engulfed foreign particle. It has been shown that the pH inside the digestive vacuoles of *Mytilus edulis* haemocytes falls continually following the phagocytosis of yeast cells which had been previously stained with a range of indicator dyes[59]. After about 25 min, the lowest pH value (4.5), which was required to activate the lysosomal enzymes, was reached.

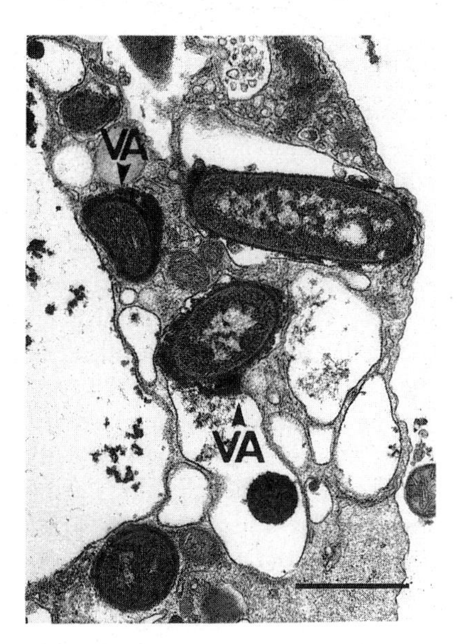

Figure 2.12. Electron micrograph of part of the cytoplasm of a granular phagocytic cell of the tunicate, *Ciona intestinalis*, containing several ingested bacteria. Note that the bacteria are surrounded by a dense halo of vanadium (VA), which may have antimicrobial activity. Scale bar = 1 μm. From reference 13 with permission.

In many cases phagocytosis stimulates the increased production of lysosomal enzymes, some of which may be released into the haemolymph during or after phagocytosis[60], where they may kill and degrade unphagocytosed particles or somehow alter their surface characteristics to make them more susceptible to recognition and ingestion. Breakdown products of bacterial digestion are probably utilized by the animal as a food source and incorporated into various body tissues. Inert, abiotic particles which cannot be degraded are usually transported and released to the exterior via various organ and tissue epithelial layers.

2.3.4 Nodule Formation

In some invertebrates, phagocytosis is unable to deal adequately with large numbers of small non-self particles such as bacteria. Here, phagocytosis is augmented by the sequestration of the particles within cellular aggregates termed nodules (Figure 2.13). The formation of these structures, as in phagocytosis, may involve cell-cell co-operative events[61].

Nodules may be formed as a result of natural infections as in, for example, annelids, insects and molluscs, or experimentally induced by injecting a variety of abiotic particles (e.g. Indian ink, carmine, turpentine) or biotic particles (e.g. bacteria, fungal spores, protozoans) into annelids, molluscs, crustaceans, insects and echinoderms[13,62], but detailed studies are few.

Generally, nodules comprise a central core of entrapped foreign particles, either intracellularly or extracellularly associated with degenerate and necrotic haemocytes, surrounding which are normal and flattened blood cells. In some cases, depending upon the invertebrate species, the central cellular aggregate may also become melanized (Figure 2.13). The attraction of haemocytes towards the non-self particles is probably, as in phagocytosis, in response to chemotactic factors released from the foreign material itself, or from degranulating haemocytes. Where nodules are

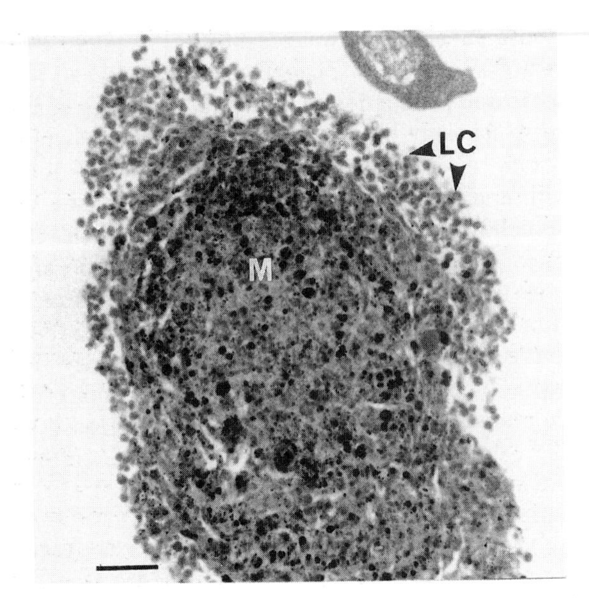

Figure 2.13. A coelomocyte aggregate (nodule) from the polychaete worm, *Arenicola marina*. The central region has formed naturally and is composed of necrotic and melanized cells (M), whilst the outer region of newly attaching cells (LC) has formed in response to bacterial injection. Scale bar = 20 μm. Reproduced by courtesy of Dr S.W. Fitzgerald from reference 200 with permission.

melanized, as in arthropods, it is probable that factors released during activation of the prophenoloxidase system are responsible for facilitating cellular aggregation[63,64]. The speed and extent of nodule formation, the size of the nodules and the fate of the entrapped foreign material, depend upon the concentration and nature of the injected test particles and the animal species involved[13].

The fate of microorganisms within nodules varies considerably. It is possible that bacteria succumb as a result of physiological stress brought about by lack of oxygen or nutrients, or, if trapped within melanized nodules, they are killed by the toxic effect of quinones and melanin generated by phenoloxidase activity[2] or by phenoloxidase itself[65]. Alternatively, proteolytic enzymes, such as described in Section 2.3.3 above, may be bacteriolytic[66]. Cellular aggregates may also be eliminated from the body either during the moult as in crustaceans[67] or, as in some polychaetes, by periodic autotomy of the tail[58]. In some cases, however, pathogens avoid sequestration in nodules or, if entrapped, break out of the nodules to reinfect the animal[68]. The mechanisms by which microorganisms evade cellular entrapment are poorly understood, although factors released from the non-self material itself may be responsible. It has been shown, for example, that destruxins, one of several secondary metabolites of various fungi, are able to inhibit or reduce β-1,3-glucans-induced activation of the haemocytes of cockroaches (*Periplaneta americana*) and locusts (*Schistocerca gregaria*), as well as suppressing β-1,3-glucans-induced nodule formation *in vivo* and *in vitro*[69].

2.3.5 Encapsulation

Foreign bodies such as cestodes, nematodes, insect parasitoids and copepods, too large to be phagocytosed, are often enclosed in multicellular sheaths called capsules (Figure 2.14), closely resembling nodules in structure and formation.

Encapsulation responses have been observed in virtually all invertebrate species and, considering the range and variety of invertebrates in which encapsulation reactions occur, it is not surprising that differences may also be found in capsule structure and the cell types involved. In the majority of cases, however, capsules are multicellular structures often consisting of two or more distinct zones or regions of variably flattened and—especially near the foreign object—necrotic and degenerate cells. Although it is often difficult to discern the cell types forming the capsules, usually more than one cell type is involved. In certain dipteran insects, however, where the blood cell numbers are low (<6000 cells/mm^3), capsules are completely acellular and apparently formed without the visible participation of blood cells, although it has been demonstrated that, in some cases, blood cell activity may indeed be required[70]. This form of encapsulation, termed humoral encapsulation, has been shown to be an adequate defence against nematodes, fungi, bacteria and microfilariae[71]. Indeed, in *Chironomus* larvae, humoral encapsulation is the main defence strategy since only low levels of phagocytic activity are observed, and cellular encapsulation, nodule for-

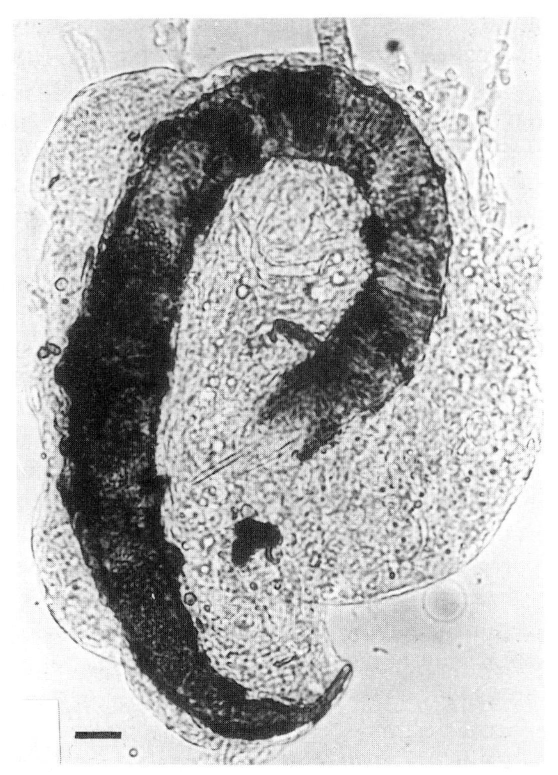

Figure 2.14. Multicellular capsule enclosing a melanized juvenile stage of the nematode, *Filipjevimermis leipsandra*, 72 h after penetration into the body cavity of a larva of the beetle, *Diabrotica* sp. Scale bar = 50 μm. Reproduced by courtesy of Professor George O. Poinar from reference 201 with permission.

mation and antibacterial activity are apparently absent[72].

Cellular encapsulation responses are possibly initiated by random contact of the circulating haemocytes with the foreign body. Subsequent events, including recruitment of other blood cells, however, are specific and probably rely on chemotactic factors[73] as well as components of the activated prophenoloxidase system[74,75]: these may constitute the 'encapsulation promotion factor' released from insect granular cells[76].

Although lectins have been shown to be opsonic in phagocytic responses, their role in encapsulation is unclear. It has, however, been demonstrated that the galactose-specific lectin from the cockroach, *Periplaneta americana*, induced the *in vivo* formation of thick cellular capsules around Sepharose beads which had previously been conjugated with galactose-rich molecules[77].

How parasites are killed within capsules is poorly understood. It is possible they may succumb due to starvation and asphyxiation, but this is not always the case since some parasites die within incomplete capsules[78]. If intact and viable haemocytes are present adjacent to the foreign body, they may infiltrate it and phagocytose the tissue as observed with encapsulated xenografts in the snail *Lymnaea stagnalis*[79]. Furthermore, the phagocytic cells may possess lytic enzymes which facilitate destruction of the foreign tissue[80]. Encapsulated parasites may also be destroyed by toxic molecules such as quinones released as by-products during melanization in arthropods[2] (Figure 2.14) as well as by peroxidase and reactive forms of oxygen present in molluscan haemocytes[56,57,81,82].

Finally, not all parasites are encapsulated. Many will survive and develop, unaffected by the host's immune system (Table 2.6). Occasionally,

Table 2.6. Possible avoidance mechanisms adopted by parasites and pathogens in their invertebrate hosts.

Passive avoidance:
Organs colonized out of reach of blood cells
Low-reactivity regions of host occupied
Immature hosts invaded
Active mechanisms:
Refractory envelopes/cell wall, etc.
Mimicry of host tissues
Acquisition of host antigens
Lysis and abrogation of leucocytes
Depression or inhibition of chemotaxis
Depression or inhibition of phagocytosis
Depression or inhibition of nodule formation/encapsulation
Provision of alternative targets
Utilization of host response
Abrogation of immune proteins
Depression or inhibition of opsonizing properties of haemolymph
Depression or inhibition of prophenoloxidase system
No obvious mechanism:
Host swamped by numbers or rate of growth of parasite/pathogen?

Modified from reference 13 with permission.

some will even survive after encapsulation[83]. How these parasites survive or escape encapsulation is a complex subject and poorly understood. For detailed discussions on this topic the reader is directed to other reviews[13,84,85].

2.3.6 Cytotoxic Reactions

In addition to phagocytosis, nodule formation and encapsulation, invertebrate blood cells are capable of spontaneous cytotoxic reactions towards foreign cells and tissues. This form of killing by non-sensitized cells is similar to so-called natural killing in the mammalian immune system and may possibly be a primordial system of immunosurveillance[86]. Encapsulated metazoan parasites may be destroyed by cytotoxic factors released from haemocytes within the capsule. Furthermore, non-fusion reactions between incompatible strains of colonial animals such as sponges, coelenterates and tunicates, also involve cytotoxic interactions[87]. Indeed, cytotoxicity may be regarded as a 'manifestation of graft rejection at the cellular level'[88]. In animals such as sipunculids, which are difficult to graft, *in vitro* cytotoxic reactions between allogeneic and xenogeneic cells may provide insight into the possible *in vivo* reactions of host cells towards grafted tissue.

Cytotoxic reactions are ubiquitous among invertebrates. In the majority of cases, contact of the effector cell with the target cell is required to initiate a cytotoxic response. Haemocytes from the mussel, *Mytilus edulis*, however, have been shown to secrete cytotoxic molecules quite independently of target cells[89].

Attempts at enhancing cytotoxic reactions by immunization have met with variable results. In sipunculids, a previous exposure to the antigen actually results in tolerance[90]. There have been some reports, however, which suggest that haemolymph factors may be able to transform non-cytotoxic cells into active ones. For example, it is possible to take non-cytotoxic haemocytes from a strain of *Biomphalaria glabrata*, susceptible towards *Schistosoma mansoni* sporocysts, and to transform them into cytotoxic cells simply by incubating them in cell-free plasma from resistant strains of *B. glabrata*[91]. Since cytotoxic responses with resistant strain haemocytes can take place in the absence of plasma, it is likely that the plasma killing factor (a lectin?) is cytophilic. Support for the involvement of lectins in cytotoxic reactions comes from the observation that cytotoxicity of blood cells from annelids, molluscs and echinoderms can be inhibited by a variety of carbohydrates[92].

Little is known about the nature of the cytotoxic factors in invertebrates, although it has been suggested that in sipunculids, a phospholipase may be responsible for target cell lysis[90]. A similar enzyme detected in haemocytes of the Asian clam, *Corbicula fluminea*, has also been implicated in haemocyte-mediated cytotoxicity of rabbit erythrocyte target cells.[93]. The possibility that components of the activated prophenoloxidase system may be cytotoxic has not yet been substantiated[94].

2.3.7 Transplant Rejection

The grafting of tissue onto allogeneic or xenogeneic hosts is an artificial situation performed mainly to determine histoincompatability and the presence of specific immunological memory. However, the ability to recognize and reject foreign tissue probably reflects the early evolution of the immune system to detect and destroy the threatening presence of mutated, neoplastic or virally transformed cells[95]. In invertebrates, there are some circumstances in which graft rejection may be considered natural, such as in the fusion/non-fusion events between adjacent members of compatible/non-compatible members of colonial sponges, coelenterates and urochordates.

In most cases, a histoincompatible union between graft and host results in the ultimate destruction of the graft by the infiltration of various types of cytotoxic and phagocytic blood cells (Figure 2.15). In different invertebrates, the type

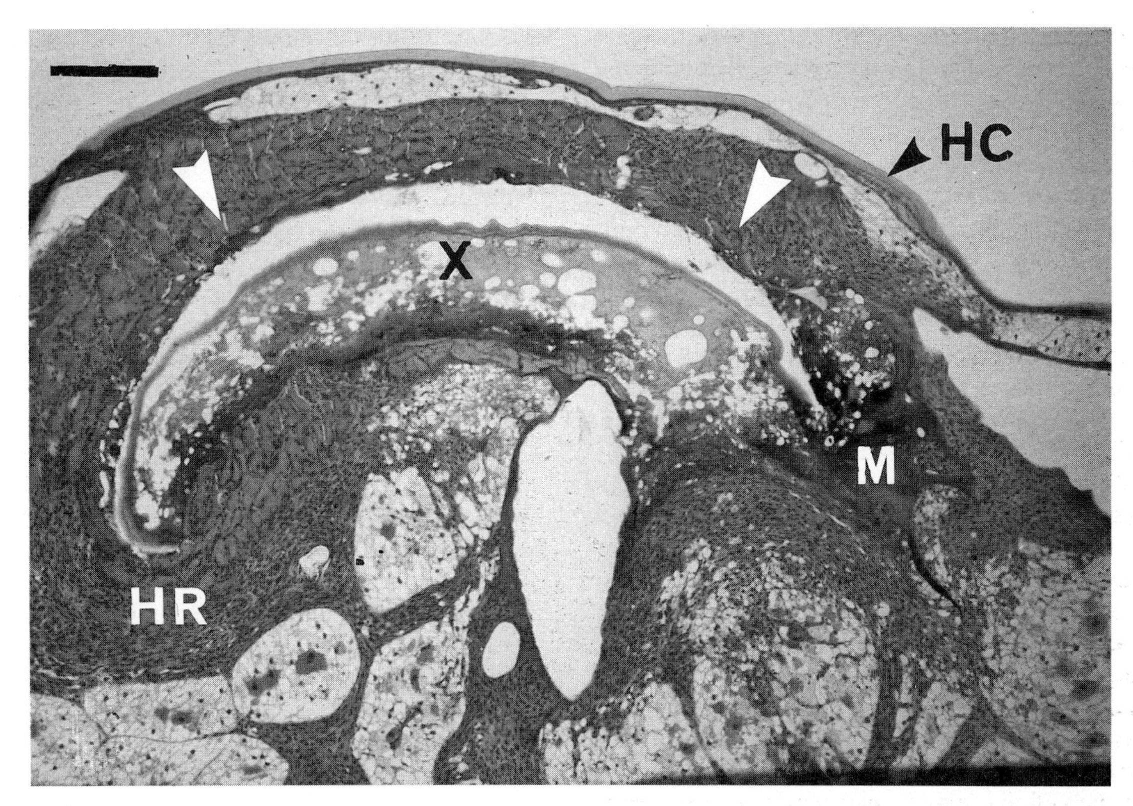

Figure 2.15. Light micrograph of a section through a cockroach, *Blaberus craniifer*, showing the rejection of a 10-day-old xenograft from the beetle, *Tenebrio molitor*. The graft (X) appears to be overlaid by host tissue (unlabelled arrowheads) due to its insertion under the overhanging margin of the previous abdominal segment. Note massive haemocytic response (HR) to the presence of the graft. Melanization (M), host cuticle (HC). Scale bar = 200 μm. From reference 97 by kind permission of Elsevier Science Ltd, The Boulevard, Langford Lane, Kidlington OX5 1GB, UK.

of blood cells involved and their commitment to graft destruction has been shown to vary[13]. Thus, in the solitary ascidian, *Styela plicata*, lymphocyte-like cells are the only cell type to specifically invade allografted tunic tissue[96], while in the insects *Blaberus craniifer* and *Extatosoma tiaratum*, the main cells infiltrating the foreign grafts are of the plasmatocyte type[97]. It is also probable that the rejection of incompatible grafts, at least in colonial ascidians, involves humoral factors as well as blood cells in provoking non-fusion reactions[98].

Transplantation experiments have revealed immunological specificity and at least a short-term memory in coelenterates, sponges, annelids, insects, echinoderms and tunicates[13,95,99], but in contrast alloreactivity has not yet been unequivocally demonstrated in nemertines or molluscs. Technical difficulties with some groups such as molluscs, sipunculids, arthropods and tunicates have exacerbated the problems associated with assessing graft rejection. It is only in recent years, for instance, that second set grafting to test for specific immunological memory has been successfully performed in tunicates[99] and allogeneic recognition determined in insects[95].

The conclusions drawn from transplantation experiments have often been disputed. For example, the report that earthworms exhibit specific memory as evidenced by accelerated second set rejection[100] has been questioned by other workers[101]. Indeed, much of the controversy arises from the numerous variables inherent in these studies which are, briefly:

1. The absence of inbred, genetically identical strains of animals.
2. The occasional application of the second graft before the primary graft has been fully rejected, which would not be acceptable in mammalian studies.
3. The temperature variable, which can affect graft survival times of invertebrate animals[102].
4. The long survival times of first set grafts (e.g. 213 days for the echinoderm, *Dermasterias imbricata*[103]) means that many recipients die before the experiments are completed.
5. The method of assessment of graft rejection, i.e. gross observation compared with histological observation, can dramatically affect results[104].

Recently, however, some effort has been made to overcome these problems and, hopefully, a better understanding of the cellular response towards foreign tissues will result. For example, experiments using inbred strains of cockroaches and a histological examination of the subcuticular epidermal layer, rather than the gross appearance of the grafts to assess rejection, have shown that previously unrecognized allograft rejection does actually occur in insects and that a memory component also exists[95].

2.4 HUMORAL IMMUNITY

Humoral components of the invertebrate immune system include lysins, agglutinins and antimicrobial agents. In addition, antibody- and complement-like factors have been described in invertebrates and these are particularly interesting from the phylogenetic viewpoint.

2.4.1 Lysins

Invertebrate lysins have often been detected by their ability to lyse vertebrate erythrocyte target cells and have been shown to be fairly uniform in nature. They are all basically proteinaceous, sensitive to proteolytic enzymes, inactivated between 50 and 60°C, susceptible to extremes of pH, operate over a wide temperature range and require divalent cations, especially calcium, for their activity[105]. Other factors included under the general heading of lysins may also be antibacterial, as in the case of lysozyme, or cytotoxic, since many cytotoxic molecules released from cells destroy their targets by lysing them[89].

2.4.2 Agglutinins

Agglutinins are ubiquitous in nature, being found in a variety of living organisms such as slime moulds, protozoans, algae, bacteria, fungi, plants and mammals as well as invertebrates.

A great deal is now known about the structure and specificity of invertebrate agglutinins, and a number have been isolated and purified to homogeneity and, in some cases[106], the amino acid sequences determined. In recent years, interest in invertebrate agglutinins has increased because of the possibility that they might function as recognition molecules[107] (see also Section 2.5.1).

Invertebrate agglutinins have usually been detected in cell-free coelomic fluid or haemolymph by their ability to agglutinate particles such as vertebrate erythrocytes, bacteria, protozoa or fungal cells. They also occur in the albumin glands of terrestrial snails[108], in mucus covering the external surface of some terrestrial and marine invertebrates[9,11,12], and on blood cells where they may be truly membrane-integrated[41] or simply adsorbed onto the cell surface[109].

Since invertebrates are a polyphyletic assemblage of animals which have evolved independently in response to different environmental pressures, it is not surprising that great heterogeneity in agglutinin structure and physicochemical properties also occurs. Differences are seen not only between agglutinins from members of diverse phyla, but also within phyla. For example, two studies[110,111] examined between them a total of 32 different ascidians for haemagglutinin activity and found great differences in the titres, dependence upon Ca^{++}, and carbohydrate specificities. Enormous variations have also been reported in the size and shape of various agglutinins as observed in the electron microscope, as well as in their molecular mass. For example, the agglutinins from the sponge, *Axinella polypoides*, have relative molecular masses of 15 and 21 kDa[112], whereas agglutinin from the snail, *Biomphalaria glabrata*[113], has been described as a protein of M_r *c.* 10^6 Da, although this may represent the aggregated forms of smaller agglutinin molecules.

Structurally, most invertebrate agglutinins are proteins or glycoproteins and possess multiple (at least two) binding sites for specific carbohydrate moieties. They may, therefore, by definition[114], be termed lectins. With some exceptions[11,113], invertebrate lectins require the presence of divalent cations, especially calcium, for their activity. The cations may function by stabilizing the conformation of the lectin molecules and/or the saccharide-binding site[115]. The lectins are usually composed of two, or multiples of two, subunits held together by non-covalent (hydrogen) or disulphide bonds. The actual number of subunits and their arrangement may vary with different agglutinin molecules. For example, the agglutinin from the sea urchin, *Anthocidaris crassispina*, comprises between 18 and 20 polypeptide chains bonded in pairs by disulphide bonds[106], whereas the agglutinin from the giant silkmoth, *Hyalophora cecropia*, contains two different M_r subunits, designated the A chain (M_r 38 kDa) and the B chain (M_r 41 kDa), which combine non-covalently in different tetrameric proportions (i.e. A_4, A_3B, A_2B_2, AB_3, B_4) to form isolectins[116]. Finally, in some instances, agglutinin subunits may have similar molecular weights but can be distinguished by their charge. The haemolymph of the octopus, *Octopus vulgaris*, for example, contains a lectin of M_r 260 kDa made up of similar subunits of M_r 30–32 kDa with differing isoelectric points (6.4, 6.6, 7.0 and 7.3)[117].

Most invertebrates possess more than one agglutinin and often different agglutinins recognize different carbohydrate moieties. Agglutinin specificity is always determined by the type of carbohydrate with which it binds. This is usually by means of competitive binding experiments in which the agglutinin is incubated with various carbohydrates (e.g. mono-, di- and polysaccharides, glycoproteins, acetylated sugars and amino sugars) prior to the addition of the test particles (red blood cells, bacteria, etc.). Nicolson[118], however, stressed that deductions made from such experiments may be an oversimplification, since the interactions of agglutinins (lectins) with simple carbohydrates are weaker than with carbohydrates isolated from the cell surface, or with complex synthetic carbohydrates. In addition, multiple lectins with differing sugar specificities may be present in invertebrate tissues at such low concentrations that they cannot be detected by crude laboratory assays without prior detailed purification techniques. If, as is now thought, invertebrate lectins are functionally important recognition molecules, then agglutinin heterogeneity is essential if diverse foreign particles, each possessing different carbohydrate determinants, are to be recognized.

In the majority of cases, invertebrate agglutinins are synthesized by and secreted from blood cells[42,84], although in insects the fat body is also involved[119]. Agglutinins present in mucus may have arisen in the haemolymph and diffused through the body wall into the mucus or, more likely, been synthesized by cells within the body wall itself.

The majority of invertebrate agglutinins are naturally occurring and attempts to stimulate the production of new proteins, or enhance existing levels, have met with varying degrees of success. Factors such as the age and sex of the animal, the nature of the immunogen and the route of administration may have an effect on the outcome of such experiments[13]. Thus, while agglutinin titres have not been enhanced following the injection of various particles such as vertebrate erythrocytes, fungal cells or bacteria into certain invertebrate species[120], some successes have been reported with others[121] (Table 2.7).

The most detailed study on induction of an agglutinin comes from work on the flesh fly, *Sarcophaga peregrina*. This lectin is present in the

Table 2.7. Some inducible immune factors in invertebrates (see text for further details).

Species	Factor	Relative molecular mass (kDa)	Activity
Sarcophaga peregrina	Agglutinin	190 32 (α subunits: $\times 4$) 30 (β subunits: $\times 2$)	Disintegration of larval tissue during metamorphosis. Recognition factor?
Lumbricus terrestris	Agglutinins	4, 15, 30, 65, 400	Antibacterial?
Hyalophora cecropia	Antibacterial proteins:		
	P4 (haemolin)	48	Major protein, complexes with 125 kDa haemolymph protein on bacterial surface. Complex initiates synthesis of further proteins and non-self recognition?
	P5 (attacins)	21–23	Narrow spectrum activity against some Gram-negative bacteria
	P7 (lysozyme)	15	Kills some Gram-positive bacteria
	P9 (cecropins)	*c.* 4	Kills Gram-positive and Gram-negative bacteria
Phormia terranovae	Antibacterial proteins:		
	Diptericin	12 (diptericin A)	Against Gram-negative bacteria
	Defensins	4	Against Gram-positive bacteria
Sarcophaga peregrina	Antibacterial proteins:		
	Sarcotoxin I	4	Against Gram-positive and Gram-negative bacteria
	Sarcotoxin II	24	Against Gram-positive and Gram-negative bacteria
	Sarcotoxin III	7	Against Gram-positive and Gram-negative bacteria
	Sapecin (insect defensin)	4	Against Gram-positive and Gram-negative bacteria
Drosophila melanogaster	Antibacterial proteins:		
	Diptericins	9	Against Gram-negative bacteria
Apis mellifera	Antibacterial proteins:		
	Apidaecins (I, II and III)	2	Inhibits Gram-negative bacteria
Periplaneta americana	Antibody-like factor (?)	700	Induces protective response against lethal doses of immunizing toxin
Lumbricus terrestris	Antibody-like factor (?)	<50	Natural function unknown
Asterias rubens	Antibody-like factor (?)	132	Natural function unknown

embryo, disappears during the larval stage, and reappears at the early pupal stage when the decomposition of larval tissue takes place[122]. The lectin can be induced in the larval stage, however, by a simple injury with a hypodermic needle[121]. The lectin has an M_r of 190 kDa and consists of four α subunits and two β subunits of M_r 32 and 30 kDa, respectively[121]. The gene responsible for the production of the lectin in the fat body codes only for the α subunit, part of which, on injury, is converted to the β subunit by partial proteolysis[123]. This results in the active lectin with the structure $\alpha_4\beta_2$. More recent work suggests that the haemolymph from an injured larva contains a mediator molecule which activates the fat body to synthesize the lectin[124]. The gene is also activated in the larval fat body when sheep red blood cells are injected, but the response is greater than when the larva is injured or injected with saline. Induction is not confined to the larval stage since injury to adult *S. peregrina* increases the haemagglutinin titre 4–8-fold within 24h[125]. However, the level of activity induced in adults is 8–16-fold less than in larvae, which may reflect different functions of the larval and adult fat body[125].

Although there have been many reports on the distribution, nature and specificity of invertebrate agglutinins, very few of them have considered their biological function. In recent years, however, the possibility that lectins may function as recognition molecules in defence reactions has been suggested[107], and experimental evidence is gradually accumulating to support this notion[41–44]. The role of agglutinins in this context is fully discussed later in the context of self-non-self recognition (Section 2.5.1).

Other ways in which invertebrate lectins may be directly or indirectly involved in defence have also been described. These include protecting the developing ova in the albumen glands of land snails against bacteria[108], possessing protease-inhibiting properties[126], facilitating the lysis of injected foreign cells[127], acting as a chemoattractant stimulator[128], as an activator of insect haemocyte coagulation[129], and as an important factor in the relationship between parasites and their vector hosts. Vector–parasite interactions, especially in insects, have received a great deal of attention in recent years[130–132], although here the role of lectins is somewhat ambiguous and requires further detailed study. This is well illustrated by the finding that midgut lectins in the tsetse fly, *Glossina morsitans morsitans*, are not only capable of preventing trypanosome establishment, possibly by lysis of the parasites[133], but may actually be involved in maturation of the parasites as well[134].

2.4.3 Other Antimicrobial Factors

Many invertebrates inhabit environments which are rich in potentially harmful microorganisms. Despite this, their internal tissues remain relatively sterile. This is due, in part, to the presence of soluble antimicrobial factors which offer protection at least against the type of microorganisms prevalent in the animal's immediate natural environment (Figure 2.16).

The majority of investigations have concentrated upon the haemolymph or coelomic fluid for the detection of antimicrobial activity. Nevertheless, external secretions such as mucus may also contain antibacterial factors which could destroy entrapped bacteria before they have a chance to invade underlying tissues or enter the body through a wound[8,10].

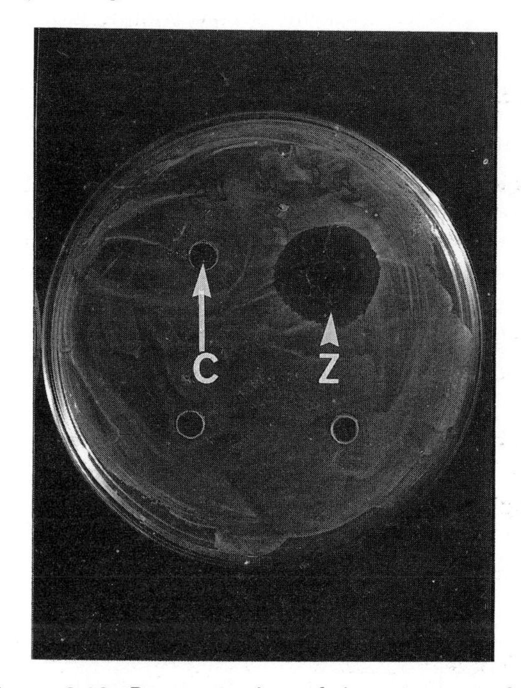

Figure 2.16. Demonstration of the presence of a natural antimicrobial factor in the blood serum of the polychaete worm, *Arenicola marina*. Note the zone of clearance (Z) in the bacterial lawn produced around the well containing the serum but not around the control well (C) inoculated with saline alone. The bacterial lawn was made from a natural isolate from the environment of the worm. Reproduced by courtesy of Dr S.W. Fitzgerald.

Although naturally-occurring and induced antibacterial activity has been detected in many invertebrate species[13], very little is known about the nature, physicochemical properties and mode of action of these factors. However, some chemical compounds with antimicrobial activity have been isolated from various marine organisms[135], and these substances are of interest as they may also have great potential clinical use as antiviral, antitumour, or antiprotozoan agents[136].

The most widely known bacteriolytic factor is the enzyme lysozyme, which is found in both vertebrates and many invertebrate species such as annelids, molluscs, insects and echinoderms, and acts by splitting the bond between *N*-acetyl-glucosamine and *N*-acetylmuramic acid present in the cell walls of some bacteria and fungi[137]. Invertebrate lysozyme (*N*-acetylmuramide glycanohydrolase), probably synthesized in and secreted from blood cells, is of relatively low M_r (*c.* 15kDa), stable at low pH and high temperature, and usually identified by its ability to lyse bacterial suspensions of *Micrococcus luteus* (formerly *M. lysodeikticus*). Although in some invertebrates haemolymph lysozyme levels are quite low, it may be possible to enhance activity by the injection of bacteria[138].

Initially, it was assumed that lysozyme was the only antibacterial factor in insects. This, however, was a misconception (see Table 2.7) since it was later observed that although the lysozyme level in the moth *Galleria mellonella* increased following the injection of *Pseudomonas aeruginosa* endotoxin, the concomitant increase in protective immunity decreased long before the level of lysozyme returned to pre-immune levels[138]. Indeed, over the last decade extensive work has been carried out on the inducible antibacterial factors in several insect species with the result that many factors (in addition to lysozyme) have been isolated, purified, and their mode of action determined (Table 2.7). Furthermore, several antibacterial proteins have now been sequenced and the genes responsible for their synthesis characterized.

The most widely used insect in these investigations has been the diapausing pupa of the giant silk moth, *Hyalophora cecropia*. This is an ideal insect to employ since, during diapause, following the injection of bacteria, or simply by wounding, only those genes responsible for synthesis of the immune proteins are switched on. This subject has been reviewed elsewhere in detail[139–141].

Following immunization of *H. cecropia* pupa, and a lag period of 12–18h, at least 15 different proteins are produced which peak at 7–8 days and then gradually decline. The main inducible proteins are a family of molecules termed P4, P5, P7 and P9. P4, also called haemolin, is the major inducible protein with an M_r of 48kDa and an isoelectric point of 8.2; until recently[142] it did not appear to have any obvious function. It is now known, however, that it is synthesized by the haemocytes and fat body, and *in vitro* is one of the first proteins which binds to the bacterial surface. Here, in the presence of divalent cations it forms a complex with another haemolymph protein with an M_r of *c.* 125kDa[141]. This complex is probably involved in initiating synthesis of the other immune proteins[141].

The next protein, P5, is actually a family of six related proteins, termed attacins A–F with M_r values of 21–23kDa and different isoelectric points. The attacins can be further divided into two subgroups depending upon their amino acid composition and N-terminal sequences. A basic group is made up of attacins A–D and an acidic group contains E and F with each group being synthesized by two different genes. The complete amino acid sequence has been determined for attacin F and it is possible that this protein is derived from attacin E by proteolytic cleavage. The antibacterial activity of the attacins is rather narrow, being effective against only *E. coli* and a few other Gram-negative bacteria, and then only on growing cells where it acts by interfering with the cell division cycle.

The third inducible protein in *H. cecropia*, P7, is a lysozyme with an M_r of 15kDa. *H. cecropia* lysozyme is lytic against only a few Gram-positive bacteria such as *Bacillus megaterium* and *Micrococcus luteus*. It is likely, however, that it acts synergistically with the attacins and

cecropins (see below) and further destroys bacteria killed by the action of these other factors.

The fourth group of induced antibacterial proteins, P9, is a family of heat-stable, strongly basic proteins termed cecropins, with M_r values of *c.* 4 kDa. Cecropin D, together with A and B, constitute the major proteins, while C, E and F are present in very low amounts. It is possible that cecropins C and E actually represent precursors of A and D, respectively. Examination of the amino acid sequences of the cecropins shows that there is a high degree of homology, especially between cecropins A, B and D, and thus they are probably products of three different genes which arose by gene duplication from a common ancestral gene. Cecropins are highly active against Gram-positive and Gram-negative bacteria and induce lysis of the target microorganism, probably by acting directly on the phospholipid layer of the cell membrane[143]. Furthermore, it appears that the amphipathic α helix structure of the cecropin molecule is important to its mode of action[143]. By using cecropin analogues with impaired α helix structure (truncated molecules or amino acid substitution), it has been shown that alterations in the molecule result in a reduced spectrum of activity[140,143].

The immune proteins in *H. cecropia* are synthesized and released from the fat body cells not only in response to injury or injection of sterile Ringer solution, but also to a greater extent following the injection of bacteria[144]. The haemocytes, which also synthesize cecropins and attacins[145], may play a part, too, in triggering the synthesis of the immune proteins from the fat body. It is possible that the foreign particles (bacteria) interact with the haemocytes (attachment or phagocytosis) which results in the release of soluble factors (lectins, components of the prophenoloxidase cascade?) which, in turn, act directly or indirectly on the fat body cells thus stimulating the production of the antibacterial proteins.

Attacins and cecropins are not specific to *H. cecropia*, since induced attacin- and/or cecropin-like proteins have been found in the pupae or larvae of other insect species[146,147]. In dipteran insects, as well as the cecropins and members of the cecropin family (sarcotoxins), other, unrelated antibacterial proteins—given the names of diptericin, defensin, sapecin—have also been investigated[148] (Table 2.7). Thus, further work on other insect species may reveal even more families of inducible antibacterial proteins, distinct in structure and mode of action from those previously mentioned, and which may also have possible uses as novel chemotherapeutic agents[1].

Although the inducible proteins of insects have been the most widely investigated antibacterial factors from invertebrates, the naturally-occurring factors in two species of annelid, the polychaete, *Glycera dibranchiata*, and the oligochaete, *Eisenia foetida andrei*, have also been characterized to some degree. The *Glycera* antibacterial factor, or GAF, is a naturally-occurring, trypsin-sensitive, heat-labile, glycoprotein of M_r 2.5–4.5 $\times$ 10^5 Da[149]. The molecule is thus very much larger than the antibacterial proteins in insects and, although difficult to purify, was found to contain bound divalent cations and at least one disulphide bond[149]. GAF is particularly bactericidal towards several Gram-negative bacteria, especially *Serratia marcescens*, and killing is a two-stage process[150]. The first stage, which is divalent cation- and slightly temperature-dependent, involves the reversible binding of GAF to the outer surface of the bacterium, and usually occurs within a few minutes. The second stage is a temperature-sensitive killing process which occurs optimally at the temperature at which the target bacteria grow. The factor is very potent: only three or four molecules of GAF may be required to kill a single bacterium[150].

Antibacterial activity of coelomic fluid from the earthworm, *Eisenia foetida andrei*, was originally measured against 23 different bacterial species from the same biotope, and found to be effective against only six Gram-positive and Gram-negative strains[151]. The coelomic fluid was later analysed by gel filtration, followed by chromatofocusing and polyacrylamide gel electrophoresis to determine the active proteins[152,153]. These techniques revealed that the antibacterial activity resided in at least four distinct proteins with relative molecular masses of 20, 40, 40–45 and 175 kDa. The mode of action of these proteins is unknown, although it has been hypothesized that they bind to the bacterial cell wall, modifying its physical state leading to immobilization or death[152]. In addition, it has recently been shown that the level of antibacterial activity and protection in *E.f. andrei* can be enhanced by bacterial injection, and that this induction shows some specificity[154].

2.4.4 Other Inducible Humoral Factors and Complement-like Activity in Invertebrates

It has been shown, above, that certain humoral components such as agglutinins, lysins and antibacterial factors may be naturally occurring or induced. In most cases induction is non-specific and may even occur simply by wounding. However, some reports have indicated that invertebrates are indeed capable of responding to antigen in a way analogous to that of vertebrates, i.e. the response is specific, transferable and shows a memory component (Table 2.7). These results, therefore, may be significant especially in relation to the phylogenetic origin of vertebrate antibody. Similarly, the possible presence of complement-like activity in some invertebrates suggests that these factors (or their precursors) may also have arisen prior to the emergence of the vertebrates.

An interesting example of a specific adaptive immune response is that observed in the cockroach, *Periplaneta americana*, following the injection of soluble protein toxins. These induce a factor which provides the cockroach with protection against subsequent, lethal doses of toxin[155]. The response, which peaks within 2 weeks and declines by 5 weeks, is specific and protection can be passively transferred to naive cockroaches by cell-free haemolymph, indicating the presence of a humoral factor. Additional injection of toxin results in a typical vertebrate-like secondary response with specific memory. This is characterized by a shorter lag time and a heightened and prolonged level of protection, which can be stimulated with sub-immunizing doses. It was further shown that the extent of response is dependent upon the sex and age of the cockroaches.

Preliminary investigations into the nature of the inducible factor have shown it to be a trypsin-sensitive protein, of *M*r 700 kDa. In addition, the factor was shown to behave like an antibody molecule since it forms precipitation lines against honey bee venom in Ouchterlony gel diffusion tests.

Specific 'antibody-like' molecules have also been induced in the earthworm, *Lumbricus terrestris*[156], and in the sea star, *Asterias rubens*[157], in response to the injection of various haptens. With earthworms, it was also shown that the extent of the response was heightened following a second injection. Thus, some invertebrates appear capable of initiating a humoral response similar to that of vertebrates. The production of short-lived, non-specific humoral factors may be adequate to counter an immediate threat, whereas in long-lived invertebrates, such as the cockroach, additional, specific, longer-lasting responses may be necessary to deal effectively with more persistent threats[155].

In contrast to antibody-like molecules, the search for soluble complement-like factors in invertebrates has been more fruitful. The fact that components of a complement-like system have been detected in a range of invertebrates[36,37] suggests that the vertebrate complement system may have its origins within the invertebrates. Many of the studies to date have employed cobra venom factor (a source of C3b) which, in the presence of factors in invertebrate haemolymph, induces a C3 convertase which splits vertebrate C3 to produce C3a and C3b. Thus, the active substances in the haemolymph are possibly analogous to vertebrate C3b or C3 proactivator factors which activate complement via the alternative pathway. The alternative pathway, which can be activated by bacterial polysaccharide, but which is independent of antibody, may, therefore, be ancestral to the antibody-mediated classical pathway which evolved much later to allow for a more efficient system.

2.5 SELF/NON-SELF RECOGNITION

From the above, invertebrates lack the refined, sophisticated immune responses characteristic of vertebrates, but nevertheless have an efficient cellular- and humoral-mediated defence system. The fundamental process common to defence systems, whether in the earthworm or in humans, is the ability to discriminate between 'self' and 'non-self' or altered 'self'. This topic has been the subject of several reviews[158].

Although recognition may, in part, be non-specific and based upon physicochemical properties such as surface charge and hydrophobicity[20], the fact that invertebrates can selectively clear various non-self particles from their circulation at different rates[159] and possess blood cells which can actively discriminate between different foreign particles *in vitro*[160], indicates that some very specific recognition factors are also involved. In vertebrates, recognition is achieved mainly through the interaction of humoral proteins, such as immunoglobulin or complement factors, with the foreign material. Invertebrates, however, probably lack immunoglobulins and although complement-like factors have been detected in some invertebrates (see above), it is only in the echinoderm, *Strongylocentrotus droebachiensis*, that an opsonic role for a complement-like factor has been described[161]. To date, the most widely proposed contenders for the role of recognition molecules in invertebrates are the agglutinins (lectins) and factors associated with prophenoloxidase activation.

2.5.1 Agglutinins as Recognition Molecules

The role of agglutinins (often referred to as lectins) as opsonins is controversial. Some workers have failed to show enhanced attachment or internalization of test particles by haemocytes in the presence of whole haemolymph[162] or purified lectin[163]. However, there are reports which unequivocally support the involvement of invertebrate agglutinins in immune recognition (see below and Table 2.8). The overall situation will only be clarified when purified lectin molecules from a large number of diverse invertebrate species are used in *in vitro* experiments under strictly optimized conditions.

Initially an opsonic role for invertebrate lectins was determined using classical 'blockade' experiments *in vivo*, which were originally employed to detect opsonizing antibody in vertebrates[164,165]. Thus, in the snail, *Helix pomatia*, it was found that preincubating human A and B erythrocytes in *H. pomatia* serum enhances their rate of clearance compared with non-treated cells[164].

Table 2.8. Evidence for role of agglutinins in invertebrate immunity (see text for details).

Species	Agglutinin function
Glossina sp.	Killing and development of trypanosome parasites
Species of land snail e.g. *Helix aspersa*	Protecting developing ova in albumin glands against bacteria
Sarcophaga peregrina	Participates in lysis of foreign cells
Manduca sexta	Triggering of haemocyte coagulation
Blaberus discoidalis	Enhancement of prophenoloxidase activation
Periplaneta americana	Induction of encapsulation
Arion empiricorum	Protease inhibition
Vivaparus malleatus	Chemoattractant stimulator
Helix pomatia	Enhanced clearance of non-self particles from circulation
Crassotrea gigas	Increased *in vitro* phagocytosis of non-self particles by haemocytes
Mytilus edulis	Increased *in vitro* phagocytosis of non-self particles by haemocytes
Spodoptera exig[illegible]	Increased *in vitro* phagocytosis of non-self particles by haemocytes
Botrylloides lea[illegible]	Increased *in vitro* adherence of foreign particles to haemocytes
Extatosoma tia[illegible]tum	Increased *in vitro* adherence of foreign particles to haemocytes

Also, a second dose of erythrocytes is cleared at the same rate as the primary dose, only if they are preincubated in either snail serum[164] or purified *H. pomatia* albumin gland agglutinin[165]. The opsonic effect is specific since erythrocytes which had been untreated or pretreated with purified concanavalin A or *Axinella polypodes* (sponge) agglutinins are cleared more slowly. Furthermore, the presence of *N*-acetylglucosamine, a specific inhibitor of *H. pomatia* serum agglutinin, significantly decreases the clearance of a primary dose of erythrocytes, whereas a non-inhibitor, fucose, has no effect[165]. The results from these experiments strongly suggest that the haemolymph opsonin and the agglutinin are the same.

The opsonic properties of invertebrate lectins have also been demonstrated *in vitro* using purified molecules from several invertebrate species including molluscs[41,166], ascidians[42] and insects[43,44]. For example, the galactose-specific lectin from the larvae of the beet army worm, *Spodoptera exigua*, enhances the *in vivo* clearance and the *in vitro* haemocyte attachment and endocytosis of fungal blastospores possessing exposed galactose residues. In contrast, blastospores from a different fungal species without galactose groups are neither agglutinated nor rapidly cleared from *S. exigua* haemolymph[43]. Studies on phagocytosis by *Mytilus edulis* haemocytes similarly showed that for effective lectin-mediated attachment, the specific carbohydrate determinant has to be on the surface of both the foreign test particles and the haemocytes[166].

There are three ways in which lectins are able to mediate attachment of foreign particles to blood cells[107]. First, the lectins may be associated with the haemocyte membrane, either as integral, membrane proteins[41], or as cytophilic haemolymph factors[109]. Here, the membrane-bound lectins bind directly with carbohydrate moieties on the non-self particle. Second, the lectin may be humoral and act as a bridge between complementary carbohydrates on both the foreign particle and the immunocompetent cell[43,166]. In the third situation, lectins present on microorganisms such as bacteria recognize carbohydrate residues on the phagocyte surface[40].

2.5.2. The Prophenoloxidase Activating System and Recognition in Arthropods

The observation that many of the immunological events in arthropods are associated with melanin production (Figures 2.10, 2.14 and 2.15) has stimulated the belief that the prophenoloxidase (PpO) system, responsible for melanization, may be an important component of the defence system of these animals. The role of the PpO activating system and melanization in the immune responses of arthropods has received considerable interest in recent years[33,167,168].

During melanization, substrates such as tyrosine and its derivatives, DOPA and dopamine, are metabolized to melanin by the enzyme, phenoloxidase (PO:*o*-diphenol:O_2 oxidoreductase; EC 1.10.3.1), which, under normal physiological conditions, occurs in its inactive precursor form—prophenoloxidase. Although PpO has been detected in a variety of invertebrate species[169], it is from the work with crustaceans and insects that most of our present knowledge has come. Detailed investigations have shown that during activation of the PpO system, various foreign materials trigger a cascade of serine proteases, terminating in the conversion of PpO to PO by limited proteolysis (Figure 2.17). Elicitors of the cascade include whole bacteria and fungi as well as minute quantities of β-1,3-glucans from fungal and algal cell walls, LPS and peptidoglycan from bacterial cell walls and also capsular polysaccharides[168].

In the first stages of activation of the PpO system, the non-self microbial elicitors interact with the haemocyte membrane, either directly or indirectly via specific receptors in the plasma, with the resulting release of the inactive components of the PpO system from the cells by exocytosis. The elicitor or elicitor/receptor complex then triggers a cascade of at least two serine proteases present as zymogens which terminates with the limited proteolysis of PpO, accompanied by the release of a 5kDa peptide and the formation of active PO (Figure 2.17). In insects, the PpO activating enzyme itself has been located within the cuticle and plasma. To prevent widespread

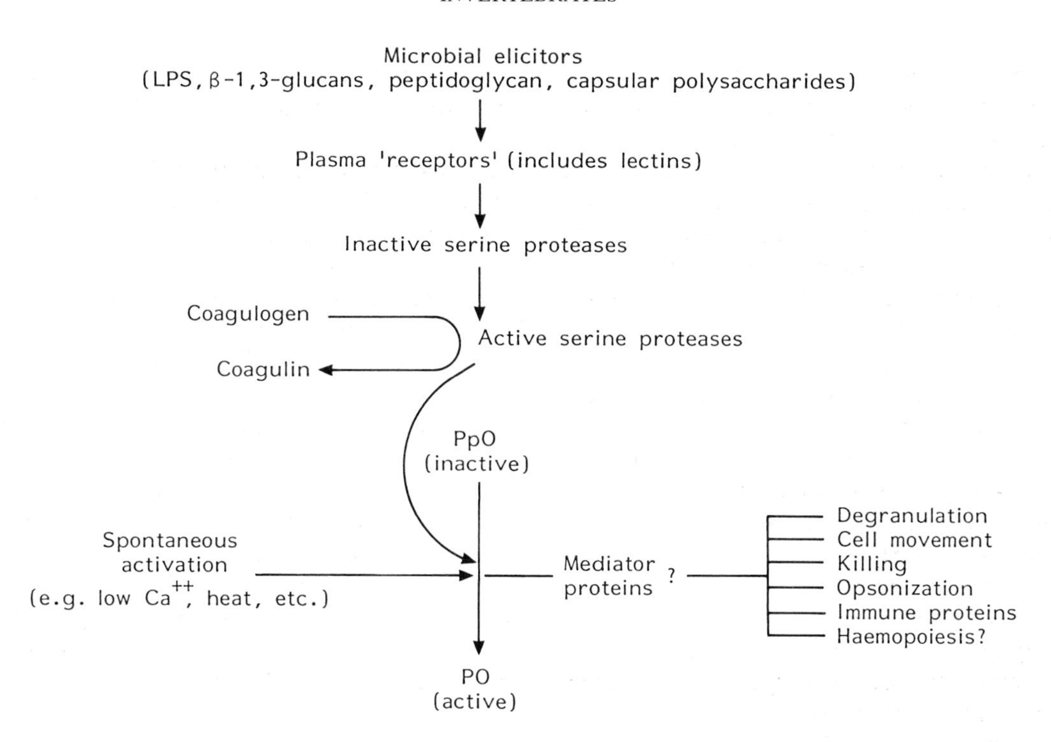

Figure 2.17. Schematic representation of prophenoloxidase (PpO) activation in arthropods.

activation of the system, inhibitors are present which act upon the serine proteases and limit the reaction[170]. There is, in addition, a self-regulating mechanism since the serine proteases, as well as PO itself, are relatively short-lived and unstable.

In addition to PO, the activated system produces at least a further four sticky proteins which adhere to foreign surfaces and which may, therefore, be the factors responsible for regulating immune responses such as phagocytosis, nodule formation and encapsulation.

Although results from several *in vivo* and *in vitro* experiments suggest a role for components of the PpO activating system in arthropod immunity (Table 2.9), the picture, as with lectin studies, is still far from clear. Nevertheless, it has been shown that substances which are known to trigger the PpO cascade, such as β-1,3-glucans, peptidoglycan or capsular material from the bacterial cell wall induce nodule formation when injected into crustaceans or insects. Similarly, fungal blastospores, coated with crayfish haemocyte lysate supernatant in which PpO had been spontaneously activated by a low concentration (5mM) of calcium, and injected into *A. astacus*, exhibit a stronger encapsulation reaction compared with blastospores treated with plasma or buffer alone[171]. However, in a similar experiment in which Sepharose beads were precoated with products of PpO activation and then injected into the locust *Schistocerca gregaria*, an encapsulation response did not occur[75].

Table 2.9. Evidence for the role of the prophenoloxidase (PpO) activating system in arthropod immunity (see text for details).

Melanin deposition associated with various immunological responses
Enhancement of haemocyte degranulation
Enhancement of cell movement
Killing of microorganisms
Plasma coagulation
Induction of nodule formation *in vivo* by PpO elicitors
Enhanced phagocytosis *in vitro* by PpO elicitors
Stimulation of *in vivo* encapsulation
Haemocyte degranulation in sequential cascade
PpO inhibition by insect parasitoids

Perhaps more convincing evidence for an opsonic role for the activated PpO system comes from *in vitro* experiments using haemocyte monolayers. For example, phagocytosis of non-self particles by crustacean and insect haemocytes can be enhanced by the presence of β-1,3-glucans or LPS, known elicitors of PpO activation[45,46]. Significantly, substances such as dextran, which have no stimulatory effect on the PpO system, also fail to promote phagocytosis[45], and the potent serine protease inhibitor, pNPGB, reduces phagocytosis to control levels[46].

Although some evidence has, thus, been provided to implicate components of the PpO activating system in immune regulation, in crustaceans and insects at least, further experimentation is required, preferably using purified molecules, before it can be stated unequivocally that the system is directly involved in self-non-self recognition. It must be pointed out, however, that the cells which contain the components of the PpO system also possess lectins which, as has been shown earlier, are also important putative recognition molecules. Indeed, it has recently been shown that lectins isolated from the haemolymph of the cockroach, *Blaberus discoidalis*, can enhance PpO activity from cockroach haemocyte lysate supernatants[172] (Table 2.10). Further, one of the cockroach lectins has been identified as the so-called plasma factor previously isolated from the plasma of this insect and found to act as a specific receptor for the microbial elicitor[173].

Therefore, immunorecognition in arthropods (and perhaps other invertebrates) is probably not a simple event, but must rely on the complex interaction of a variety of factors such as surface charge and hydrophobicity, lectins and components of the PpO system.

Table 2.10. Enhancement of phenoloxidase (PO) activity in the cockroach, *Blaberus discoidalis*, haemocytes by endogenous and exogenous lectins[a].

Incubation mixture	PO activity ($\Delta A490_{max}$/10 min)
HLS[b] + buffer	0.04 ± 0.01
HLS + laminarin	0.11 ± 0.02
HLS + BDL1[c]	0.10 ± 0.02
HLS + BDL1 + laminarin	0.18 ± 0.03
HLS + p.f.[d]	0.12 ± 0.02
HLS + p.f. + laminarin	0.21 ± 0.03
HLS + Con A[e]	0.09 ± 0.02
HLS + Con A + laminarin	0.16 ± 0.04

[a] Data from reference 172.
[b] HLS = haemocyte lysate supernatant.
[c] BDL1 = α-D-glucose/D-galactose/D-mannose specific lectin from *B. discoidalis*.
[d] p.f. = β-1,3-glucan specific lectin from *B. discoidalis*.
[e] Con A = concanavalin A, exogenous phytolectin.
Note the increase above control levels (with buffer) of PO activity in the presence of the endogenous lectins BDL1 or p.f. as well as with Con A, with and without the specific elicitor of the PpO cascade, laminarin (a β-1,3-glucan).

2.6 CONCLUDING REMARKS: PHYLOGENETIC RELATIONSHIPS BETWEEN INVERTEBRATE AND VERTEBRATE IMMUNE SYSTEMS

This subject has intrigued comparative immunologists for a considerable time[174]. The main problem when speculating on the origin and evolution of the vertebrate immune system is that the immediate ancestors of vertebrates are now extinct. Thus, it is inevitable that comparisons have to be made from studies with existing species which means that the topic is, at best, conjectural, but always controversial.

Although most invertebrates are of relatively small size, of simple morphology, have short life spans and produce large numbers of progeny to ensure the survival of the species, the need for a specialized immune system probably arose in these animals to counter threats imposed by certain evolutionary pressures. The development of cells and molecules committed to prevent infection by potential pathogens and parasites and to recognize somatic mutations which could lead to cancer, are essential for survival. In addition, a discriminatory system evolved to prevent fusion between incompatible species of colonial animals such as sponges, coelenterates and tunicates, thus maintaining heterogeneity and polymorphism in the genome.

As we have shown above, invertebrates have a

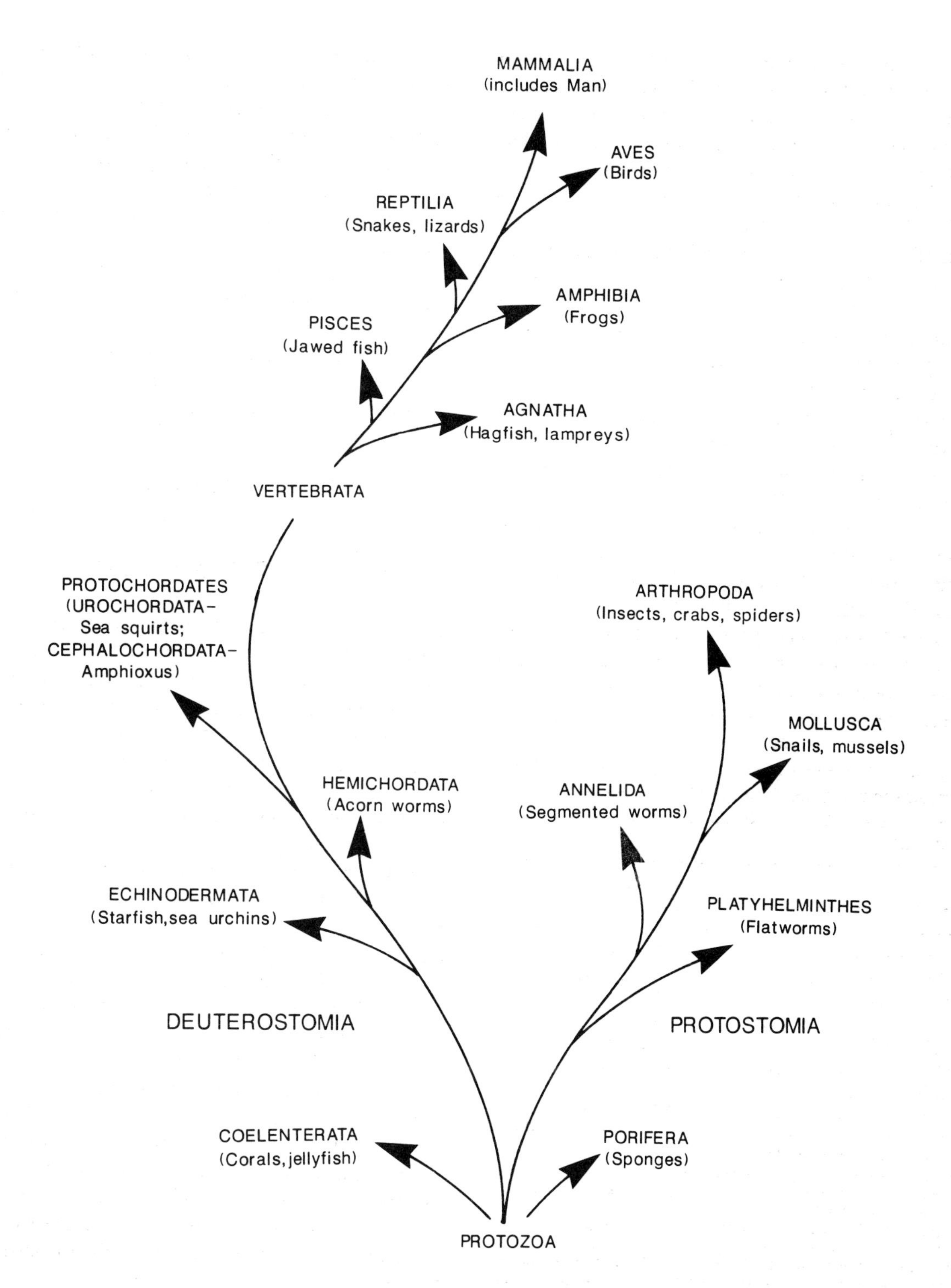

Figure 2.18. Phylogenetic relationships between the main groups in the animal kingdom.

variety of effective defence strategies which, when compared with the sophisticated, finely-tuned immune system of vertebrates, may appear rather 'primitive'. However, the main difference between invertebrate and vertebrate immune systems is that the latter, but possibly not the former, are capable of what Klein[175] terms an 'anticipatory' immune response. Klein argues that an anticipatory response which is inducible, based on lymphocytes and antibody, and effectively deals with previously unencountered antigen by the production of a large number of primed cells with specific receptors and memory, simply does not occur in invertebrates. He also maintains that if anticipatory responses do occur in invertebrates, then they are based upon mechanisms different from those in vertebrates. Therefore, as distinct lymphocytes and immunoglobulin-producing plasma cells are recognizable for the first time in cyclostome fish (hagfish and lampreys), are these components of the immune system purely vertebrate inventions or may precursors of vertebrate immunoreactive cells and molecules be found in advanced invertebrates?

To consider this further, it is first necessary to examine the phylogenetic relationships between the main groups in the animal kingdom. The traditional pathway (Figure 2.18) essentially divides the animal kingdom into two main lineages—Protostomia and Deuterostomia—based upon differences in embryological development of the mouth and anus. The bifurcation of the Protostomia and Deuterostomia probably occurred at the level of the sponges and coelenterates more than 500 million years ago. Thus in the protostome lineage, invertebrates such as the arthropods, annelids and molluscs, evolved separately but over a similar time period as members of the deuterostome lineage—the echinoderms, hemichordates, protochordates and chordates. It may, therefore, be fruitless to search for origins of the vertebrate immune system in protostome invertebrates, which are obviously not ancestral to the vertebrates[175]. However, it is possible that the ancestors of vertebrate blood cells and precursors of vertebrate humoral recognition molecules may be found in deuterostome invertebrates. In other words, cells and molecules from these invertebrates and from vertebrates should be homologous (i.e. share a common ancestor), while any structural and functional similarities between cells of protostomate invertebrates and those from vertebrates are purely analogous and have arisen as a result of convergent evolution.

The hemichordates and cephalochordates have simple body plans, lack circulating cells resembling lymphocytes[176] and are unlikely to possess the rudiments of a vertebrate immune system (Millar and Ratcliffe, unpublished observations). In contrast, as will be seen below, studies with echinoderms and urochordates have revealed some very interesting facts which may be of significance in the phylogeny of immunity.

2.6.1 Origin of Blood Cells

The first blood cells probably evolved from a free-living wandering protozoan-like ancestor[177]. With the increase in body size and complexity, together with the formation of a coelom, a blood vascular system developed for the transport of food, gases and waste material. Eventually, phagocytic cells migrated from the connective tissue into the circulatory system[178] and took on a role specifically involved in immune surveillance. Within different species, this phagocytic cell subsequently evolved different receptors and intracellular enzymes, and is probably the only blood cell type which has persisted throughout evolution and corresponds to the mammalian granulocyte, monocyte or tissue and organ macrophage[177].

Cells morphologically resembling lymphocytes have been found in many invertebrate species[13], although such cells in protostome invertebrates are most likely to be analogous. Such is the case with lymphocyte-like cells in earthworms, which exhibit functions similar to those cells in vertebrates responsible for tissue graft rejection and responses to specific mitogens[88,179]. More compelling evidence for homology comes from work on the coelomocytes of the star fish, *Asterias rubens*. Cells from the axial organ of this animal can be separated on nylon wool columns into a non-adherent population which, as with vertebrate T-lymphocytes, responds to concanavalin A, and an adherent cell population which can be stimulated by lipopolysaccharide, a B-lymphocyte mitogen[180]. Furthermore, the non-adherent cells induced angiogenesis in irradiated mice and splenomegaly in chick embryos[181,182], reactions similar to T-cell-induced graft-versus-host reactions in mammals. In addition, the axial organ cells secrete 'antibody-like' molecules in response to an injected hapten[157]. Interestingly, the production of these molecules requires the co-operation of the adherent and non-adherent cells as well as phagocytic cells[183] and is, thus, similar to antibody production in vertebrates.

In tunicates, too, lymphocyte-like cells have been shown to respond to allografts, form rosettes with sheep red blood cells and proliferate in response to T- and B-cell mitogens[96,184,185]. Whether tunicate lymphocyte-like cells and true vertebrate lymphocytes are homologous, as proposed by Wright[186], remains to be elucidated. There are those, however, who argue against homology[187] and suggest that the cells morphologically resembling lymphocytes in tunicates may simply be undifferentiated stem cells which give rise to other cell types[188].

2.6.2 Receptors and Recognition Molecules

The fact that allorecognition occurs in most invertebrate species suggests the presence of recognition molecules which react with specific receptors on the cell surface. In invertebrates, such molecules probably include lectins and components of the activated PpO system, as discussed earlier (Section 2.5). In vertebrates, lectins may also act as recognition factors, but specificity lies with immunoglobulin (Ig) molecules, either free in the circulation or membrane-bound as receptors on B-lymphocytes. Thus, phagocytosis in vertebrates is facilitated by receptors for the Fc region of Ig and the C3 component of complement. Systems resembling the antibody-independent pathway of complement have been described in invertebrates[13] (Section 2.4.4), which, together with the discovery of C3 receptors on echinoderm coelomocytes[189], suggest that the basic components of the complement system may have arisen quite early on within the invertebrates.

Evidence from serological cross-reactivity has suggested that members of a superfamily of vertebrate recognition molecules, including C-reactive protein, β_2-microglobulin, Ig and products of the major histocompatibility complex (MHC), also occur in invertebrates[190]. For example, molecules related to C-reactive protein (an acute-phase protein synthesized during infection and tissue damage), which is unrelated to Ig but which can bind to phagocytic cells and fix complement, have been found in tunicates[191] and horseshoe crabs[192], where they may function in defence[190].

In vertebrates, allorecognition is controlled by a single cluster of gene loci termed the MHC. The ancestral form of this complex may be the histocompatibility (H) system of cell-mediated immunity which Hildemann[193] proposed arose in the coelenterates to prevent colony fusion. This system is distinct from the Ig system of vertebrates, which may have been added later in phylogeny to provide a finely-tuned recognition system. Ig and modern MHC, characteristic of vertebrates, may have arisen from a common ancestral gene which also coded for β_2-microglobulin (β_2-m), Thy-1 and MHC antigens[194]. Marchalonis *et al.*[195] have, however, expressed doubt that MHC products and Thy-1 share a common ancestry with Ig. β_2-m may be the common link between invertebrates and vertebrates since it is related to both the MHC and Ig systems[194], and β_2-m-like molecules have been detected on cells from a variety of invertebrate animals[196].

It is possible, therefore, that the genes encoding for various vertebrate recognition molecules emerged early in evolution. Marchalonis and Schluter[190] believe that the most likely time for emergence of genes specific for Ig is between the forms ancestral to the tunicates and the true vertebrates, an assumption based upon the ability to detect Ig-related molecules in tunicates[197], but not in echinoderms.

Acknowledgements

We gratefully acknowledge the financial support kindly provided by the Royal Society and the Science and Engineering Research Council, grant numbers GR/G40224 and GR/F17421.

2.7 REFERENCES

1. Casteels, P. (1990). Possible applications of insect antibacterial peptides. *Res. Immunol.* **141,** 940–942.
2. Söderhäll, K. and Ajaxon, R. (1982). Effect of quinones and melanin on mycelial growth of *Aphanomyces* spp. and extracellular protease of *Aphanomyces astaci* a parasite on crayfish. *J. Invert. Pathol.* **39,** 105–109.
3. Häll, L. and Söderhäll, K. (1983). Isolation and properties of a protease inhibitor in crayfish (*Astacus astacus*) cuticle. *Comp. Biochem. Physiol.* **76B,** 699–702.
4. Stagner, J.I. and Redmond, J.R. (1975). The immunological mechanisms of the horseshoe crab, *Limulus polyphemus. Mar. Fish. Rev.* **37,** 11–19.
5. Wright, R.K. and Cooper, E.L. (1983). Inflammatory reactions of the Protochordata. *Amer. Zool.* **23,** 205–211.
6. Denny, M.W. (1989). Invertebrate mucous secretions: functional alternatives to vertebrate paradigms. In *Mucus and Related Topics. Symp. Soc. Exp. Biol. No. 43*, Chantler, E. and Ratcliffe, N.A. (eds). The Company of Biologists Ltd., Cambridge, pp. 337–366.
7. Valembois, P., Roch, P. and Lassègues, M. (1988). Evidence of plasma clotting system in earthworms. *J. Invert. Pathol.* **51,** 221–228.
8. Iguchi, S.M.M., Aikawa, T. and Matsumoto, J.J. (1982). Antibacterial activity of snail mucus mucin. *Comp. Biochem. Physiol.* **72A,** 571–574.
9. Iguchi, S.M.M., Momoi, T., Egawa, K. and Matsumoto, J.J. (1985). An *N*-acetylneuraminic acid-specific lectin from the body surface mucus of African giant snail. *Comp. Biochem. Physiol.* **81B,** 897–900.
10. Millar, D.A. and Ratcliffe, N.A. (1987). Antibacterial activity of the hemichordate, *Saccoglossus ruber* (Enteropneusta). *J. Invert. Pathol.* **50,** 191–200.
11. Millar, D.A. and Ratcliffe, N.A. (1987). Activity and preliminary characterization of a hemagglutinin from the hemichordate *Saccoglossus ruber. Dev. Comp. Immunol.* **11,** 309–320.
12. Möck, A. and Renwrantz, L. (1987). Lectin activity and a protease inhibitor in mucus from the skin of *Branchiostoma lanceolatum. J. Invert. Pathol.* **49,** 221–222.
13. Ratcliffe, N.A., Rowley, A.F., Fitzgerald, S.W. and Rhodes, C.P. (1985). Invertebrate immunity—basic concepts and recent advances. *Int. Rev. Cytol.* **97,** 183–350.
14. Ratcliffe, N.A. and Rowley, A.F. (eds) (1981). *Invertebrate Blood Cells* vols 1 and 2. Academic Press, London.
15. Ratcliffe, N.A. and Götz, P. (1990). Functional studies on insect haemocytes, including non-self recognition. *Res. Immunol.* **141,** 919–923.
16. Ratcliffe, N.A. and Rowley, A.F. (1979). A comparative synopsis of the structure and function of the blood cells of insects and other invertebrates. *Dev. Comp. Immunol.* **3,** 189–243.
17. Ratcliffe, N.A. (1985). Invertebrate immunity—a primer for the non-specialist. *Immunology Letters* **10,** 253–270.
18. Smith, V.J. (1981). Echinoderms. In *Invertebrate Blood Cells, Vol. 2*, Ratcliffe, N.A. and Rowley, A.F. (eds). Academic Press, London, pp. 513–562.
19. Yoshino, T.P. (1986). Surface membrane components of circulating invertebrate blood cells and their role in internal defense. In *Immunity in Invertebrates. Cells, Molecules and Defense Reactions*, Brehélin, M. (ed.). Springer-Verlag, Berlin, pp. 12–24.
20. Lackie, A.M. (1988). Haemocyte behaviour. In *Advances in Insect Physiology, Vol. 21*, Evans, P.D. and Wigglesworth, V.B. (eds). Academic Press, London, pp. 85–178.
21. Söderhäll, K., Smith, V.J. and Johansson, M.W. (1986). Exocytosis and uptake of bacteria by isolated haemocyte populations of two crustaceans. Evidence for cellular co-operation in the defence reactions of arthropods. *Cell. Tiss. Res.* **245,** 43–49.
22. Anggraeni, T. and Ratcliffe, N.A. (1991). Studies on cell-cell co-operation during phagocytosis by purified haemocyte populations of the Wax moth, *Galleria mellonella. J. Insect Physiol.* **37,** 453–460.
23. Ratcliffe, N.A., White, K.N., Rowley, A.F. and Walters, J.B. (1982). Cellular defense systems of the Arthropoda. In *The Reticuloendothelial System. A Comprehensive Treatise, Vol. 3:*

Phylogeny and Ontogeny, Cohen, N. and Sigel, M.M. (eds). Plenum Press, New York, pp. 167–255.

24. Sminia, T., Pietersma, K. and Scheerboom, J.E.M. (1973). Histological and ultrastructural observation on wound healing in the freshwater pulmonate *Lymnaea stagnalis*. *Z. Zellforsch. mikrosk. Anat.* **141,** 561–573.
25. Cherbas, L. (1973). The induction of an injury reaction in cultured haemocytes from saturniid pupae. *J. Insect Physiol.* **19,** 2011–2023.
26. Bohn, H. (1975). Growth promoting effect of haemocytes on insect epidermis *in vitro*. *J. Insect Physiol.* **21,** 1283–1293.
27. Prendergast, R.A. and Suzuki, M. (1970) Invertebrate protein simulating mediators of delayed hypersensitivity. *Nature* **227,** 277–279.
28. Bohn, H. (1986). Hemolymph clotting in insects. In *Immunity in Invertebrates: Cells, Molecules and Defense Reactions*, Brehélin, M. (ed.). Springer-Verlag, Berlin, pp. 188–207.
29. Durliat, M. (1985). Clotting processes in crustacean Decapoda. *Biol. Rev.* **60,** 473–498.
30. Graham, D.G., Tiffany, S.M. and Vogel, F.S. (1978). The toxicity of melanin precursors. *J. Invest. Dermatol.* **70,** 113.
31. Marks, D.H., Stein, E.A. and Cooper, E.L. (1979). Chemotactic attraction of *Lumbricus terrestris* coelomocytes to foreign tissue. *Dev. Comp. Immunol.* **3,** 277–285.
32. Howland, K.H. and Cheng, T.C. (1982). Identification of bacterial chemoattractants for oyster (*Crassostrea virginica*) hemocytes. *J. Invert. Pathol.* **39,** 123–132.
33. Söderhäll, K. and Smith, V.J. (1986). The prophenoloxidase activating system: The biochemistry of its activation and role in arthropod cellular immunity, with special reference to crustaceans. In *Immunity in Invertebrates. Cells, Molecules, and Defense Reactions*, Brehélin, M. (ed.). Springer-Verlag, Berlin, pp. 209–223.
34. Smith, V.J. and Söderhäll, K. (1986). Cellular immune mechanisms in the Crustacea. *Symp. Zool. Soc. Lond.* **56,** 59–79.
35. Takle, G.B. and Lackie, A.M. (1986). Chemokinetic behaviour of insect haemocytes *in vitro*. *J. Cell. Sci.* **85,** 85–94.
36. Day, N.K.B., Gewurz, H., Johannsen, R., Finstad, J. and Good, R.A. (1970). Complement and complement-like activity in lower vertebrates and invertebrates. *J. Exp. Med.* **132,** 941–950.
37. Phipps, D., Menger, M., Chadwick, J.S. and Aston, W.P. (1987). A cobra venom factor (CVF)-induced C3 convertase activity in the hemolymph of *Galleria mellonella*. *Dev. Comp. Immunol.* **11,** 37–46.
38. Bertheussen, K. (1982). Receptors for complement on echinoid phagocytes II. Purified human complement mediates echinoid phagocytosis. *Dev. Comp. Immunol.* **6,** 635–642.
39. Laulan, A., Lestage, J., Bouc, A.M. and Châteaureynaud-Duprat, P. (1988). The phagocytic activity of *Lumbricus terrestris* leukocytes is enhanced by the vertebrate opsonins: IgG and complement C3b fragment. *Dev. Comp. Immunol.* **12,** 269–277.
40. Ofek, I. and Sharon, N. (1988). Lectinophagocytosis. A molecular mechanism of recognition between cell sugars and lectins in the phagocytosis of bacteria. *Infection and Immunity* **56,** 539–547.
41. Renwrantz, L. and Stahmer, A. (1983). Opsonizing properties of an isolated haemolymph agglutinin and demonstration of lectin-like recognition molecules on the surface of hemocytes from *Mytilus edulis*. *J. Comp. Physiol.* **149,** 535–546.
42. Coombe, D.R., Ey, P.L. and Jenkin, C.R. (1984). Particle recognition by haemocytes from the colonial ascidian *Botrylloides leachii*: Evidence that the *B. leachii* HA-2 agglutinin is opsonic. *J. Comp. Physiol.* **154B,** 509–521.
43. Pendland, J.C., Heath, M.A. and Boucias, D.G. (1988). Function of a galactose-binding lectin from *Spodoptera exigua* larval haemolymph: Opsonization of blastospores from entomogeneous hyphomycetes. *J. Insect Physiol.* **34,** 533–540.
44. Richards, E.H. and Ratcliffe, N.A. (1990). Direct binding and lectin-mediated binding of erythrocytes to haemocytes of the insect, *Extatosoma tiaratum*. *Dev. Comp. Immunol.* **14,** 269–281.
45. Smith, V.J. and Söderhäll, K. (1983). β-1,3 Glucan activation of crustacean hemocytes *in vitro* and *in vivo*. *Biol. Bull.* **164,** 299–314.
46. Ratcliffe, N.A., Leonard, C.M. and Rowley, A.F. (1984). Prophenoloxidase activation, non-self recognition and cell cooperation in insect immunity. *Science* **226,** 557–559.
47. Johansson, M.W. and Söderhäll, K. (1988). Isolation and purification of a cell adhesion factor from crayfish blood cells. *J. Cell Biol.* **106,** 1795–1803.
48. Johansson, M.W. and Söderhäll, K. (1989). A peptide containing the cell adhesion sequence RGD can mediate degranulation and cell adhesion of crayfish granular haemocytes *in vitro*. *Insect Biochem.* **19,** 573–580.
49. Rantamaki, J., Durrant, H., Liang, Z., Ratcliffe, N.A., Duvic, B. and Söderhäll, K. (1991).

Isolation of a 90kD protein from haemocytes of *Blaberus craniifer* which has similar functional and immunological properties to the 76kD protein from crayfish haemocytes. *J. Insect Physiol.* **37,** 627–634.
50. Grimm-Jørgensen, Y. (1987). Somatostatin and calcitonin modulate gastropod internal defense mechanisms. *Dev. Comp. Immunol.* **11,** 487–499.
51. Anderson, R.S. (1976). Expression of receptors by insect macrophages. In *Phylogeny of Thymus and Bone Marrow—Bursa Cells*, Wright, R.K. and Cooper, E.L. (eds). Elsevier/North-Holland Biomedical Press, Amsterdam, pp. 27–34.
52. Anderson, R.S., Holmes, B. and Good, R.A. (1973). Comparative biochemistry of phagocytosing insect hemocytes. *Comp. Biochem. Physiol.* **46B,** 595–602.
53. Cheng, T.C. (1976). Aspects of substrate utilization and energy requirement during molluscan phagocytosis. *J. Invert. Pathol.* **27,** 263–268.
54. Chakravarti, B. and Chakravarti, D.N. (1987). Phagocytosis: An overview. *Pathol. Immunopathol. Res.* **6,** 316–342.
55. Shozawa, A., Suto, C. and Kumada, N. (1989). Superoxide production by the haemocytes of the freshwater snail *Biomphalaria glabrata*, stimulated by miracidia of *Schistosoma mansoni. Zool. Science* **6,** 1019–1022.
56. Dikkeboom, R., van der Knaap, W.P.W., van der Bovenkamp, W., Tijnagel, J.M.G.H. and Bayne, C.J. (1988). The production of toxic oxygen metabolites by hemocytes of different snail species. *Dev. Comp. Immunol.* **12,** 509–520.
57. Nakamura, M., Mori, K., Inooka, S. and Nomura, T. (1985). *In vitro* production of hydrogen peroxide by the amoebocytes of the scallop *Patinopecten yessoensis* (Jay). *Dev. Comp. Immunol.* **9,** 407–417.
58. Pilgrim, M. (1965). The coelomocytes of the maldanid polychaetes *Clymenella torquata* and *Euclymene oerstedi. J. Zool.* **147,** 30–37.
59. Kroschinski, J. and Renwrantz, L. (1988). Determination of pH values inside the digestive vacuoles of hemocytes from *Mytilus edulis. J. Invert. Pathol.* **51,** 73–79.
60. Cheng, T.C. (1983). The role of lysosomes in molluscan inflammation. *Amer. Zool.* **23,** 129–144.
61. Ratcliffe, N.A. and Gagen, S.J. (1977). Studies on the *in vivo* cellular reactions of insects: An ultrastructural analysis of nodule formation in *Galleria mellonella. Tissue and Cell* **9,** 73–85.
62. Bayne, C.J. (1983). Molluscan immunobiology. In *The Mollusca, Vol. 5*, Saleuddin, A.S.M. and Wilbur, K.M. (eds). Academic Press, San Diego, pp. 407–486.
63. Smith, V.J., Söderhäll, K. and Hamilton, M. (1984). β-1,3-glucan induced cellular defence reactions in the shore crab *Carcinus maenas. Comp. Biochem. Physiol.* **77A,** 635–639.
64. Gunnarsson, S.G.S. and Lackie, A.M. (1985). Hemocytic aggregation in *Schistocerca gregaria* and *Periplaneta americana* as response to injected substances of microbial origin. *J. Invert. Pathol.* **46,** 312–319.
65. Zlotkin, E., Gurevitz, M. and Shulov, A. (1973). The toxic effects of phenoloxidase from the haemolymph of tenebrionid beetles. *J. Insect Physiol.* **19,** 1057–1065.
66. Walters, J.B. and Ratcliffe, N.A. (1980). A comparison of the immune response of the wax moth, *Galleria mellonella*, to pathogenic and non-pathogenic bacteria. In *Aspects of Developmental and Comparative Immunology—1*, Solomon, J.B. (ed.). Pergamon Press, Oxford, pp. 147–152.
67. Fontaine, C.T., Bruss, R.G., Sanderson, I.A. and Lightner, D.V. (1975). Histopathological response to turpentine in the white shrimp *Penaeus setiferus. J. Invert. Pathol.* **25,** 321–330.
68. Walters, J.B. and Ratcliffe, N.A. (1983). Studies on the *in vivo* cellular reactions of insects: Fate of pathogenic and non-pathogenic bacteria in *Galleria mellonella* nodules. *J. Insect Physiol.* **29,** 417–424.
69. Huxham, I.M., Lackie, A.M. and McCorkindale, N.J. (1989). Inhibitory effects of cyclodepsipeptides, destruxins, from the fungus *Metarhizium anisopliae*, on cellular immunity in insects. *J. Insect. Physiol.* **35,** 97–105.
70. Christensen, B.M., Sutherland, D.R. and Gleason, L.N. (1984). Defense reactions of mosquitoes to filarial worms: Comparative studies on the response of three different mosquitoes to inoculated *Brugia pahangi* and *Dirofilaria immitis* microfilariae. *J. Invert. Pathol.* **44,** 267–274.
71. Götz, P. (1986). Mechanisms of encapsulation in dipteran hosts. *Symp. Zool. Soc. Lond.* **56,** 1–19.
72. Götz, P., Enderlein, G. and Roettgen, I. (1987). Immune reactions of *Chironomus* larvae (Insecta: Diptera) against bacteria. *J. Insect Physiol.* **33,** 993–1004.
73. Nappi, A.J. and Stoffolano, J.G. Jr. (1971). *Heterotylenchus autumnalis.* Hemocytic reactions and capsule formation in the host, *Musca domestica. Exp. Parasitol.* **29,** 116–125.
74. Persson, M., Vey, A. and Söderhäll, K. (1987). Encapsulation of foreign particles *in vitro* by separated blood cells from crayfish *Astacus leptodactylus. Cell Tissue Res.* **247,** 409–415.
75. Dularay, B. and Lackie, A.M. (1985). Haemocytic encapsulation and the prophenol-

oxidase-activation pathway in the locust *Schistocerca gregaria* Forsk. *Insect Biochem.* **15,** 827–834.
76. Ratner, S. and Vinson, S.B. (1983). Phagocytosis and encapsulation: Cellular immune responses in Arthropoda. *Amer. Zool.* **23,** 185–194.
77. Lackie, A.M. and Vasta, G.R. (1988). The role of galactosyl-binding lectin in the cellular immune response of the cockroach *Periplaneta americana* (Dictyoptera). *Immunology* **64,** 353–357.
78. Loker, E.S. (1979). Pathology and host responses induced by *Schistosomatium douthitti* in the freshwater snail *Lymnaea catascopium. J. Invert. Pathol.* **33,** 265–273.
79. Sminia, T., Borghart-Reinders, E. and Van de Linde, A.W. (1974). Encapsulation of foreign materials experimentally introduced into the freshwater snail *Lymnaea stagnalis. Cell Tiss. Res.* **153,** 307–326.
80. Cheng, T.C. and Garrabrant, T.A. (1977). Acid phosphatase in granulocytic capsules formed in strains of *Biomphalaria glabrata* totally and partially resistant to *Schistosoma mansoni. Int. J. Parasitol.* **7,** 467–472.
81. Dikkeboom, R., Tijnagel, J.M.G.H., Mulder, E.C. and van der Knaap, W.P.W. (1987). Hemocytes of the pond snail *Lymnaea stagnalis* generate reactive forms of oxygen. *J. Invert. Pathol.* **49,** 321–331.
82. Sminia, T., van der Knaap, W.P.W. and Boerrigter-Barendsen, L.H. (1982). Peroxidase-positive blood cells in snails. *J. Reticuloendothel. Soc.* **31,** 399–404.
83. Klühspies, G. (1983). Angeborene und erworbene Resistenz von *Lymnaea stagnalis* und *Radix ovata* gegen Miracidien-Invasion. *Z. Parsitenk.* **69,** 591–611.
84. Whitcomb, R.F., Shapiro, M. and Granados, R.R. (1974). Insect defence mechanisms against microorganisms and parasitoids. In *The Physiology of Insecta, Vol. 5*, Rockstein, M. (ed.). Academic Press, New York, pp. 447–536.
85. Lackie, A.M. (1980). Invertebrate immunity. *Parasitology* **80,** 393–412.
86. Valembois, P., Roch, P. and Boiledieu, D. (1982). Cellular defense systems of the Platyhelminthes, Nemertea, Sipunculida, and Annelida. In *The Reticuloendothelial System. A Comprehensive Treatise, Vol. 3: Phylogeny and Ontogeny*, Cohen, N. and Sigel, M.M. (eds). Plenum Press, New York, pp. 89–139.
87. Johnston, I.S. and Hildemann, W.H. (1982). Cellular defense systems of the Porifera. In *The Reticuloendothelial System. A Comprehensive Treatise. Vol. 3: Phylogeny and Ontogeny*, Cohen, N. and Sigel, M.M. (eds). Plenum Press, New York, pp 37–57.
88. Cooper, E.L. (1982). Invertebrate defense systems. An overview. In *The Reticuloendothelial System. A Comprehensive Treatise. Vol. 3: Phylogeny and Ontogeny*, Cohen, N. and Sigel, M.M. (eds). Plenum Press, New York, pp. 1–35.
89. Leippe, M. and Renwrantz, L. (1988). Release of cytotoxic and agglutinating molecules by *Mytilus* hemocytes. *Dev. Comp. Immunol.* **12,** 297–308.
90. Valembois, P., Roch, P. and Boiledieu, D. (1980). Natural and induced cytotoxicities in sipunculids and annelids. In *Phylogeny of Immunological Memory*, Manning, M.J. (ed.). Elsevier/North-Holland Biomedical Press, Amsterdam, pp. 47–55.
91. Bayne, C.J., Buckley, P.M. and DeWan, P.C. (1980). *Schistosoma mansoni*: cytotoxicity of hemocytes from susceptible snail hosts for sporocysts in plasma from resistant *Biomphalaria glabrata. Exp. Parasit.* **50,** 409–416.
92. Decker, J.M., Elmholt, A. and Muchmore, A.V. (1981). Spontaneous cytotoxicity mediated by invertebrate mononuclear cells towards normal and malignant vertebrate targets: Inhibition by defined mono- and disaccarides. *Cell. Immunol.* **59,** 161–170.
93. Yoshino, T.P. (1988). Phospholipase C-like activity in phagocytic cells of the Asian clam, *Corbicula fluminea*, and its possible role in cell-mediated cytolytic reactions. *J. Invert. Pathol.* **51,** 32–40.
94. Söderhäll, K., Wingren, A., Johansson, M.W. and Bertheussen, K. (1985). The cytotoxic reaction of hemocytes from the freshwater crayfish *Astacus astacus. Cell. Immunol.* **94,** 326–332.
95. Karp, R.D. (1990). Transplantation immunity in insects: Does allograft responsiveness exist? *Res. Immunol.* **141,** 923–927.
96. Raftos, D.A., Tait, N.N. and Briscoe, D.A. (1987). Cellular basis of allograft rejection in the solitary urochordate, *Styela plicata. Dev. Comp. Immunol.* **11,** 713–725.
97. Thomas, I.G. and Ratcliffe, N.A. (1982). Integumental grafting and immunorecognition in insects. *Dev. Comp. Immunol.* **6,** 643–654.
98. Taneda, Y. (1985). Simultaneous occurrence of fusion and non fusion reaction in two colonies in contact of the compound ascidian, *Botryllus primigenus. Dev. Comp. Immunol.* **9,** 371–375.
99. Raftos, D.A., Tait, N.N. and Briscoe, D.A. (1987). Allograft rejection and alloimmune memory in the solitary urochordate, *Styela plicata. Dev. Comp. Immunol.* **11,** 343–351.

100. Cooper, E.L. (1968). Transplantation immunity in annelids. I. Rejection of xenografts exchanged between *Lumbricus terrestris* and *Eisenia foetida*. *Transplantation* **6,** 322–337.
101. Dales, R.P. (1978). The basis of graft rejection in the earthworms *Lumbricus terrestris* and *Eisenia fetida*. *J. Invert. Pathol.* **32,** 264–277.
102. Johnston, I.S., Jokiel, P.L., Bigger, C.H. and Hildemann, W.H. (1981). The influence of temperature on the kinetics of allograft reactions in a tropical sponge and a reef coral. *Biol. Bull.* **160,** 280–291.
103. Karp, R.D. and Hildemann, W.H. (1976). Specific allograft reactivity in the sea star *Dermasterias imbricata*. *Transplantation* **22,** 434–439.
104. Varadarajan, J. and Karp, R.D. (1983). Histological versus morphological assessment of graft rejection in invertebrates. *Transplantation* **35,** 629–631.
105. Canicatti, C. (1990). Hemolysins: Pore-forming proteins in invertebrates. *Experientia* **46,** 239–244.
106. Giga, Y., Ikai, A. an Takahashi, K. (1987). The complete amino acid sequence of echinoidin, a lectin from the coelomic fluid of the sea urchin *Anthocidaris crassispina*. *J. Biol. Chem.* **262,** 6197–6203.
107. Renwrantz, L. (1983). Involvement of agglutinins (lectins) in invertebrate defense reactions: The immunobiological importance of carbohydrate-specific binding molecules. *Dev. Comp. Immunol.* **7,** 603–608.
108. Uhlenbruck, G., Pardoe, G.I., Prokop, O. and Ishiyama, I. (1972). The serological specificity of snail agglutinins (protectins). *Anim. Blood Grps. Biochem. Genet.* **3,** 125–139.
109. van der Knaap, W.P.W., Sminia, T., Schutte, R. and Boerrigter-Barendsen, L.H. (1983). Cytophilic receptors for foreignness, and some factors which influence phagocytosis by invertebrate leukocytes: *In vitro* phagocytosis by amoebocytes of the snail *Lymnaea stagnalis*. *Immunology* **48,** 377–383.
110. Coombe, D.R., Ey, P.L. and Jenkin, C.R. (1984). Ascidian haemagglutinins: Incidence in various species, binding specificities and preliminary characterisation of selected agglutinins. *Comp. Biochem. Physiol.* **77B,** 811–819.
111. Vasta, G.R., Warr, G.W. and Marchalonis, J.J. (1982). Tunicate lectins: Distribution and specificity. *Comp. Biochem. Physiol.* **73B,** 887–900.
112. Bretting, H. and Kabat, E.A. (1976). Purification and characterization of the agglutinins from the sponge *Axinella polypoides* and a study of their combining sites. *Biochemistry* **15,** 3228–3236.
113. Boswell, C.A. and Bayne, C.J. (1984). Isolation, characterization and functional assessment of a hemagglutinin from the plasma of *Biomphalaria glabrata*, intermediate host of *Schistosoma mansoni*. *Dev. Comp. Immunol.* **8,** 559–568.
114. Goldstein, I.J., Hughes, R.C., Monsigny, M., Osawa, T. and Sharon, N. (1980). What should be called a lectin? *Nature* **285,** 66.
115. Finstad, C.L., Good, R.A. and Litman, G.W. (1974). The erythrocyte agglutinin from *Limulus polyphemus* hemolymph: Molecular structure and biological function. *Ann. N.Y. Acad. Sci.* **234,** 170–180.
116. Castro, V.M., Boman, H.G. and Hammarström, S. (1987). Isolation and characterization of a group of isolectins with galactose/*N*-acetylgalactosamine specificity from hemolymph of the giant silk moth *Hyalophora cecropia*. *Insect Biochem.* **17,** 513–523.
117. Rögener, W., Renwrantz, L. and Uhlenbruck, G. (1985). Isolation and characterization of a lectin from the hemolymph of the cephalopod *Octopus vulgaris* (Lam.) inhibited by α-D-lactose and *N*-acetyl-lactosamine. *Dev. Comp. Immunol.* **9,** 605–616.
118. Nicolson, G.L. (1974). The interactions of lectins with animal cell surfaces. *Int. Rev. Cytol.* **30,** 89–190.
119. Komano, H., Nozawa, R., Mizuno, D. and Natori, S. (1983). Measurement of *Sarcophaga peregrina* lectin under various physiological conditions by radioimmunoassay. *J. Biol. Chem.* **258,** 2143–2147.
120. Pendland, J.C. and Boucias, D.G. (1986). Characteristics of a galactose-binding hemagglutinin (lectin) from hemolymph of *Spodoptera exigua* larvae. *Dev. Comp. Immunol.* **10,** 477–487.
121. Komano, H., Mizuno, D. and Natori, S. (1980). Purification of lectin induced in the hemolymph of *Sarcophaga peregrina* larvae on injury. *J. Biol. Chem.* **255,** 2919–2924.
122. Takahashi, H., Komano, H. and Natori, S. (1986). Expression of the lectin gene in *Sarcophaga peregrina* during normal development and under conditions where the defence mechanism is activated. *J. Insect Physiol.* **32,** 771–779.
123. Takahashi, H., Komano, H., Kawaguchi, N., Kitamura, N., Nakanishi, S. and Natori, S. (1985). Cloning and sequencing of cDNA of *Sarcophaga peregrina* humoral lectin induced on injury of the body wall. *J. Biol. Chem.* **260,** 12228–12233.
124. Shiraishi, A. and Natori, S. (1988). Humoral mediator-dependent activation of the

Sarcophaga lectin gene. *FEBS Letts.* **232,** 163–166.
125. Kubo, T., Komano, H., Okada, M. and Natori, S. (1984). Identification of hemagglutinating protein and bactericidal activity in the hemolymph of adult *Sarcophaga peregrina* on injury of the body wall. *Dev. Comp. Immunol.* **8,** 283–291.
126. Habets, L., Vieth, U.C. and Hermann, G. (1979). Isolation and new biological properties of *Arion empiricorum* lectin. *Biochim. Biophys. Acta* **582,** 154–163.
127. Komano, H. and Natori, S. (1985). Participation of *Sarcophaga peregrina* humoral lectin in the lysis of sheep red blood cells injected into the abdominal cavity of larvae. *Dev. Comp. Immunol.* **9,** 31–40.
128. Schmid, L.S. (1975). Chemotaxis of hemocytes from the snail, *Viviparus malleatus. J. Invert. Pathol.* **25,** 125–131.
129. Minnick, M.F., Rupp, R.A. and Spence, K.D. (1986). A bacterial-induced lectin which triggers hemocyte coagulation in *Manduca sexta. Biochem. Biophys. Res. Comm.* **137,** 729–735.
130. Molyneux, D.H., Takle, G., Ibrahim, E.A.R. and Ingram, G.A. (1986). Insect immunity to Trypanosomatidae. *Symp. Zool. Soc. Lond.* **56,** 117–144.
131. Ratcliffe, N.A. and Rowley, A.F. (1987). Insect responses to parasites and other pathogens. In *Immune Responses in Parasitic Infections: Immunology, Immunopathology and Immunoprophylaxis, Vol. 4*, Soulsby, E.J.L. (ed.). CRC Press, Boca Raton, Florida, pp. 271–332.
132. Kaaya, G.P. (1989). A review of the progress made in recent years on research and understanding of immunity in insect vectors of human and animal diseases. *Insect. Sci. Applic.* **10,** 751–769.
133. Welburn, S.C., Maudlin, I. and Ellis, D.S. (1989). Rate of trypanosome killing by lectins in mid guts of different species and strains of *Glossina. Med. Vet. Entomol.* **3,** 77–82.
134. Welburn, S.C. and Maudlin, I. (1990). Haemolymph lectins and the maturation of trypanosome infections in tsetse. *Med. Vet. Entomol.* **4,** 43–48.
135. Azumi, K., Yokosawa, H. and Ishii, S. (1990). Halocyamines: Novel antimicrobial tetrapeptide-like substances isolated from the hemocytes of the solitary ascidian *Halocynthia roretzi. Biochem.* **29,** 159–165.
136. Rinehart, K.L. Jr., Glover, J.B., Hughes, R.G. Jr., Renis, H.E., McGovren, J.P., Swynenberg, E.B., Stringfellow, D.A., Kuentzel, S.L. and Li, L.H. (1981). Didemnins: Antiviral and antitumor depsipeptides from a Caribbean tunicate. *Science* **212,** 933–935.
137. Anderson, R.S. (1984). Lysozyme—an inducible protective agent in invertebrate serum. In *Comparative Pathobiology, Vol. 6: Invertebrate Blood Cells and Serum Factors*, Cheng, T.C. (ed.). Plenum Press, New York, pp.173–185.
138. Chadwick, J.S. (1970). Relation of lysozyme concentration to acquired immunity against *Pseudomonas aeruginosa* in *Galleria mellonella. J. Invert. Pathol.* **15,** 455–456.
139. Boman, H.G., Fay, I., Hofsten, P.V., Kockum, K., Lee, J.-Y., Xanthopoulos, K.G., Bennich, H., Engström, Å., Merrifield, B.R. and Andreu, D. (1986). Antibacterial immune proteins in insects—a review of some current perspectives. In *Immunity in Invertebrates, Cells, Molecules, and Defense Reactions*, Brehélin, M. (ed.). Springer-Verlag, Berlin, pp.63–73.
140. Boman, H.G. and Hultmark, D. (1987). Cell-free immunity in insects. *Ann. Rev. Microbiol.* **41,** 103–126.
141. Faye, I. (1990). Acquired immunity in insects. The recognition of nonself and the subsequent onset of immune protein genes. *Res. Immunol.* **141,** 927–932.
142. Sun, S.-C., Lindström, I., Boman, H.G., Faye, I. and Schmidt, O. (1990). Hemolin: an insect immune protein belonging to the immunoglobin super family. *Science* **250,** 1729–1732.
143. Steiner, H., Andreu, D. and Merrifield, R.B. (1988). Binding and action of cecropin and cecropin analogues: antibacterial peptides from insects. *Biochim. Biophys. Acta* **939,** 260–266.
144. Trenczek, T. and Faye, I. (1988). Synthesis of immune proteins in primary cultures of fat body from *Hyalophora cecropia. Insect Biochem.* **18,** 299–312.
145. Trenczek, T. (1986). Injury and immunity in insects: *In vitro* studies with fat body and haemocyte primary cultures from *Hyalophora cecropia* pupae. *Dev. Comp. Immunol.* **10,** 627.
146. Spies, A.G., Karlinsey, J.E. and Spence, K. (1986). The immune proteins of the darkling beetle, *Eleodes* (Coleoptera: Tenebrionidae). *J. Invert. Pathol.* **47,** 234–235.
147. Morishima, I., Suginaka, S., Ueno, T. and Hirano, H. (1990). Isolation and structure of cecropins, inducible antibacterial peptides, from the silkworm, *Bombyx mori. Comp. Biochem. Physiol.* **95B,** 551–554.
148. Hoffmann, J.A. and Hoffmann, D. (1990). The inducible antibacterial peptides of dipteran insects. *Res. Immunol.* **141,** 910–918.
149. Chain, B.M. and Anderson, R.S. (1983). Antibacterial activity of the coelomic fluid from

the polychaete *Glycera dibranchiata*. II. Partial purification and biochemical characterization of the active factor. *Biol. Bull.* **164,** 41–49.
150. Chain, B.M. and Anderson, R.S. (1983). Antibacterial activity of the coelomic fluid of the polychaete, *Glycera dibranchiata*. I. The kinetics of the bactericidal reaction. *Biol. Bull.* **164,** 28–40.
151. Valembois, P., Roch, P., Lassègues, M. and Cassand, P. (1982). Antibacterial activity of the hemolytic system from the earthworm *Eisenia fetida andrei*. *J. Invert. Pathol.* **40,** 21–27.
152. Vaillier, J., Cadoret, M.A., Roch, P. and Valembois, P. (1985). Protein analysis of earthworm coelomic fluid. III. Isolation and characterization of several bacteriostatic molecules from *Eisenia fetida andrei*. *Dev. Comp. Immunol.* **9,** 11–20.
153. Valembois, P., Roch, P. and Lassègues, M. (1986). Antibacterial molecules in annelids. In *Immunity in Invertebrates. Cells, Molecules, and Defense Reactions*, Brehélin, M. (ed.). Springer-Verlag, Berlin, pp. 74–93.
154. Lassègues, M., Roch, P. and Valembois, P. (1989). Antibacterial activity of *Eisenia fetida andrei* coelomic fluid: Specificity of the induced activity. *J. Invert. Pathol.* **54,** 28–31.
155. Karp, R.D. (1990). Inducible humoral immunity in insects: does an antibody-like response exist in invertebrates? *Res. Immunol.* **141,** 932–934.
156. Laulan, A., Morel, A., Lestage, J., Delaage, M. and Châteaureynaud-Duprat, P. (1985). Evidence of synthesis by *Lumbricus terrestris* of specific substances in response to an immunization with a synthetic hapten. *Immunology* **56,** 751–758.
157. Brillouet, C., Leclerc, M., Binaghi, R.A. and Luquet, G. (1984). Specific immune response in the sea star *Asterias rubens*: Production of 'antibody-like' factors. *Cell. Immunol.* **84,** 138–144.
158. Coombe, D.R., Ey, P.L. and Jenkin, C.R. (1984). Self/non-self recognition in invertebrates. *Quart. Rev. Biol.* **59,** 231–255.
159. Renwrantz, L., Schäncke, W., Harm, H., Erl, H., Liebsch, H. and Gercken, J. (1981). Discriminative ability and function of the immunobiological recognition system of the snail *Helix pomatia*. *J. Comp. Physiol.* **141,** 477–488.
160. Fitzgerald, S.W. and Ratcliffe, N.A. (1982). Evidence for the presence of subpopulations of *Arenicola marina* coelomocytes identified by their selective response towards Gram +ve and Gram −ve bacteria. *Dev. Comp. Immunol.* **6,** 23–34.
161. Bertheussen, K. (1983). Complement-like activity in sea urchin coelomic fluid. *Dev. Comp. Immunol.* **7,** 21–31.
162. Rowley, A.F. and Ratcliffe, N.A. (1980). Insect erythrocyte agglutinins. *In vitro* opsonization experiments with *Clitumnus extradentatus* and *Periplaneta americana* haemocytes. *Immunology* **40,** 483–492.
163. Bradley, R.S., Stuart, G.S., Stiles, B. and Hapner, K.D. (1989). Grasshopper haemagglutinin: Immunocytochemical localization in haemocytes and investigation of opsonic properties. *J. Insect Physiol.* **35,** 353–361.
164. Renwrantz, L. and Mohr, W. (1978). Opsonizing effect of serum and albumin gland extracts on the elimination of human erythrocytes from the circulation of *Helix pomatia*. *J. Invert. Pathol.* **31,** 164–171.
165. Harm, H. and Renwrantz, L. (1980). The inhibition of serum opsonins by a carbohydrate and the opsonizing effect of purified agglutinin on the clearance of nonself particles from the circulation of *Helix pomatia*. *J. Invert. Pathol.* **36,** 64–70.
166. Mullainadhan, P. and Renwrantz, L. (1986). Lectin-dependent recognition of foreign cells by hemocytes of the mussel, *Mytilus edulis*. *Immunobiol.* **171,** 263–273.
167. Ashida, M. and Yamazaki, H.I. (1990). Biochemistry of the phenoloxidase system in insects: with special reference to its activation. In *Moulting and Metamorphosis*, Ohnishi, E. and Ishizaki, H. (eds). Japan Sci. Soc. Press, Tokyo/Springer-Verlag, Berlin, pp. 239–265.
168. Ratcliffe, N.A. (1991). The prophenoloxidase system and its role in arthropod immunity. In *Phylogenesis of Immune Functions*, Warr, G.W. and Cohen, N. (eds). CRC Press, Boca Raton, Florida, pp. 46–65.
169. Smith, V.J. and Söderhäll, K. (1991). A comparison of phenoloxidase activity in the blood of marine invertebrates. *Dev. Comp. Immunol.* **15,** 251–261.
170. Saul, S.J. and Sugumaran, M. (1986). Protease inhibitor controls prophenoloxidase activation in *Manduca sexta*. *FEBS Letts.* **208,** 113–116.
171. Söderhäll, K., Vey, A. and Ramstedt, M. (1984). Hemocyte lysate enhancement of fungal spore encapsulation by crayfish hemocytes. *Dev. Comp. Immunol.* **8,** 23–29.
172. Durrant, H.J., Ratcliffe, N.A., Chen, C. Lectin enhancement of prophenoloxidase activation in the cockroach, *Blaberus discoidalis*. Unpublished observations.
173. Söderhäll, K., Rögener, W., Söderhäll, I., Newton, R.P. and Ratcliffe, N.A. (1988). The properties and purification of a *Blaberus crani-*

ifer plasma protein which enhances the activation of haemocyte prophenoloxidase by a β-1,3-glucan. *Insect. Biochem.* **18,** 323–330.
174. Ratcliffe, N.A. and Millar, D.A. (1988). Comparative aspects and possible phylogenetic affinities of vertebrate and invertebrate blood cells. In *Vertebrate Blood Cells*, Rowley, A.F. and Ratcliffe, N.A. (eds). University Press, Cambridge, pp. 1–17.
175. Klein, J. (1989). Are invertebrates capable of anticipatory immune responses? *Scand. J. Immunol.* **29,** 499–505.
176. Rowley, A.F., Rhodes, C.P. and Ratcliffe, N.A. (1984). Protochordate leucocytes: a review. *Zool. J. Linn. Soc.* **80,** 283–295.
177. Cooper, E.L. (1976). Evolution of blood cells. *Ann. Immunol.* (*Inst. Pasteur*) **127c,** 817–825.
178. Stang-Voss, C. (1974). On the ultrastructure of invertebrate hemocytes: An interpretation of their role in comparative hematology. In *Contemporary Topics in Immunobiology, Vol. 4*, Cooper, E.L. (ed.). Plenum Press, London, pp. 65–76.
179. Cooper, E.L. (1981). Immunity in invertebrates. *CRC Crit. Rev. Immunol.* **2,** 1–32.
180. Brillouet, C., Leclerc, M., Panijel, J., and Binaghi, R.A. (1981). *In vitro* effect of various mitogens on starfish (*Asterias rubens*) axial organ cells. *Cell. Immunol.* **57,** 136–144.
181. Leclerc, M., Redziniak, G., Panijel, J. and El Lababidi, M. (1977). Reactions induced in vertebrates by invertebrate cell suspensions. I. Specific effects of sea star axial organ cells injection. *Dev. Comp. Immunol.* **1,** 299–310.
182. Leclerc, M., Redziniak, G., Panijel, J. and El Lababidi, M. (1977). Reactions induced in vertebrates by invertebrate cell suspensions. II. Non adherent axial organ cells as effector cells. *Dev. Comp. Immunol.* **1,** 311–320.
183. Leclerc, M., Brillouet, C., Luquet, G. and Binaghi, R.A. (1986). Production of an antibody-like factor in the sea star *Asterias rubens*: Involvement of at least three cellular populations. *Immunology* **57,** 479–482.
184. Hildemann, W.H. and Uhlenbruck, G. (1974). Invertebrate immunology. *Progr. Immunol.* **2,** 292–317.
185. Ermak, T.H. (1976). The hematogenic tissue of tunicates. In *Phylogeny of Thymus and Bone Marrow-Bursa Cells*, Wright, R.K. and Cooper, E.L. (eds). Elsevier/North-Holland Biomedical Press, Amsterdam, pp. 45–56.
186. Wright, R.K. (1976). Phylogenetic origin of the vertebrate lymphocyte and lymphoid tissue. In *Phylogeny of Thymus and Bone Marrow-Bursa Cells*, Wright, R.K. and Cooper, E.L. (eds). Elsevier/North-Holland Biomedical Press, Amsterdam, pp. 57–70.
187. Warr, G.W. and Marchalonis, J.J. (1978). Specific immune recognition by lymphocytes: An evolutionary perspective. *Q. Rev. Biol.* **53,** 225–241.
188. Wright, R.K. and Ermak, T.H. (1982). Cellular defense systems of the Protochordata. In *The Reticuloendothelial System. A Comprehensive Treatise, Vol. 3*, Cohen, N. and Sigel, M.M. (eds). Plenum Press, New York, pp. 283–320.
189. Bertheussen, K. and Seljelid, R. (1982). Receptors for complement on echinoid phagocytes. I. The opsonic effect of vertebrate sera on echinoid phagocytosis. *Dev. Comp. Immunol.* **6,** 423–431.
190. Marchalonis, J.J. and Schluter, S.F. (1990). On the relevance of invertebrate recognition and defence mechanisms to the emergence of the immune response of vertebrates. *Scand. J. Immunol.* **32,** 13–20.
191. Vasta, G.R., Hunt, J., Marchalonis, J.J. and Fish, W.W. (1986). Galactosyl-binding lectins from the tunicate *Didemnum candidum.* Purification and physicochemical characterization. *J. Biol. Chem.* **261,** 9174–9181.
192. Robey, F.A. and Liu, T-Y. (1981). Limulin: A C-reactive protein from *Limulus polyphemus. J. Biol. Chem.* **256,** 969–975.
193. Hildemann, W.H. (1977). Specific immunorecognition by histocompatibility markers. The original polymorphic system of immunoreactivity characteristic of all multicellular animals. *Immunogenetics* **5,** 193–202.
194. Shalev, A., Pla, M., Ginsburger-Vogel, T., Echalier, G., Lögdberg, L., Björck, L., Colombani, J. and Segal, S. (1983). Evidence for β_2-microglobulin-like and H-2-like antigenic determinants in *Drosophila. J. Immunol.* **130,** 297–302.
195. Marchalonis, J.J., Vasta, G.R., Warr, G.W. and Barker, W.C. (1984). Probing the boundaries of the extended immunoglobulin family of recognition molecules: jumping domains, convergence and minigenes. *Immunol. Today* **5,** 133–142.
196. Shalev, A., Lögdberg, L. and Björck, L. (1987). β_2-microglobulin, histocompatibility and H-Y antigen homologues in invertebrates and early vertebrates. In *Invertebrate Models, Cell Receptors and Cell Communication*, Greenberg, A.H. (ed.). Karger, Basel, pp. 49–77.
197. Rosenshein, I.L., Schluter, S.F., Vasta, G.R. and Marchalonis, J.J. (1985). Phylogenetic conservation of heavy chain determinants of vertebrates and protochordates. *Dev. Comp.*

Immunol. **9,** 783–795.

198. Rowley, A.F. (1977). The role of the haemocytes of *Clitumnus extradentatus* in haemolymph coagulation. *Cell Tiss. Res.* **182,** 513–524.
199. Rowley, A.F. and Ratcliffe, N.A. (1978). A histological study of wound healing and hemocyte function in the wax-moth *Galleria mellonella. J. Morphol.* **157,** 181–200.
200. Fitzgerald, S.W. and Ratcliffe, N.A. (1989). *In vivo* cellular reactions and clearance of bacteria from the coelomic fluid of the marine annelid, *Arenicola marina* L. (Polychaeta). *J. Exp. Zool.* **249,** 293–307.
201. Poinar, G.O. Jr., Leutenegger, R. and Gotz, P. (1968). Ultrastructure of the formation of a melanotic capsule in *Diabrotica* (Coleoptera) in response to a parasitic nematode (Mermithidae). *J. Ultrastruct. Res.* **25,** 293–306.
202. Mullett, H., Ratcliffe, N.A. and Rowley, A.F. (1993). The generation and characterization of anti-insect blood cell monoclonal antibodies. *J. Cell Sci.* **105,** 93–100.

3

FISHES

M.J. Manning

Department of Biological Sciences, University of Plymouth, UK

3.1 INTRODUCTION

Jawless vertebrates (Agnatha) were the first vertebrates to appear in the fossil record. They date from the Ordovician and their present-day descendants, the hagfish and lampreys, retain certain features characteristic of this primitive level. The earliest jawed vertebrates radiated in the Palaeozoic and by the Devonian period had already given rise to three main lines. One line, that of cartilaginous fish (Chondrichthyes), became a successful group, mainly of marine carnivores, as represented today by elasmobranchs such as the sharks and rays. The other two lines were bony fish (Osteichthyes) of which the fleshy-finned fish (Sarcopterygii) have only a few modern representatives (Dipnoi, Coelacanthini) but are the group which gave rise to the tetrapods. A different line of bony fishes (the Actinopterygii) form the majority of present-day fish, mainly owing to the success of one large group, the order Teleostei (Figure 3.1)[1,2].

There are over 20000 species of teleost fish, representing almost half the number of vertebrate species in existence today. Indeed the teleosts, as masters of the aquatic environment, are as diverse and highly specialized in the water as the higher tetrapods are on land[3]. Furthermore, they include the majority of food fishes. Thus the interests of aquaculture and fisheries provide a major impetus to the study of fish immunology, for a better understanding of the host response to pathogens and to improve methods for vaccinating fish against disease[4,5]. Several families of teleost fish are represented amongst commercially farmed fish: salmonids (trout and salmon), carp, catfish, tilapia, eels, sea bass, sea bream, marine flatfish, etc. Thus a good range has been examined, the most widely studied species being the rainbow trout, *Oncorhynchus mykiss* (formerly known as *Salmo gairdneri*).

Immunology: A Comparative Approach. Edited by R.J. Turner

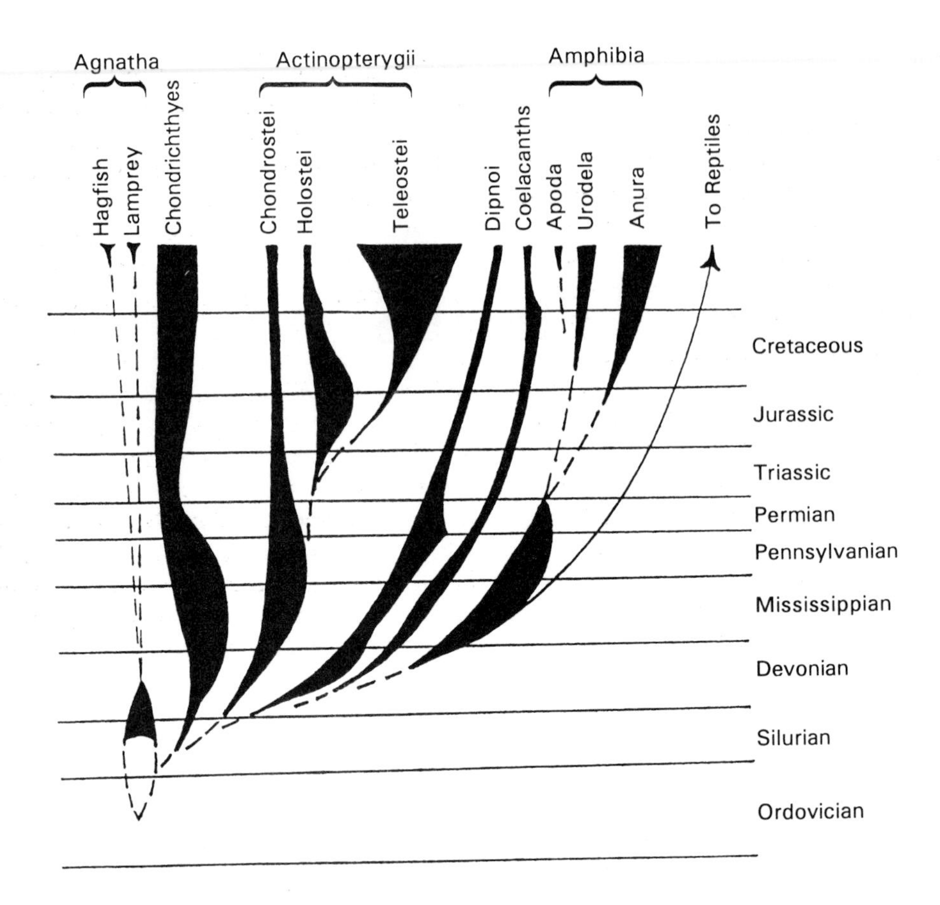

Figure 3.1. Phylogenetic relationships in the fishes and amphibians: distribution through time of the different groups. From reference 2, by permission.

3.2 LYMPHOID TISSUES: STRUCTURE AND FUNCTION

Agnathan lymphoid tissue lacks organization and appears to be intimately admixed with general haemopoietic (blood-forming) areas. In all jawed vertebrates, on the other hand, there is special provision of a microenvironment (the thymus) for the differentiation of T-cells. Separate secondary lymphoid organs provide the architectural arrangements for trapping antigen and ensuring that it is presented in suitable form to cells capable of reacting to it. In teleost fish the secondary organs are the spleen, the head-kidney (pronephros) and the trunk kidney (Figure 3.2)[6].

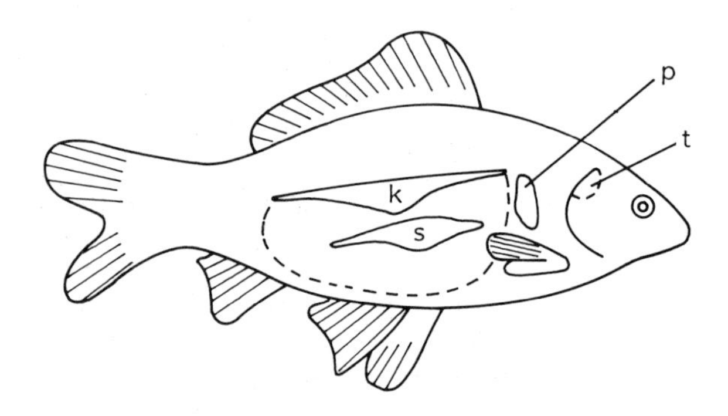

Figure 3.2. Lymphoid organs of carp: k = trunk kidney (opisthonephros); p = head kidney (pronephros), s = spleen; t = thymus.

3.2.1 Thymus

The thymus starts its embryological development as a thickening of the pharyngeal epithelium but in most fish, in contrast to tetrapod vertebrates, instead of budding off from the pharyngeal epithelium (e.g. *Xenopus*, see Chapter 4) it retains contact with the pharynx (Figure 3.3). The superficial position is particularly noticeable in species like the rainbow trout where, even in adult fish, the thymic cells are separated from the pharyngeal lumen by only a single layer of cells[7]. Furthermore, in young trout, pores a few micrometres in diameter have been demonstrated on the thymic surface[8], leading to the suggestion that, in immature fish, the thymus might be amenable to direct stimulation from the surrounding water. In other species the thymus becomes more internalized[9], although it may still be linked to the pharyngeal surface by strands of tissue[10]. In the Dipnoi, thymic development more closely resembles that of amphibians[2].

The thymus in fish, as in all jawed vertebrates, is composed of lymphocytes and lymphoblasts contained within a network of reticular epithelial cells. A distinction between cortex and medullary regions occurs in the thymus of some cartilaginous and bony fish; for example, amongst teleosts, in the tilapia[11], the viviparous blenny[12] and in mullet (Figure 3.4). This distinction seems to be lacking in other species, such as the rainbow trout[8]. Derivatives of the thymic epithelium such as Hassall's corpuscles, myoid cells and cysts, although occasionally described in fish[9], are generally less conspicuous than in tetrapods[13]. Involution of the thymus occurs in fish and may be related to age, season and hormones[14]. In rainbow trout, signs of involution begin to appear at

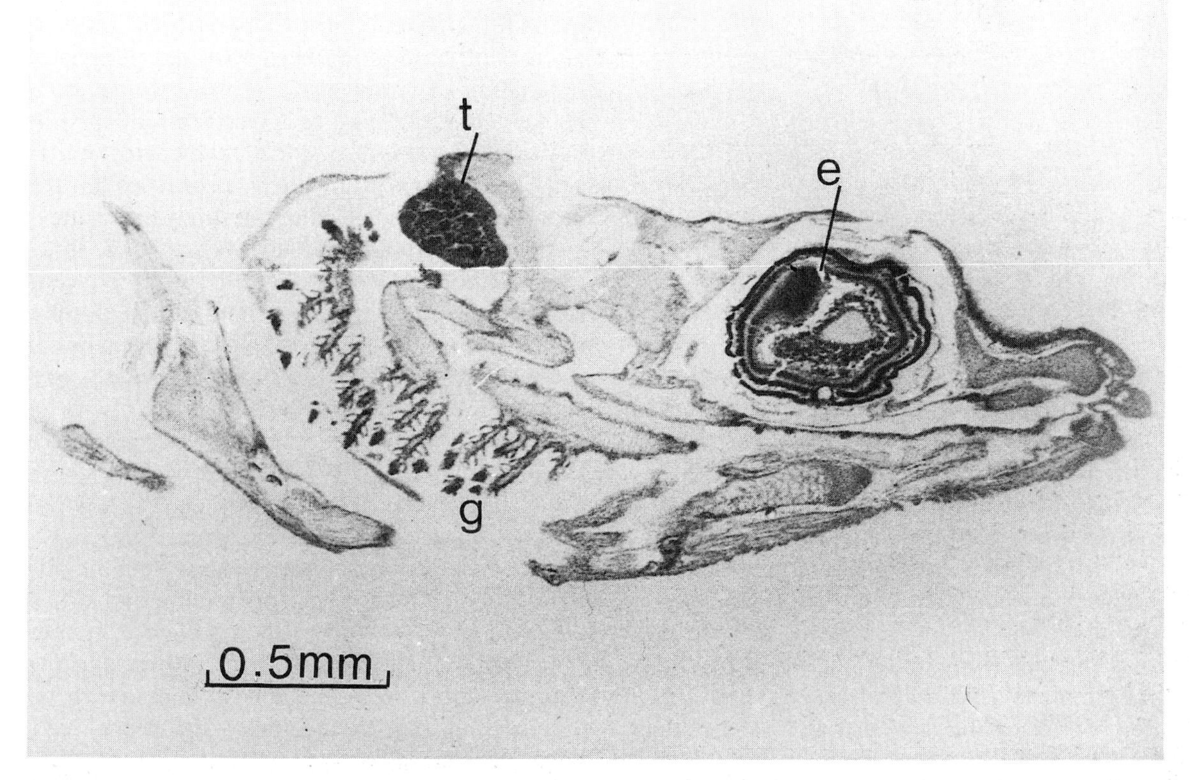

Figure 3.3. Longitudinal section through the head of a 2-week old carp, *Cyprinus carpio*, to show the superficial position of the thymus in relation to the pharyngeal cavity: e = eye; g = gills; t = thymus.

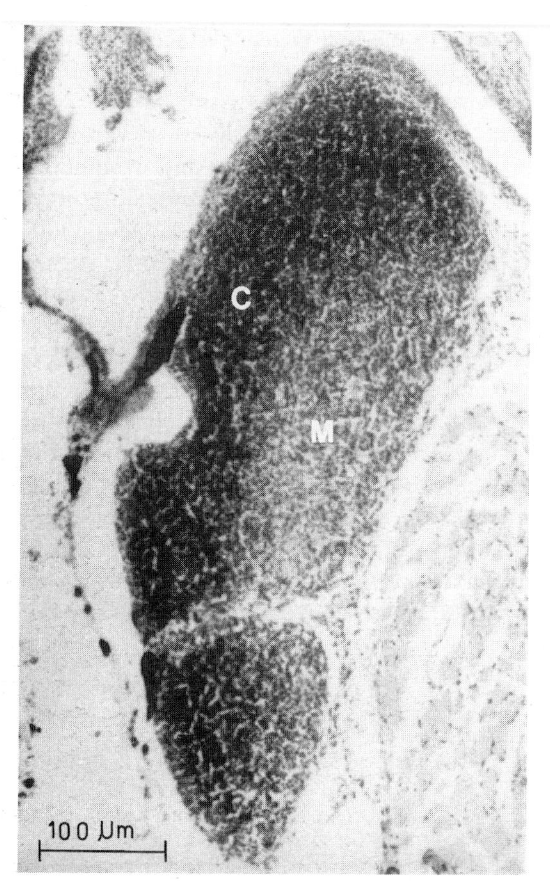

Figure 3.4. Transverse section through the thymus of a mullet, *Chelon labrosus*, approximately 6 months old, showing distinct cortex (C) and medulla (M).

15 months of age[15]. The thymus is often difficult to find in adult fish and has been described as being fully involuted in a number of species[16]. In the teleost *Astyanex mexicanus*, however, the thymus could be identified histologically even when it was not visible with the naked eye[17].

The thymus plays a primary, non-executive role in fish as in mammals. Its importance in ontogeny is apparent from the fact that it is the first organ to acquire mature lymphocytes during histogenesis of the lymphoid tissues. This has been shown for a variety of freshwater and marine fish including tilapia[18,19], salmon[20], carp[21], rainbow trout[8,22], the rock fish *Sebasticus marmoratus*[23], the antarctic fish *Harpagifer antarcticus*[24], sea bream[25] and dogfish[26]. Moreover in carp and rainbow trout, the onset of alloimmune reactivity to foreign skin grafts can be correlated with maturation of the thymus, along with lymphocytic differentiation in the kidney and the presence of small lymphocytes in the peripheral blood[21,22]. In sea bream it was noted that, during the course of development, the thymus and head kidney grew closer together and an apparent migration of cells was seen between them[25]. Furthermore, using a mouse anti-carp thymocyte monoclonal antibody, positively labelled thymocytes could be identified in the developing carp thymus as early as day 4 post-fertilization, with positively stained lymphocytes appearing in the head kidney some 3 days later[27]. Also in older fish (rainbow trout), thymocytes labelled with isotope (tritiated thymidine) by intra-thymic injection could be traced to the kidney and spleen[28]. These experiments suggest that cells are exported from the thymus to the periphery as in mammals.

There is strong evidence, both on functional grounds and from the use of monoclonal antibodies as cell surface markers, that the peripheral population of lymphocytes in fish includes both B-like cells and T-like cells (see Section 3.3.3). It has further been shown that thymocytes are T-cell-like in respect of these properties in the channel catfish[29]. Despite this body of indirect evidence the crucial case, as to whether the functional dichotomy into T-like cells and B-like cells is based upon a thymic origin of the T-cells, remains to be proven for fish. The problem lies in the technical difficulty of achieving early and complete ablation of the thymus. This is in marked contrast to amphibians (*Xenopus*) where the operation is easy (see Chapter 4). The anatomy of the head in bony fish is such that the thymus lies quite close to the ear (Figure 3.5) and is wrapped around muscles which pass from the otic region of the skull to the operculum and which are used in respiration[25]. The problem is greater for some species of fish than for others and the thymus is more accessible in trout than in carp[30].

In young fish thymectomy operations have been performed on tilapia[18] and on rainbow trout[30,31]. Removal of the thymus in trout fry aged 4 to 24 days post-hatching had no noticeable

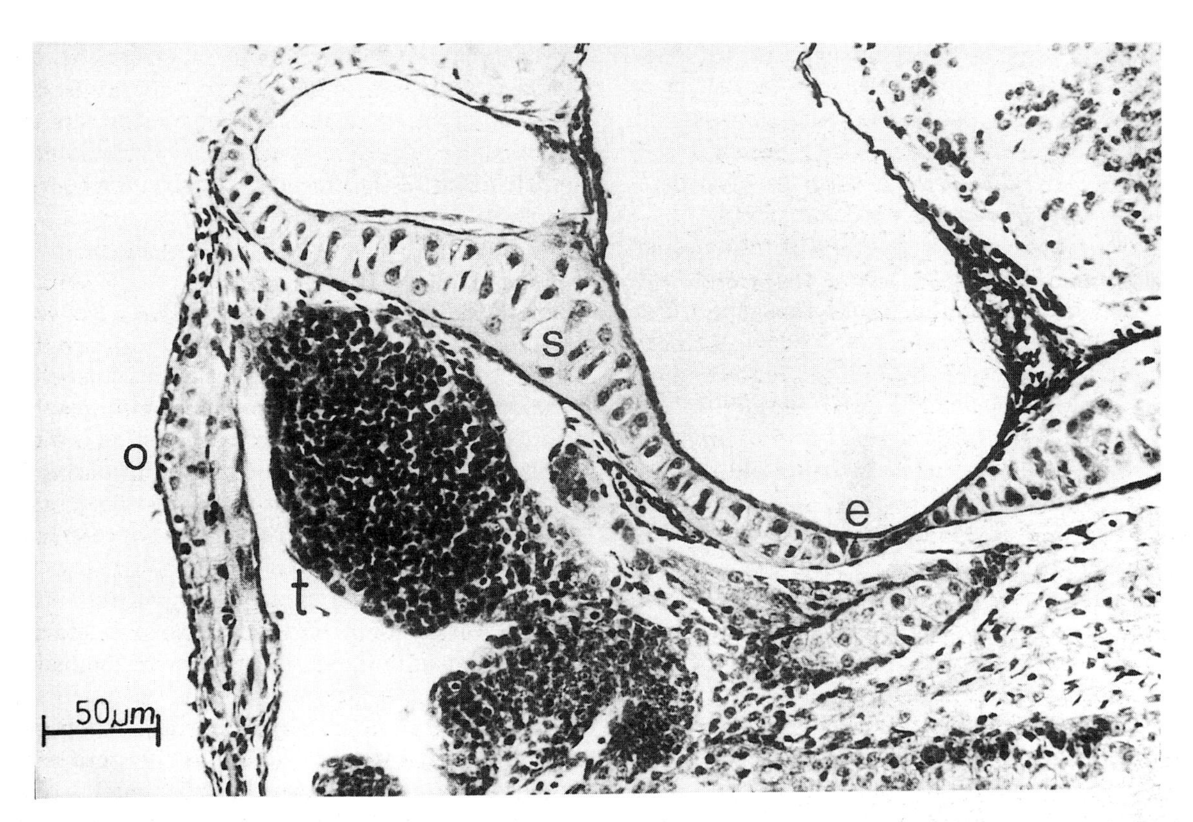

Figure 3.5. Transverse section through head of a 7-day-old carp, *Cyprinus carpio*, showing the position of the thymus. Respiratory muscles (the adductor muscles of the operculum, not seen in this section) pass from the otic region of the skull to the operculum, crossing the region of the developing thymus: e = ear; o = operculum; s = otic region of skull; t = thymus.

effect on the lymphocytic population of the kidney, but decreased the numbers of lymphoid cells in the spleen. In trout thymectomized at day 14, the lymphoid cells of both spleen and kidney demonstrated a proliferative response to antigenic stimulation when the fish were immunized with a soluble protein antigen 1 month after thymectomy and killed 4–6 weeks later. This response was, however, less intense than in sham-thymectomized fish. Similarly, a lymphocytic response to allografts occurs in young thymectomized fish but is less intense than in controls.

Thymectomy experiments on slightly older fry or on adult fish have been performed to determine whether the thymus is a source of helper T-cells and/or of suppressor T-cells. In fish, soluble proteins and cells such as foreign erythrocytes, are putative thymus-dependent antigens, while bacterial cells and lipopolysaccharides are believed to be thymus-independent (capable of directly stimulating B-cells)[32–34]. In tilapia, thymectomy of 2-month-old fry[18] or of adult fish[35] reduced the antibody response to sheep erythrocytes. In rainbow trout thymectomized at 1 or 2 months of age, antibody production against a different thymus-dependent antigen (human gamma globulin, HGG) was diminished in the secondary response[34]. In a similar experiment on adult rainbow trout, a reduced response to HGG occurred (in a primary response) but only if immunization was postponed until 9 months after thymectomy[36]. One possible explanation for these findings is that peripheral T-cells are quite long lived, so if thymectomy is performed after the cells have already been seeded to the periphery, any effect of thymus removal may become apparent only in the longer term, as has been suggested for mammals[37]. In general, thymectomy has less effect on the antibody response to putative thymus-independent antigens such as killed bacterial cells (*Aeromonas salmonicida*), which in most cases remains similar to that in the control fish[30,34,36].

Some tentative evidence that in certain circumstances the thymus may have a suppressive role in antibody production comes from short-term thymectomy experiments on adult fish. Thus, when adult carp were thymectomized then immunized 4 weeks later, the fish showed elevated serum antibody titres at day 7 to *Aeromonas salmonicida* vaccine, while in 1-year-old rainbow trout, the antibody response to HGG was markedly increased in the thymectomized fish[6,30].

Overall, the outcome of removal of the thymus in fish appears to depend on the completeness of the operation, the stage of development of the fish, the interval between thymectomy and immunization and doubtless many other factors, including the type and dose of antigen used. The results are compatible with the hypothesis that the role of the thymus in fish is the same as that in higher vertebrates, i.e. that it is a source of T-cells, but the case has yet to be proven unequivocally. In addition the fish thymus may play a minor role as a residence for a small population of B-cells. This view arises from the identification of plasma cells in the thymus[9,38], antibody production by the thymus in culture[39] and the detection of antibody-forming cells in the thymus of tilapia after immunization[11]. This need not, however, detract from the conclusion that the fish thymus is a typical primary lymphoid organ since B-cell elements occur in the thymus of many vertebrate groups[13].

Finally, in relation to the evolution of the thymus, it has been argued that the thymus is unlikely to have arisen *de novo* in the jawed fishes, even though it is doubtful whether a true thymus exists in the agnathans[2]. No such organ could be identified in the pharynx of hagfish[40]. The ammocoete larva of the lamprey possesses a number of lymphocytic accumulations in the pharynx but their structure is somewhat dissimilar to that usually associated with the thymus[41] and there is no evidence as to their function. However, the ammocoete larva uses its pharynx for filter feeding, as did the vertebrate ancestor, and the constant trapping of microorganisms in the food stream of the pharynx may have provided an evolutionary impetus which led to further specialization of the pharyngeal lymphoid cell populations[2].

3.2.2 Secondary Lymphoid Organs

The secondary lymphoid organs show considerable diversity in vertebrates, which to some extent can be correlated with the evolution of other organ systems. In poikilotherms, much of the lymphoid tissue occurs in organs with sinusoidal blood flow where lymphoid cells can cluster and respond to the presence of antigens[2]. In fish these organs include the kidney and spleen. Elasmobranchs also have a variety of other such organs, including Leydig's organ in the oesophageal wall; the liver; and the epigonal organ of the gonads[42].

The kidney of fish is provided with a low-pressure renal portal system. The sluggish blood supply ramifies into sinusoids with a haemopoietic stroma whose framework of reticular cells and phagocytes resembles that of the bone marrow of tetrapods[43]. Indeed, the kidney probably plays a role similar to that of bone marrow as a blood-forming organ and source of stem cells. In addition, however, it has an important function as a lymphoid organ. In response to antigenic stimulation both the head kidney (pronephros) and the trunk kidney (opisthonephros) show a proliferative response involving cells of the haemopoietic parenchyma. The cytoplasm of these cells contains RNA and stains deep pink with pyronin (pyroninophilic cells). In carp which were injected with a soluble protein antigen (HGG) in Freund's complete adjuvant, this response peaked at week 3 with the formation of clusters of pyroninophilic cells, especially in the pronephros[44]. Antibody-forming cells have been identified in both the head and trunk kidney of a variety of fish species using either haemolytic plaque-forming assays[45–49] or, more recently, methods based upon solid phase immunoenzyme techniques[50]. These studies suggest that the kidney is an important antibody-producing organ in teleost fish.

Pigment-containing cells (melanomacrophages) occur in the kidney and spleen. They increase in number following immunization, particularly in the spleen but also in the pronephros and opisthonephros. They are particularly conspicuous in the response to particulate antigens and are often seen in areas where pyroninophilic cells had been prominent a few weeks earlier[44]. These pigment-containing cells form aggregates (melanomacrophage centres[51]) with which lymphocytes and antibody-containing cells may also be associated. Small circulatory lymphocytes have been reported to home onto the melanomacrophage aggregates[52]. Most of the pigment is melanin but various amounts of lipofuscin and haemosiderin may also be present[53]. The melanomacrophages have various functions as scavengers of unwanted material or as metabolic depots (e.g. for iron). In addition, the melanin may have an important role in protection against free radical damage[54]. It has also been suggested that melanomacrophages function as primitive germinal centres in the production of memory cells[44,55], especially as long-term retention of antigen (up to a year) has been reported in the melanomacrophage centres of carp[56].

The important role of the kidney in haemopoiesis is apparent early in the ontogeny of teleost fish. The pronephros is the first organ to contain haemopoietic tissue based on histological criteria, although erythrocytes and macrophages are present before differentiation of the kidney so the yolk sac is probably the earliest organ of limited haemopoiesis[57]. The pronephros forms the first excretory kidney of the larval fish with haemopoietic tissue interspersed between the excretory tubules (Figure 3.6). Later in development the

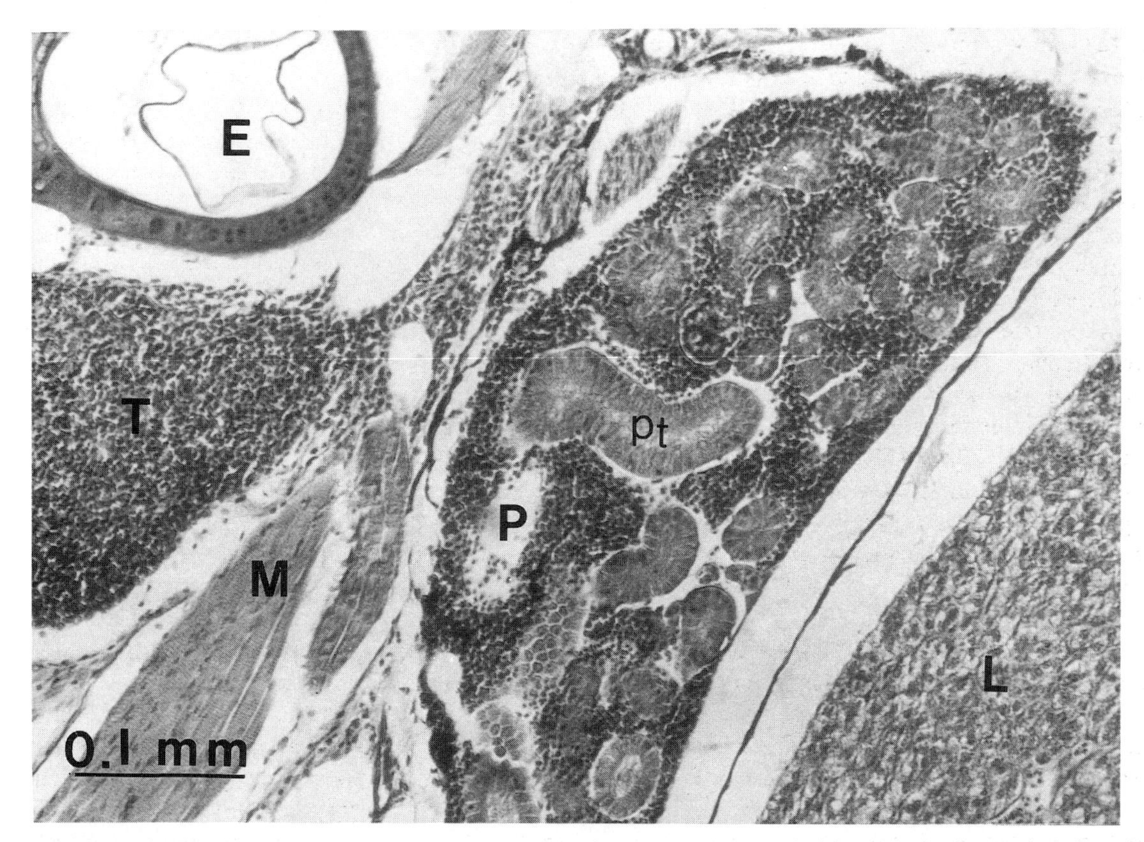

Figure 3.6. Longitudinal section through the head of a young guppy, *Lebistes reticulatus*, showing the head kidney (pronephros) with excretory tubules and intertubular lymphoid/haemopoietic tissue: E = ear; L = liver; M = respiratory muscle; P = pronephros; pt = pronephric tubule; T = thymus.

excretory role is assumed by the larval mesonephros and the adult opisthonephros (trunk kidneys). The pronephros then loses its excretory function. However, the pronephros is retained in teleost fish as a haemopoietic/lymphoid organ, while the trunk kidney also becomes important in this respect in its intertubular regions.

The spleen is the last organ to develop during histogenesis of the lymphoid system in teleost fish[20-22]. Like the kidney, it plays a role both in haemopoiesis and in immune reactivity. Melanomacrophage centres are often prominent, typically with a higher iron content than those of the kidney. This is perhaps due to the splenic breakdown of effete erythrocytes. In many teleosts, unlike tetrapods, there is no clear distinction between splenic red pulp and white pulp. Moreover, the proliferative response to antigenic stimulation takes place in a diffuse pattern throughout the organ and no thymus-dependent and thymus-independent areas have as yet been distinguished (Figure 3.7).

Ellipsoids are conspicuous in the spleen of most, but not all, teleost fish[58]. These are terminal capillaries with a thin endothelial cell layer surrounded by a sheath of reticular fibres and macrophages. They open into the red pulp sinuses which drain into the venous system. Injected carbon particles rapidly accumulate in the ellipsoid walls (Figure 3.8). In the case of antigens, the fates of injected soluble antigens and of particulate forms appear to be somewhat different. When killed bacteria (*Aeromonas*) are injected, immunoreactivity first appears in the ellipsoids and in solitary phagocytes in the red pulp. Later, after about 2 weeks, when specific antibodies are

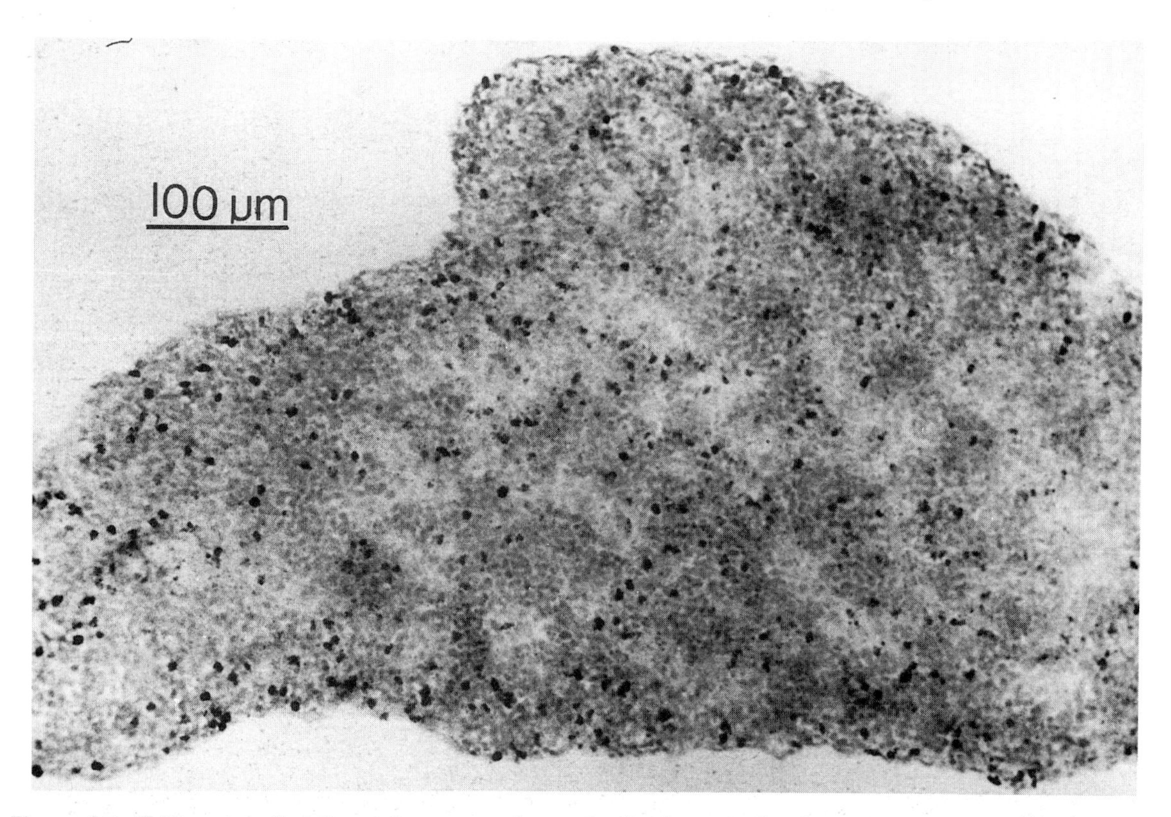

Figure 3.7. Tritium-labelled thymidine autoradiograph of spleen section from a young carp, *Cyprinus carpio*, 3 weeks after injection of antigen (HGG in Freund's complete adjuvant). Proliferating cells are indicated by black silver grains above labelled nuclei. Note the scattered distribution of the labelling with lack of distinct white pulp and red pulp areas.

produced, immunoreactivity disappears from the ellipsoids and is subsequently observed at the surface of cells in and around melanomacrophage centres, where it is retained for at least one year[59]. When soluble antigens are injected (HGG in carp[60] and bovine serum albumin in plaice[61]), they can be detected, after an initial delay, in the ellipsoids, where they appear to be associated with the reticulin fibres of the ellipsoid walls (Figure 3.9). Homologous immunoglobulin can also be detected in these areas in both control and immunized fish, suggesting that Fc receptors for homologous immunoglobulin may be present. In the primary response, the peak specific serum antibody titre coincides with the period of increased antigen trapping[44]. It has also been shown that in secondary antibody responses, or when antigen is administered in the form of immune complexes, the antigen-trapping in the ellipsoids takes place much more rapidly[62]. These findings, taken together, have led to the suggestion that antigen trapping in teleosts occurs in a manner similar to that of birds and mammals, i.e. it is localized in the form of immune complexes and retained extracellularly, as it is in the germinal centres of higher vertebrates. In some secondary immune responses, pyroninophilic cells collect within the ellipsoid sheaths in large numbers and form nodules (Figure 3.10). Immunoglobulin-containing cells have been identified in the pyroninophilic cell clusters of immunized fish[63]. These collections of pyroninophilic cells and of cells associated with melanomacrophage centres are therefore considered to be analogues of the homoiotherm germinal centre. True germinal centres do not occur in lower vertebrates, however.

The spleen is lacking in agnathans although in the lamprey the typhlosole of the gut accommodates haemopoietic tissue and is sometimes

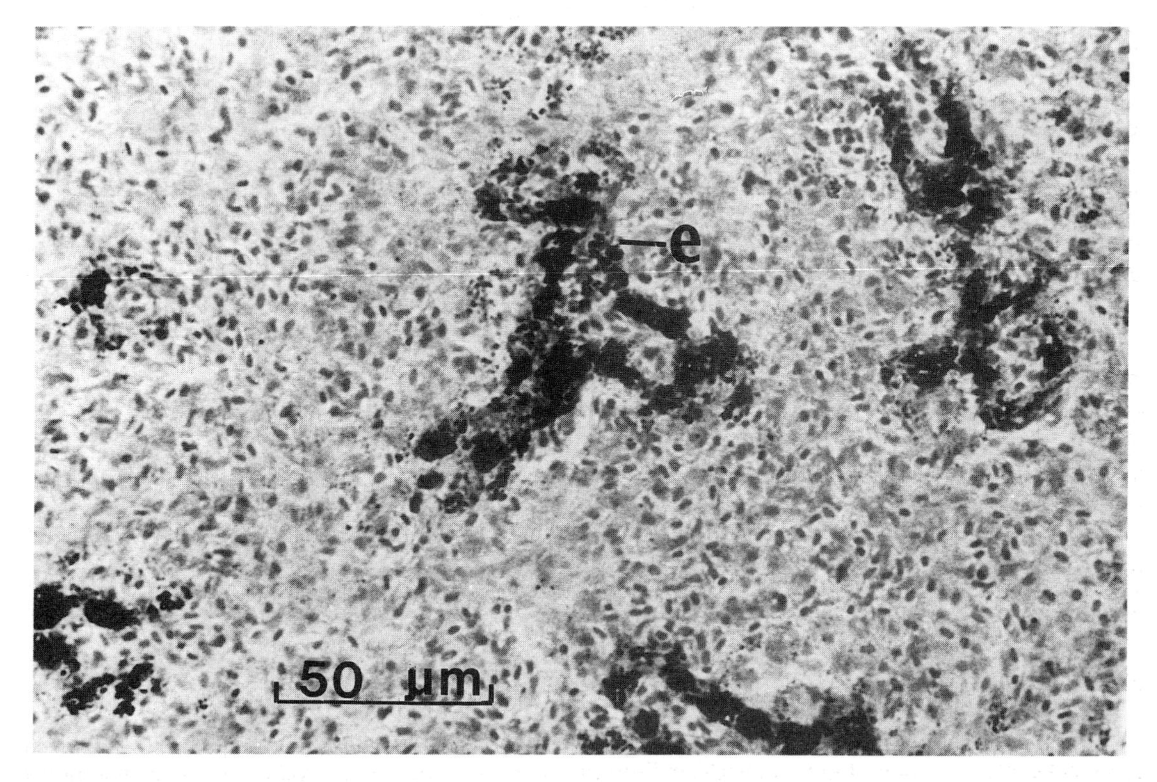

Figure 3.8. Spleen from carp, *Cyprinus carpio,* killed 6 h after injection of colloidal carbon. Carbon particles are localized in the splenic ellipsoids (e).

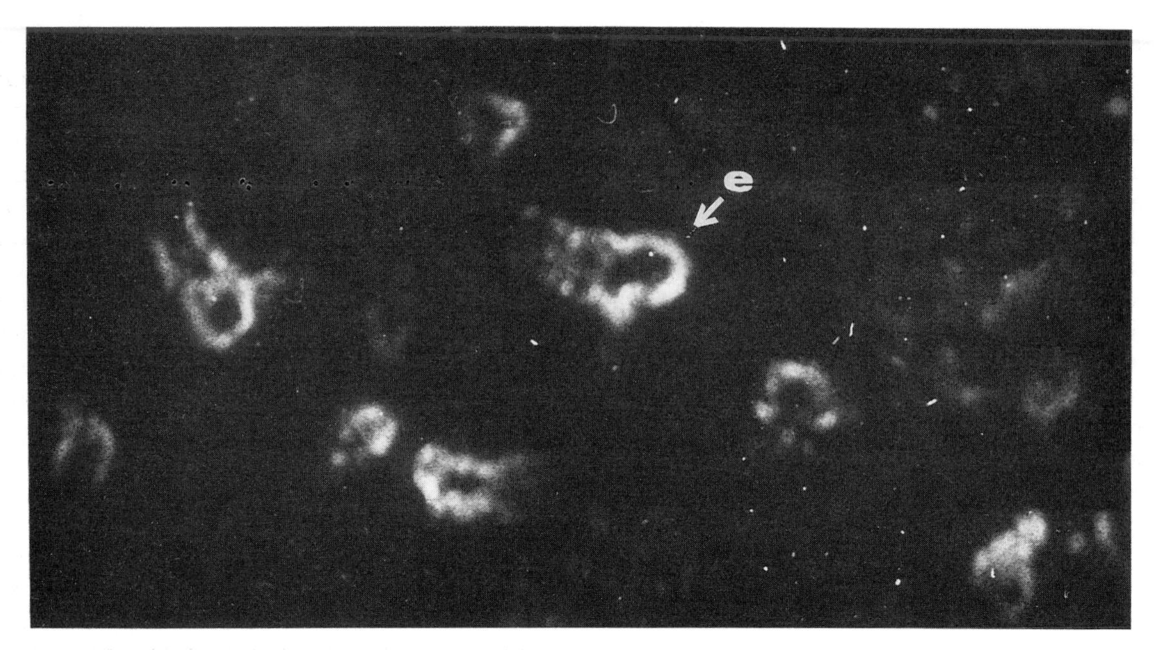

Figure 3.9. Spleen from carp, *Cyprinus carpio*. Immunofluorescence preparation to show antigen localization 3 weeks after injection of HGG in Freund's complete adjuvant. Fluorescence is seen in the ellipsoids following incubation of the section with a fluorescein-labelled sheep anti-HGG antiserum. e = splenic ellipsoids.

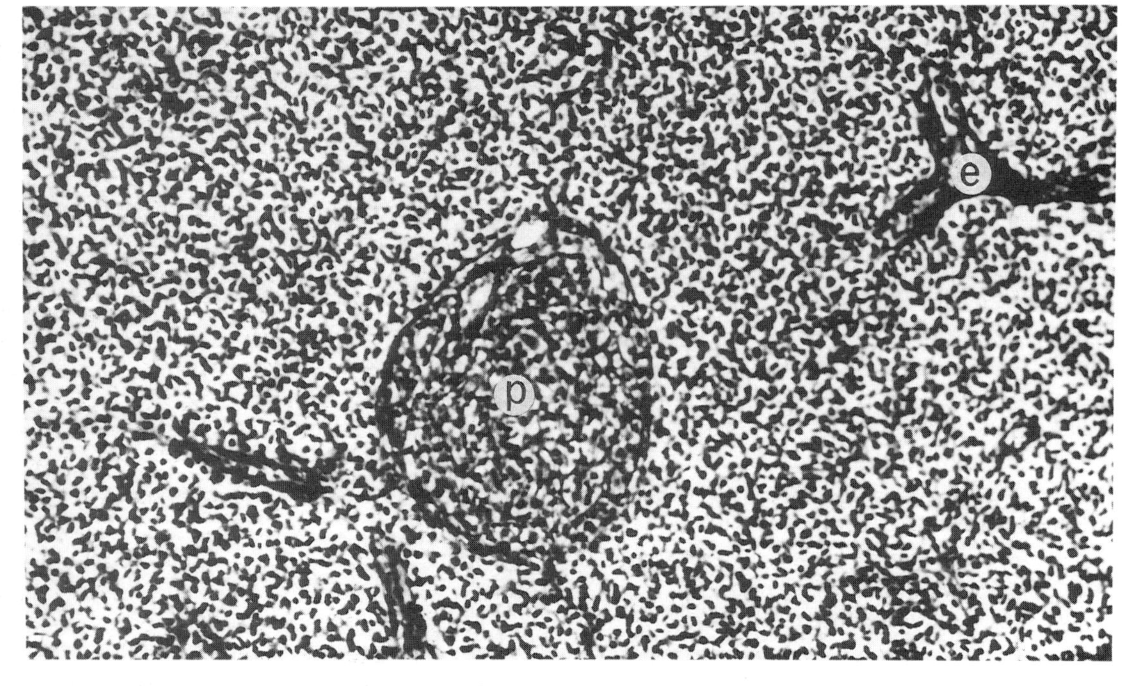

Figure 3.10. Silver-stained preparation of carp, *Cyprinus carpio*, given a booster injection of the same antigen 2 months after primary immunization, and killed 1 week later. A mass of pyroninophilic cells is contained within a metalophilic sheath which appears to be an expansion from the reticulum of the ellipsoid. e = splenic ellipsoids; p = area of pyroninophilic cells.

regarded as a primitive spleen. A true spleen is present in all groups of jawed fishes. In the Chondrichthyes, it is a large organ with some differentiation into red pulp and white pulp. As in teleosts, it contains ellipsoids and is a haemopoietic organ. Plasma cells have been described in the spleen of a variety of cartilaginous fishes[64] and the spleen is believed to be the major antibody-producing organ in this group of fish. Both in the bullhead shark, *Heterodontus japonicus*[65], and in the skate, *Raja kenojeii*[66], considerably more immunoglobulin-containing cells were found in the spleen than in the other lymphoid organs (Leydig's organ, epigonal organ and spiral valve of the intestine). In the skate there were two populations of cells, one containing high molecular weight immunoglobulin (IgM), the other an immunoglobulin of a lower molecular weight class[66].

3.2.3 Mucosa-associated Lymphoid Tissues

The mucosal surfaces of gut, skin and gills are protected by both humoral and cellular mechanisms. In addition, the mucus itself forms a physical barrier which is constantly being secreted and shed. Non-specific protective agents such as lysozyme are found in the mucus, together with immunoglobulin and specific antibody, as has been shown, for example, in plaice[67], channel catfish[68], and rainbow trout[69]. In all fish groups, the gut wall is infiltrated by lymphocytes and other leucocytes. Lymphocytes also occur in the skin[70] and gills[71]. In the skin, immunoglobulin-containing lymphocytes have been found in the epidermis of rainbow trout[70]. Also, in the same species, specific antibody-producing cells were detected in subepidermal sites after immunization[69].

Although mammalian-type Peyer's patches are lacking in fishes, various scattered leucocytes or aggregates of lymphoid cells are found in all groups[72,73]. In the hagfish, haemopoietic sites have been described in the lamina propria of the gut wall[74]. In several groups of jawed fishes, including cartilaginous fish[26,72], chondrostean bony fish (paddlefish and sturgeons)[73] and sarcopterygian lungfish (Dipnoi)[42], the internal surface of the gut is increased by means of a spiral valve. Substantial accumulations of lymphoid cells are present in this region. In teleost fish, lymphoid cells occur in all parts of the gut, most extensively in the intestine, where lymphocytes, granulocytes, macrophages and plasma cells are present in and under the intestinal epithelium, either scattered individually or in small groups. Two distinct regions of the intestine can be identified in teleosts, the anterior first segment and the posterior second segment. In carp, many more large intraepithelial macrophages are found in the second segment than in the first segment, and this posterior part of the intestine is considered to be an important immunological site in teleost fish[75,76].

A number of current studies on the gut-associated immune system of teleost fish relate to the commercial interests of aquaculture, particularly with the aim of improving methods for immunizing fish against disease. Procedures which involve administration of the antigen by injection are reasonably effective. Other routes of administration are considerably more practical for the fish farmer but remain somewhat problematic although they produce worthwhile results with some vaccines[5]. These alternative routes include direct immersion of the fish in an antigen bath or administration of the vaccine orally. Uptake of antigen across the gut epithelium is probably involved in both techniques since it has been shown that during immersion some of the vaccine is swallowed by the fish[5]. A number of current studies are therefore concerned with research on the fate of antigens administered enterically; also with attempts to improve the efficacy of vaccination by the use of adjuvants, for example detergents, lipid preparations, saponins, and agents which add three-dimensionality to soluble antigenic preparations or that act as protective vesicles[4,77]. In addition, there is considerable interest in knowing whether fish have distinct systemic and local (mucosal) immune compartments, as is the case in higher vertebrates.

Recent studies on gut immunology have included the use of soluble antigens (e.g. ferritin and HGG) and particulate antigens such as killed cells of bacteria pathogenic to fish (e.g. *Vibrio anguillarum* vaccines). In some experiments the antigen has been administered orally, in others by anal intubation. The latter route brings the antigen directly into contact with the immunologically reactive second segment of the intestine without prior passage through the anterior digestive regions of the gut.

Antigens reaching the second gut segment, anally or orally, are transported from the lumen to large intraepithelial macrophages. Finally the antigen (or determinants thereof) appears on the surface of these macrophages, suggesting an antigen-processing and presenting function[78,79]. Surface-immunoglobulin-positive lymphocytes (putative B-cells) also occur in the second gut segment although the proportion of surface-immunoglobulin-positive to surface-immunoglobulin-negative lymphocytes was found to be lower than that in the pronephros[79]. The demonstration of these immune components in the second gut segment of carp, together with the finding that antigen-binding and antibody-producing cells can be induced in the posterior intestinal epithelium of rainbow trout by antigen administered orally[80], support the interpretation that an intestinal secretory immune system is present in fish.

When particulate antigens such as *V. anguillarum* vaccines were administered orally or anally in studies on several different species of agastric (stomachless) or gastric, marine or freshwater species, the antigen could be detected in the second gut segment. In some but not all cases[81,82], protective immunity ensued. In the ayu, *Pleuroglossus altivelia*, oral vaccination with *V. anguillarum* led to agglutinating antibody activity in the skin mucus, but not in the serum. Furthermore, this anti-*Vibrio* activity prevented the colonization of viable *V. anguillarum* on the skin[83].

Other studies in which antigen was administered via the gut or by immersion have also noted the presence of antigen-specific antibodies in mucus or bile, but not in the serum[84–86]. This indicates that mucosal antibody may have been produced locally, a suggestion borne out by the finding that radiolabelled homologous immunoglobulin injected intravenously was not transferred to the mucus or bile[87]. Furthermore, a monoclonal antibody directed against carp serum immunoglobulin reacted with relatively few of the antibody-producing cells of the gut compared with many of those of the pronephros, suggesting that the gut cells may contain a different (secretory) form of immunoglobulin[79]. Although a number of studies on teleost serum and mucus immunoglobulins have documented the similarity between the two forms (in their antigenicity and in their tetrameric structure)[69], an analysis of the bile immunoglobulin of a marine teleost, the sheepshead, showed that biliary immunoglobulin exists as a dimer in physiological buffer and that the molecular weight of its heavy chain differs from that of the serum immunoglobulin[88]. Again this suggests a difference between mucosal and serum antibodies.

When soluble antigens are administered enterically, the intact molecule may be taken up from the gut lumen and transported as antigen or antigenic fragments through the enterocytes of the second gut segment into the bloodstream. Alternatively, they may be transferred to intraepithelial macrophages, some of which possibly migrate to the spleen or kidney[78]. Under these circumstances antigen-specific antibody can appear in the serum and may, or may not, also occur in the mucus. The presence of antibodies in the serum, but not in the mucus, has been reported after regular oral administration in carp. This may subsequently lead to immunosuppression[86]. The enteric uptake of soluble antigens can be enhanced by the use of adjuvants[77]. Both the ability of the gastrointestinal tract of fishes to absorb intact macromolecules[89] and the occurrence of orally induced immunosuppression indicate some similarities with the gut uptake of proteins and the phenomenon of orally-induced tolerance shown by higher vertebrates.

3.3 NON-SPECIFIC AND SPECIFIC HUMORAL IMMUNITY

3.3.1 Non-specific Immune Factors and Complement

A number of non-immunological humoral factors and body secretions are thought to contribute to the natural resistance of fish[90–92]. These include:

1. *Lectins.* Lectins are involved in the agglutination of microorganisms or the precipitation of soluble substances. They are proteins or glycoproteins of non-immune origin and characterized by their interaction with carbohydrates, for which they are considered recognition molecules.
2. *Lytic enzymes.* Cell lysis can be mediated via the complement system, also by single enzymes such as lysozyme. Lysozyme is found in phagocytic cells, serum and mucus. It acts upon the peptidoglycan component of the cell wall of microorganisms.
3. *Transferrin.* Transferrin is an iron-binding glycoprotein. Iron is an essential growth element for microorganisms. Their growth is therefore inhibited in the presence of transferrin owing to its high iron-binding power.
4. *Enzyme inhibitors.* Inhibitors such as anti-proteinases and anti-trypsinases act by neutralizing the activity of pathogen exoenzymes.
5. *Interferon.* The production of interferon can be readily induced upon infection. Its production has been demonstrated in a number of fish species following infection by various pathogenic viruses[93].
6. *C-reactive protein.* C-reactive protein is a common serum component that may increase significantly upon exposure to bacterial endotoxin. It is characterized by its Ca^{++}-dependent binding to phosphocholine-containing molecules. Such molecules are commonly found in the cell walls or surface structure of invading microorganisms.

Fish therefore appear to be well endowed with a battery of innate mechanisms of resistance to disease. Many of these non-specific defences precede specific immunity in phylogeny and have analogues in the invertebrates (Chapter 2).

Complement

The complement system is an important part of the organism's defence against microorganisms. The third complement component (C3) is the key element in this system. It is present in all vertebrates. Activation of C3 can occur either by the alternative (antibody-independent) pathway or by the classical (antibody-dependent) pathway. In lampreys only the alternative pathway is used and the lamprey complement system seems to function primarily to promote phagocytosis (opsonization) rather than in cytolysis[94]. Studies have been carried out by several groups on the isolation and characterization of cDNA clones for the so-called IgM of hagfish. These predict an amino acid sequence which bears close similarity to mammalian complement C3, rather than to immunoglobulin[95]. Lamprey C3 has now been isolated and shown by its physicochemical characteristics to be homologous to mammalian C3[96]. It seems likely therefore that the complement system had an early phylogenetic origin.

With the evolution of the jawed fishes and the emergence of the immunoglobulins, the classical (antibody-dependent) pathway of complement activation becomes evident. As in higher vertebrates, the jawed fishes use complement to mediate target cell lysis, as well as to cause opsonization. A series of studies on the nurse shark, *Ginglyostoma cirratum*, clearly established that elasmobranch immunoglobulin and complement have the capacity to interact in a functional way. Of the six components identified in the shark complement system, three were functionally compatible with their mammalian counterparts (C1, C8 and C9)[97,98]. The ability of the shark complement to induce a cytolytic membrane attack could be demonstrated with electron microscopy by the appearance of 'pores' in the target cell membrane similar to those produced by the mammalian complement system[97].

In teleost fish the complement system is essentially similar to that of higher vertebrates. Rainbow trout complement possesses both classical and alternative pathway activity and contains components which are structurally and functionally similar to mammalian C3 and C5[98–100]. Teleost complement forms a terminal membrane attack complex to produce lysis[99]. It also has an opsonizing function in phagocytosis. In some studies opsonization was found to be effected by complement alone, in others antibody was also required[101].

3.3.2 Immunoglobulins

In the Agnatha immunoglobulin is present only in small amounts and is not well characterized. The antibody induced in the hagfish, described as IgM, now appears to be a component of complement (see Section 3.3.1). Thus the status of agnathan immunoglobulin remains problematic. In the jawed fishes, the immunoglobulin is of the IgM class with a heavy chain basically similar to the μ-chain of mammals. In Chondrichthyes and Dipnoi it is a pentameric molecule as in most other vertebrates. In Actinopterygii it is tetrameric. Fish IgM can also occur in monomeric or dimeric forms[102,103]. It seems likely that these low molecular weight versions are not simply precursor or breakdown molecules but possibly have a physiological purpose, for example where a more readily diffusible type of immunoglobulin is required.

A number of studies have revealed heterogeneity in immunoglobulin structure. In teleosts the immunoglobulin of the channel catfish is represented by different populations of molecules which vary in their covalent structure as well as in their component light chains[104], while the sheepshead shows differences between the immunoglobulins of the serum and the bile[88] (see Section 3.2.3). Amongst the variants of fish immunoglobulin are some that are perhaps sufficiently different, structurally or antigenically, to be consigned to a separate subclass or class, other than IgM. Such variants have been described in Dipnoi (the low molecular weight `IgN' immunoglobulin[105]), in skates[65,66], in sharks[106] and in teleosts[79]. Little is known about the functional significance of these different forms, although in teleosts some are associated with mucosal surfaces[78,88]. The conspicuous switch from IgM to a low molecular weight antibody class (IgG in mammals) that occurs during the course of an immune response in higher vertebrates, is not, however, manifested by fish.

There are a number of studies on the immunoglobulins of sharks, including some using modern techniques such as screening of genomic or cDNA libraries. The interest in sharks lies in their phylogenetic status as survivors from ancient stocks. The two groups that have received most attention are the horned sharks, which belong to a genus (*Heterodontus*) dating back to the Jurassic period, and sharks of the family Caracharhinidae, first found in the Cretaceous[1].

In terms of homology with the immunoglobulins of higher vertebrates, the immunoglobulin light chains of sharks can be identified as lambda (λ) and the heavy chains as mu (μ). However, the individual segments are not arranged as in higher vertebrates but instead form clusters, with many such clusters occurring throughout the shark genome. In summary, whereas in the heavy chain variable region of bony fishes the sets V (variable), D (diversity), and J (joining) are assembled to C (the constant region) in a similar way to that in higher vertebrates, as VVVDDJJC (multiple, V, D and J segments respectively, associated with a single C), in sharks the segments are arranged in multiple VDJC, VDJC, VDJC... (etc.) clusters[107,108]. Similarly in the shark light chains, the multiple light chain genomic clones contain individual, closely-linked, $V_LJ_LC_L$ clusters[109].

In poikilothermic vertebrates, including teleosts and elasmobranchs, the antibody repertoire is very restricted compared with that of mammals. In the latter, a single antigenic determinant can stimulate the immune system to produce several hundreds of different antibodies. In fish, on the other hand, although the variable region of the antibody molecule appears to possess framework and hypervariable regions as in higher vertebrates, there is only a low level of heterogeneity in the antibodies produced in response to a single defined hapten[110]. In *Heterodontus*, the antibody produced is of low affinity. It varies very little between individuals and does not increase in affinity after immunization[107]. In other elasmobranchs, such as the nurse shark, *Ginglyostoma cirratum*, and in teleost fish, some variation exists between individuals and each individual produces more than one antibody per antigenic determinant. Nevertheless the number of different antibodies elicited by a single antigen remains very limited and there is an apparent lack of[107], or relatively low expression of[112], affinity maturation in response to immunization. There remains the possibility that the various mechanisms for generating antibody diversity are similar in all vertebrates and the restricted antibody repertoire of poikilotherms results from other factors. These could include, among other possibilities, the mode of ontogenetic development in lower vertebrates, reduced rates of cell division in poikilotherms, or lack of the antigen-dependent memory B-cell lineages responsible for affinity maturation[107,110].

3.3.3 Antibody Production: B-cells and Cellular Co-operation

It is now well established that fish possess lymphocyte populations analogous to the B-cells and T-cells of mammals. This was first suggested by the demonstration of a hapten-carrier effect *in vivo*, indicating that the production of specific antibody against a hapten requires cellular co-operation[57]. Also fish lymphocyte populations respond in a way similar to higher vertebrates when stimulated *in vitro* with either B-cell (lipopolysaccharide, LPS) or T-cell (concanavalin A) mitogenic probes[112]. More recently, monoclonal antibodies directed against homologous immunoglobulin (Ig) have become available for several fish species. Since B-lymphocytes are characterized by the presence of membrane-bound immunoglobulin, these monoclonal antibodies can be used to differentiate between surface Ig-positive (sIg^{+ve}) and surface Ig-negative (sIg^{-ve}) lymphocytes. By this means, an sIg^{+ve} lymphocyte subpopulation analogous to the mammalian B-lymphocyte can be clearly demonstrated in fish, e.g. rainbow trout[113,114], carp[115] and channel catfish[116].

The use of monoclonal antibodies and the separation of sIg^{+ve} lymphocytes from sIg^{-ve} cells by panning techniques, in conjunction with relevant *in vitro* functional assays, has considerably furthered the understanding of cellular co-operation in the channel catfish. These studies demonstrate that teleosts have the functional equivalents of T- and B-cells, and that monocytes serve as accessory cells[116–118]. Thus the *in vitro* anti-hapten antibody responses of channel catfish peripheral blood leucocytes to haptenated-protein carrier conjugates (considered to be thymus-dependent in mammals) requires at least three cell types: carrier-specific sIg^{-ve} lymphocytes, hapten-specific sIg^{+ve} lymphocytes and monocytes (as accessory cells). With antigens considered to be thymus independent, the requirement is for hapten-specific sIg^{+ve} lymphocytes and monocytes (again as accessory cells). The accessory cells could, however, be replaced by a monocyte culture supernatant which had cytokine-like properties (possibly interleukin-1)[117].

An impetus to studies on antigen-presenting cells came from the serendipitous discovery that in 40% of channel catfish tested, long-term spontaneous leucocyte lines could be established from peripheral blood cells without any type of stimulation[119]. When cells from long-term monocyte-like lines were used, pulsed, fixed and co-cultured with autologous peripheral blood leucocytes from antigen-primed fish, specific secondary *in vitro* proliferative and antibody responses to various thymus-dependent antigens were elicited. These responses failed to occur when allogeneic peripheral blood leucocytes were employed as responders. In addition, the use of homologous alloantisera against these cell lines inhibited antigen presentation. These results have been interpreted as providing evidence that MHC (major histocompatibility complex)-like molecules govern antigen presentation and that putative MHC restriction of teleost immune responses occurs, reminiscent of that found in mammals[120,121]

3.3.4 Antibody Production *In Vivo*: Immunological Memory and Tolerance

The secondary antibody response in teleosts is often more rapid and reaches a higher peak than the primary response. The extent of the increase varies with the conditions of the experiment but at best there is only some 10-fold to 50-fold improvement over the primary response, as compared to the increases of a hundred-fold or more which are commonly found in mammals[122]. The kinetics of the response have been recorded both by measuring serum antibody levels and by counting antibody-forming cells[123]. The induction of memory depends on factors such as route, dose and temperature[122].

Clear secondary antibody responses have been reported to a variety of antigens including erythrocytes, bacterial antigens, virus particles, proteins, lipopolysaccharides and haptens. In some instances the elevated responses are quite long-lasting (at least 1 year)[124]. In some of these experiments, however, the true anamnestic nature of the results has been questioned[125]: there has not always been sufficient account taken of individual variability among the fish used; and the specificity of memory induction has frequently remained unchecked. Thus in some of the shorter term experiments, the enhanced antibody production might have been influenced by non-specific mechanisms such as the polyclonal expansion of B-cell populations or the mobilization and activation of phagocytes. The apparent absence of a secondary Ig class switch and affinity maturation has already been noted. Nevertheless some changes in antibody quality have been reported. Thus carp immunized with antigen-antibody complexes formed at equivalence, produced better secondary responses than with antigen alone, including a shift to the formation of precipitating antibody. This could perhaps be attributed to an affinity change[127]. In rainbow trout *in vivo* and *in vitro* analyses confirmed the existence of immunological memory but with no affinity maturation. In this species limiting dilution analyses of the responder spleen cells indicated that the memory population was not functionally different from the primary population with respect to its proliferative potential. Possibly memory in rainbow trout is due to a simple expansion of the antigen-specific precursor pool without many of the qualitative changes in antibody or B-cell function associated with the expression of memory in mammals[128].

Table 3.1 Factors influencing immunity versus tolerance.

Immaturity of the recipient
Physical state of the antigen
Antigen status:
T-independent or T-dependent
More or less foreign
Route of antigen administration
Antigen dose
Use of adjuvants
Ambient temperature

Some of the factors which may influence the outcome of immunization are shown in Table 3.1. It is likely that many newly hatched fish larvae rely on non-specific defence mechanisms such as phagocytosis for their early protection[31]. However, in fish, protective immunity following vaccination, and antibody production to bacterial antigens have been shown to develop quite rapidly[31,33], the ability to respond to thymus-independent antigens preceding that to thymus-dependent antigens[129]. Indeed, several authors have shown that early exposure to thymus-dependent antigens can induce tolerance[6,30]. Thus young rainbow trout were unable to produce an antibody response against human gamma globulin (HGG) when this was injected at 21 days post-hatching, whereas they could respond at this age to a thymus-independent bacterial vaccine. The trout developed a state of tolerance to HGG, for they did not respond to a second injection 8 weeks later at an age when normal fish (or controls injected at 21 days with bovine serum albumin instead of HGG) produced a positive anti-HGG response. Similarly, 4-week-old carp injected with sheep erythrocytes (SRBC) and re-immunized 3 months later failed to respond with antibody production, whereas normal fish receiving their first SRBC injection at 4 months old showed a good anti-SRBC response[130]. The tolerance effect could last for up to 1 year. Antibody tolerance can also be induced in 4-week-old carp following injection of HGG. However, in these experiments, although tolerance was induced for antibody production, the tolerance was incomplete inasmuch as proliferation of lymphoid cells in response to re-injection of the antigen (HGG) occurred in the spleen and kidney of the antibody-tolerant fish[131].

Antibody tolerance to thymus-dependent antigens can also be elicited in adult fish. The outcome of priming adult carp with soluble BSA (positive memory or tolerance) has been shown to depend on the physicochemical state of the antigen, the route of immunization and the environmental temperature[132–134]. The suggestion that this cold-induced tolerance might result from a modification of the immunoregulatory balance between helper cells and suppressor cells has recently been re-examined using the more precise *in vitro* methods developed for the channel catfish[118,135,136]. These latter studies indicate that the immunosuppression induced by low temperature does not involve T-suppressor cells or the induction of true immunological tolerance, but is the result of specific inhibition of the generation/activation of virgin T-helper cells. Leucocytes from fish immunized with a thymus-dependent hapten-carrier antigen while immunosuppressed at 11°C produced poor secondary antibody responses *in vitro*. However, if these fish were raised back to a water temperature of 23°C their cells subsequently exhibited good *in vitro* secondary responses, indicating that immunization at low temperatures did not induce true antigen-specific tolerance. The implication of this finding for prophylactic vaccination in aquaculture is that immunization conducted over the winter months may, nonetheless, prime the fish, so that secondary responses rather than tolerance can ensue if the antigen is encountered in the warmer seasons[135]. Thus vaccination on a farm could take place over winter and still provide good protection in the following spring/summer when the natural challenges appear[137].

3.3.5 Transfer of Immunity from Mother to Young

There is some interest in the possibility of vaccinating hen fish in the hope that protective antibody may be transferred to the eggs. Passive immunization from the mother would be particularly beneficial in the case of pathogens which are transmitted vertically or to which the young are particularly susceptible, for example the IPN (infectious pancreatic necrosis) virus. Non-specific humoral factors such as C-reactive protein and lectin-like agglutinins, as well as immunoglobulins, have been found in fish ova[4]. In addition, the transfer of immune substances from an immunized female to her ova has been reported in plaice, tilapia and channel catfish[138–140] while in an ovoviviparous guppy, immunity was transferred to the newborn fry[141]. In tilapias, the specificity of the transferred antibody when the mother was immunized with various protein antigens, has been established[139]. In the channel catfish IgM was found both in the yolk and in the egg membranes, suggesting that it provides an immune barrier at the external surface of the egg, as well as protection for the developing fry[140].

3.4 NON-SPECIFIC AND SPECIFIC CELL-MEDIATED IMMUNITY

3.4.1 Non-specific Cellular Immunity

Inflammatory reactions involving granulocytes, monocyte/macrophages and lymphocytes occur in fish in response to a variety of insults including bacterial, viral, mycotic, protozoan and parasitic infections[71]. Granulocytes are found in agnathans and in the various groups of jawed fishes. Their functions have recently been reviewed[73,142]. Although the morphological heterogeneity of fish granulocytes is well established, the respective roles of the different populations is less clear and may vary from species to species. At least one type is functionally equivalent to the neutrophil.

The first stage of an inflammatory reaction involves the accumulation of granulocytes. This peaks at about 12–24h and is followed by an influx of monocytes and lymphocytes. Leucocyte migration is enhanced by a variety of host-derived chemoattractants including serum factors, possibly complement fragments[143,144], cytokines[145] and eicosanoids[78,146]. Eicosanoids such as leukotrienes and lipoxins are produced by macrophages. The lipoxins are generated more slowly than the leukotrienes. However, in fish, unlike mammals, more lipoxins are produced overall[147,148]. Chemoattractants may also be derived from the pathogen itself. Bacterial products and culture supernatants elicited a chemokinetic response in plaice neutrophils[149], while in rainbow trout parasitized by *Diphyllobothrium dendriticum*, a parasite-derived factor was chemokinetic and a host-derived factor chemotactic[150].

Phagocytes play an important role in the defence mechanisms of all vertebrates. In fish it is generally assumed that circulating monocytes and tissue macrophages belong to the same cell line. The clearance of particulate matter from fish blood has been described as 'biphasic'. In the first phase most of the material, typically over 90%, is cleared within half an hour, the rest is cleared more slowly[134,151]. Macrophages are the predominant phagocyte of fish, being highly phagocytic for both inert and antigenic materials. Fish neutrophils also participate in phagocytosis, at least under some circumstances[151,152].

Phagocytosis in fish is enhanced by a number of opsonins including lectins, C-reactive protein, complement and antibodies. Opsonization can be demonstrated in all groups, including agnathans. In the Arctic lamprey the uptake of sheep erythrocytes (SRBC) was enhanced by opsonization with homologous anti-SRBC antibody[153]. There is good evidence for Fc receptors in sharks[154], while both complement and antibody have been implicated in opsonization in several teleost species for both neutrophils[101,152] and macrophages[155,156]. The fact that the presence of complement and/or Fc receptors on fish cells has sometimes been disputed is probably due to the source of the cells, their stage of development and their state of activation, as well as to experimental circumstances. Circulating monocytes, resident macrophages and peritoneal exudate cells have been used in various studies, sometimes in mixed populations, sometimes after Percoll separation.

The events which occur when fish phagocytes are activated are very similar to those in higher vertebrates. Phagocytosis proceeds through the stages of recognition, attachment, engulfment, killing and digestion in a process which, as in mammals, involves the generation of reactive oxygen species including superoxide anion, hydrogen peroxide, singlet oxygen and hydroxyl radicals. The biochemical pathway, termed the respiratory burst, is oxygen dependent but is not inhibited by azide/cyanide. The respiratory burst, detected by chemiluminescence[157] or by measurement of reactive oxygen products, can be elicited both from neutrophils[158,159] and from macrophages[160,161]. Activation products are released to the extracellular environment. They are known to be bactericidal[162] and to have toxic effects on helminth parasites[163].

Another type of fish cell which can effect the destruction of foreign cells is the non-specific cytotoxic cell (NCC). These are small cells found in the peripheral blood, spleen and pronephros of teleost fish. They are cytotoxic towards a range of normal and transformed cell lines of both fish and mammalian origin[73,164–166], also towards parasitic protozoa[167]. NCC are non-adherent, non-phagocytic cells, which require physical contact and a high ratio of NCC to target cell to cause killing. Their role in the non-specific defence mechanisms of fish seems quite similar to that of natural killer (NK) cells of mammals. Indeed, it has been suggested that the NCC represents a progenitor of the NK cell, but with greater diversity of function. The NCC of channel catfish lack the cytoplasmic granules of the NK cell, indicating that the mechanisms of target cell lysis might be different. The kinesis of cell killing also differs from that of the NK cell in that the NCC effect is rapid and is almost complete after 1 hour of contact with the target cell. In sharks, the peripheral blood contains an adherent cell, macrophage-like in appearance, which is spontaneously cytotoxic. The activities of this cell are down-regulated by a non-adherent cell population which can suppress the activity of the cytotoxic cell when in direct contact with it[168].

In considering the inflammatory reactions of fish, the question arises whether fish possess any cell which is functionally equivalent to the mast cell of higher vertebrates. The teleost eosinophilic granular cell has been investigated in this respect. These cells are uncommon in the blood but occur in quite high numbers in regions of antigen encounter such as the gut and gills. They can be degranulated experimentally. When rainbow trout were injected with extracellular products from pathogenic bacteria, *Aeromonas salmonicida*, there was a marked increase in the levels of histamine in the blood and a degranulation and disappearance of eosinophilic granular cells in the stratum compactum of the gut wall. This response occurred rapidly and bore some resemblance to the mammalian Type 1 anaphylactic hypersensitivity reaction[169]. Anaphylactic hypersensitivity reactions in fish have been described by several authors but no fish homocytotropic antibody has been identified and the involvement of antigen as an initiating factor is problematical[170].

3.4.2 Cytokine Activities

Cytokines are soluble products which play a regulatory or enhancing role in cell-mediated reactions. Cytokines putatively present in fish include interleukin-1 (IL-1), interleukin-2 (IL-2), macrophage activating factor (MAF), interferon-gamma (IFN-γ), chemotactic factor (CF) and macrophage migration inhibition factor (MIF).

Interleukin-1 appears to be a conserved molecule which evolved early in the phylogeny of the vertebrates. In mammals IL-1 serves as a starting point for a number of cascade reactions including its action on T-cells which leads to an increase in their receptors for IL-2. Interleukin-1 is known to cross species barriers within the Mammalia. Furthermore, channel catfish peripheral blood lymphocytes can recognize and respond to human IL-1; also fish cells can themselves produce an IL-1-like substance. Interleukin-1 that stimulated proliferation in catfish peripheral blood lymphocytes, could be obtained from carp epithelial cell lines[171]. An IL-1-like agent was also demonstrated in catfish monocyte culture supernatants, as described in Section 3.3.3[117].

Interleukin-2 is a polypeptide which induces proliferation of any T-cells that have the appropriate receptors. In mammals these IL-2 receptors are synthesized and expressed by activated T-cells which have encountered antigen (or mitogen) and macrophage-derived IL-1. Mammalian IL-2 is primarily produced by T-helper cells although other subsets of T-cells may also be involved. In fish (carp), supernatants from mitogen-stimulated cultures of peripheral blood leucocytes induced proliferation of homologous (putatively T-cell) lymphoblasts. These cultures also responded to mammalian IL-2[172]. A similar result was obtained using mitogen-stimulated carp pronephric cells. Supernatants from these cultures caused proliferation of blast cells but had little effect on virgin leucocytes. A similar but less reactive supernatant was obtained from mixed leucocyte cultures[173]. These lymphocyte growth-promoting factors resemble the IL-2 of higher vertebrates although more needs to be discovered about their regulatory role in fish.

In mammals antigenic or mitogenic stimulation of T-lymphocytes also induces the production of MAF, particularly IFN-γ, which is a potent macrophage activating factor. MAF has been demonstrated in fish. Rainbow trout macrophages responded to supernatants from mitogen (concanavalin A)-stimulated homologous lymphocytes with increased respiratory burst activity and increased bacterial killing[174]. Panning experiments utilizing a monoclonal anti-trout Ig have demonstrated that only sIg^{-ve} lymphocytes secrete this MAF and that accessory cells are also necessary[175]. The MAF-containing supernatants were found to confer viral resistance on a rainbow trout epithelial cell line challenged with infectious pancreatic necrosis virus[175]. Fish, like mammals, may therefore gain protection not only from the broader interferons (α and β-like), which are induced in cells infected with virus (and probably most cell types can produce them), but also from IFN-γ-like cytokines produced by stimulated T-cells.

Cytokines which affect the movements of leucocytes are less well characterized but probably play a role in the inflammatory responses of fish. A macrophage migration inhibition factor (MIF) can be elicited from carp pronephric cells by stimulation with mitogens such as concanavalin A. MIF has been demonstrated in elasmobranchs as well as teleosts[31]. In carp, mitogens[145] or antigens[176] also elicit a chemotactic factor, which mainly seems to attract granulocytes.

3.4.3 Specific Cell-mediated Immunity: T-cell Functions

There is strong evidence, based on functional criteria, that fish possess a lymphocyte population with many of the attributes of mammalian T-cells. In teleosts, these cells can be separated from sIg^{+ve} B-cells and investigated for their ability to perform in a T-cell-like manner (see Section 3.3.3). In such experiments the sIg^{-ve} lymphocytes were able to respond to T-cell mitogens[116], to act as helper cells in antibody production against thymus-dependent antigens[117,118,177], to respond in mixed leucocyte reactions (MLR)[178], and to play a role in the production of macrophage-activating lymphokines[179]. In MLR the sIg^{-ve} cells were found to be stimulators as well as responder cells. This differs from mammalian T-cells but could be explained by the probability that not all fish sIg^{-ve} cells are putative T-cells: a small proportion could be other leucocyte types which might act as potent stimulators in MLR.

Monoclonal antibodies against the sIg^{-ve} cells have been made for carp[115] and for channel catfish[177]. These antibodies reacted with most homologous thymocytes and with peripheral sIg^{-ve} lymphocytes including, in the catfish, lymphocytes which were helper cells in antibody production. The monoclonal antibodies appear to be pan-T-cell reagents which also react with other cell types such as neutrophils and some brain cells. There are no markers to date for any T-cell subpopulations in fish, for example for any possible teleost equivalents to the mammalian CD4 and CD8 co-receptor-bearing T-cells.

The requirements for the T-cell activities outlined above included the presence of accessory cells, in this case monocytes. These presumably functioned by phagocytosis or pinocytosis of the stimulant, followed by degradation of the internalized molecules, coupled with re-expression on the accessory cell surface. Appropriate presentation of the antigen to the putative T-cell therefore appears to be an initiating factor in the specific immune responses of fish, as in higher vertebrates.

Simple *in vivo* tests for T-cell activity were among the earliest methods used in studies on specific cell-mediated immunity in fish. These include skin or scale transplantation and tests for delayed-type hypersensitivity reactions. The latter (DTH) is based upon the interaction between antigen and primed T-cells. The antigen (usually killed mycobacteria) is injected locally into a primed fish which then responds within 1 or 2

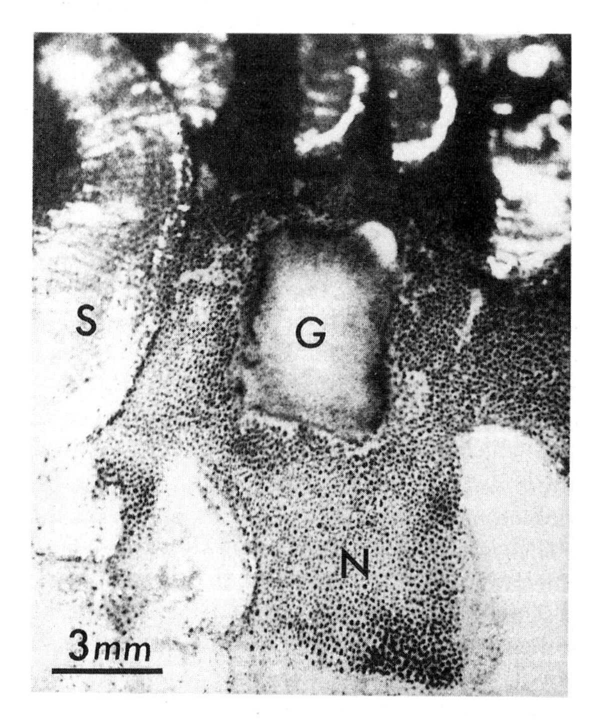

Figure 3.11. Allograft of skin 12 days after transplantation on to a mirror carp, *Cyprinus carpio*, showing the size and position of the skin graft in relation to the host scales. Note the lack of normal pigmentation in the allograft: G = graft; N = normal skin; S = large scale.

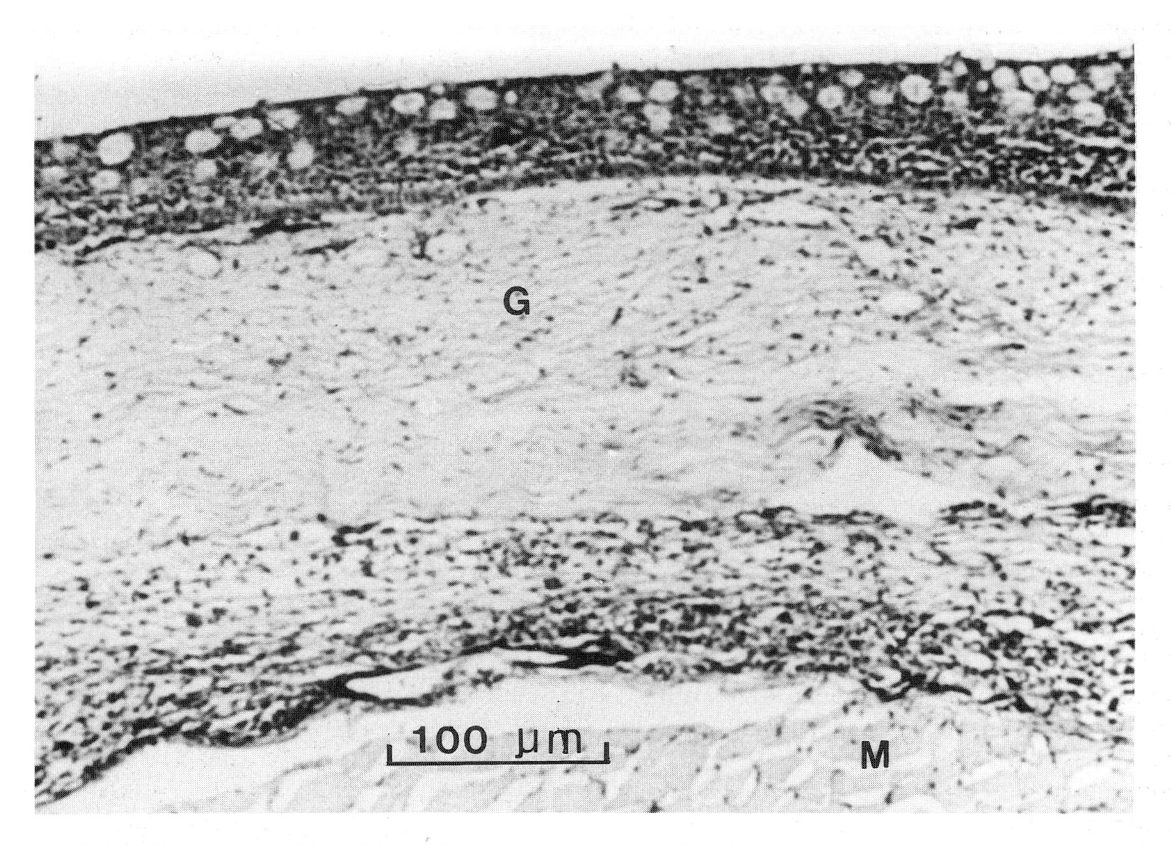

Figure 3.12. Autograft of skin 12 days after grafting on carp, *Cyprinus carpio*. Section to show the healthy appearance of the graft: G = graft; M = muscle below the graft.

days with an accumulation of lymphocytes and macrophages at the site. DTH has been demonstrated in lampreys[180] as well as in teleost fish[181,182]. The reaction has not as yet been sufficiently well characterized, however, for the nature of the cytotoxic/inflammatory or lymphokine-producing cells to be ascertained in fish.

The ability to recognize and respond to allogeneic tissue transplants with specific accelerated second-set memory is shown in all fish groups studied, including agnathans (hagfish and lampreys), elasmobranchs (stingrays and sharks) and a wide variety of teleost fish. In the more primitive groups allogeneic grafts are rejected slowly, in the chronic manner presumed to be due to cumulative minor histoincompatibility reactions. In contrast, teleost fish reject allografts more rapidly in acute responses usually associated with the presence of a major histocompatibility complex[2,183].

A first-set skin allograft on an adult carp is shown in Figure 3.11. As early as day 4 at 22°C, the graft becomes swollen and the melanin pigmentation begins to break up, the mean survival time at 22°C of a first-set graft being 14 days; for second-set grafts it is 7 days. Autografts, on the other hand, heal in well and are eventually indistinguishable from the surrounding tissue; compare Figure 3.12 (autograft) and Figure 3.13 (allograft)[21]. Skin transplantation reactions provide an easy way to monitor the onset of specific cell-mediated immunity in ontogenetic studies because it is possible to place grafts on delicate fish fry from an early stage of development. Such experiments indicate that the cellular immune component matures rapidly in teleosts. Thus

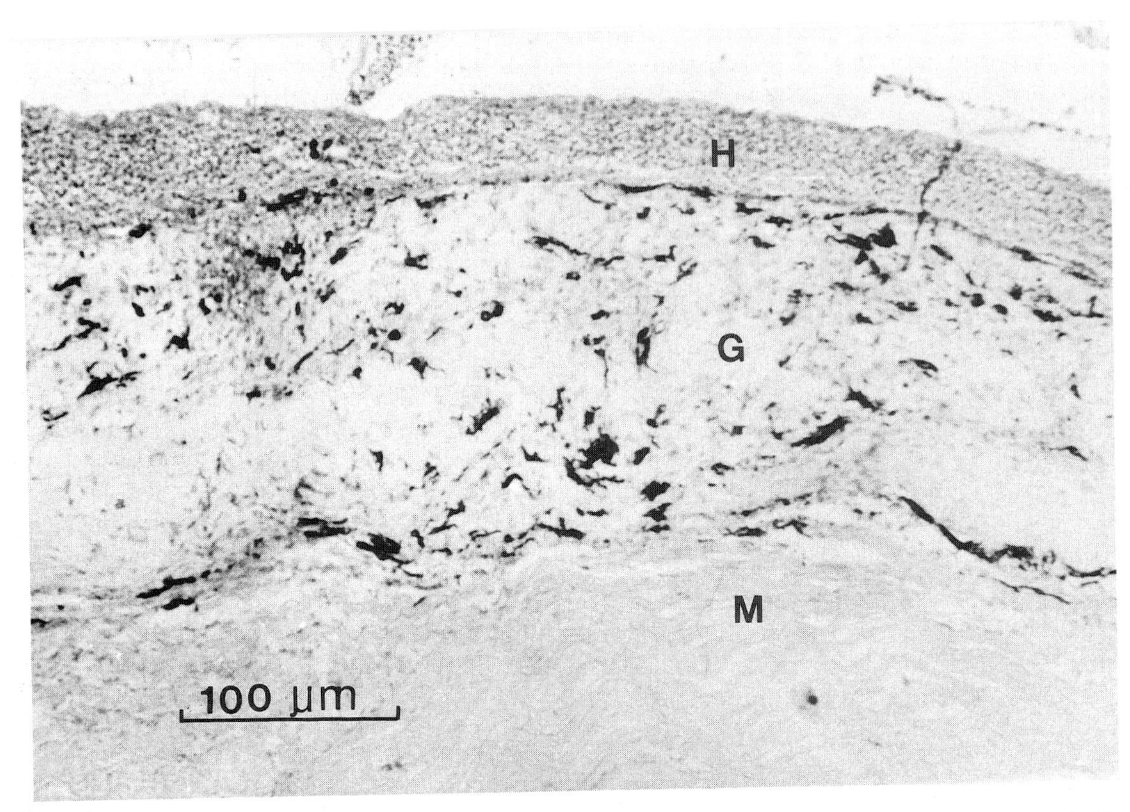

Figure 3.13. Allograft of skin 18 days after grafting on carp, *Cyprinus carpio*. Section to show graft disruption: G = graft; H = host epidermis which has overgrown the graft; M = host muscle beneath the graft.

young carp are capable not only of mounting an allograft response by day 16 post-hatching (albeit less efficiently than in adult fish) but also of developing a memory component to the response[31].

3.4.4 The Major Histocompatibility Complex (MHC)

It has been assumed for some time that teleosts possess an MHC. This is based on indirect evidence which includes the acute rejection of allografts, *in vitro* MLR and, more recently, the demonstration of an MHC-like restriction of antigen presentation (see Section 3.3.3[120,121]). There is now a new thrust towards formally identifying and characterizing the MHC class I and class II genes and/or their gene products. In higher vertebrates these MHC products are cell membrane glycoproteins which show a high degree of polymorphism and whose functional importance lies in the regulation and control of T-lymphocyte activation.

Recent advances in research on the fish MHC have become possible through the use of modern techniques of molecular genetics, together with the employment of histocompatible lines of fish made available through artificial or naturally occurring gynogenesis[184]. Previous studies on the immunogenetics of transplantation reactions were performed on inbred lines of fish established by sibling mating, for example in platyfish and swordtails, guppies and meduka. In the latter species, the histocompatibilities could be assigned to one or two major (MHC) loci[185]. The production of histocompatible lines of carp by gynogenesis has allowed specific alloantisera to be raised against the allelic products of a putative MHC locus. This has been achieved by immunizing carp with peripheral blood leucocytes from gynogenetic siblings. These antisera identify two co-dominantly expressed allelic products of a single locus which appear to incorporate MHC class I properties[186,187].

Strategies using polymerase chain reactions to amplify the MHC gene-encoding DNA segments of the carp genome have succeeded in isolating two putative MHC-antigen encoding sequences. One yielded reasonable homology to the MHC class I heavy chains of birds and mammals, while the other was homologous to the higher vertebrate MHC class II β chain[188]. Using similar polymerase chain reaction techniques with elasmobranchs, a sequence has recently been identified in the shark, *Triakis scyllia*, that resembles the MHC class I $\alpha 3$ domain[189].

3.5 DISCUSSION AND FUTURE DIRECTIONS

The MHC of teleosts is currently being established at the molecular level. The way is also opening up for the use of similar molecular techniques to investigate the origins and homologies of the molecules of the immunoglobulin superfamily in more primitive fish and in their invertebrate relatives. In particular, the T-cell antigen receptor, which operates in the context of the MHC, has yet to be identified in fish. In addition, at the functional level, a specific cytotoxic (T-like) cell has not yet been described at the fish stage in phylogeny. We do not know, either, whether MHC-restricted antigen-presenting cells of the dendritic type occur in fish.

The recent development of histocompatible lines of fish, in species which are of a size large enough to yield sufficient cells and serum for experimental purposes, should lend impetus to investigations hitherto limited by logistical problems. For example, reconstitution experiments and cell migration studies (ecotaxis or 'cell-trafficking'[190]) should now be easier to perform. Investigations of the stem cell and progenitor sources of the effector cells of the fish immune system should also be more feasible, along with characterization of any cytokines or other substances which may drive the differentiation pathways. The microenvironment where these events take place is also poorly understood at the present time.

The ancestor of the higher vertebrates would have been aquatic, poikilothermic and probably relatively long-lived, with external fertilization and a free-living larval stage. It would have had a much more extensive mucosal system than that of a terrestrial animal. Mucus covers the external surfaces of fish, as well as the internal ones, and is important both for streamlining the fish for swimming and for protection, including the sloughing off of unwanted material. Relationships between the mucosal and systemic components of the fish immune system are therefore of interest. The fact that blood flows through the body at relatively low pressure in the 'single circulatory' vascular system of fish may obviate the requirement for specialized architectural arrangements in the lymphoid organs, or for the extensive production of a diffusible low molecular weight class of immunoglobulin. Low molecular weight antibodies may sometimes be required, however, and have been described in fish (Section 3.3.2).

Temperature probably had a major influence on the early evolution of the vertebrate immune system, especially as it is often subject to seasonal and/or other fluctuations[136]. The lipids of fish tissues are mainly composed of polyunsaturated fatty acids, probably as an adaptation to low temperatures. These are prone to oxidation by free-radical reactions. Melanin protects against these, which may explain its prominence in the defences of fish and other poikilotherms[57].

Free-living larvae need to have some protective mechanisms in place from a very early stage in development. At first they seem to rely on non-specific immunity[31]. Later, as the specific immune system develops, it must operate at first with relatively low numbers of cells. This may be one of the factors restricting antibody diversity[110]. Indeed, there is still a lot to be learned about immunological memory and tolerance in fish. An important aspect of the vertebrate immune system is that the first event after an appropriate antigen encounter is the triggering of proliferation of the specifically reactive lymphocytes. In fish there is enhanced sensitivity to antigen upon priming and an increase in the size of the antigen-specific precursor pool. Qualitative and quantitative refinement of the specific secondary response, although incipient in fish, may be further emphasized in higher vertebrates, perhaps driven by the demands of homoiothermy, this being reflected morphologically by the presence of germinal centres and physiologically by the increasing proliferative potential of individual precursor cells in higher vertebrates, as well as their switch to a low molecular weight class of antibody[128].

Finally, we should acknowledge the extensive body of work being carried out on the applied aspects of fish immunology. It has not been possible to cover, in this chapter, the many excellent studies on host/pathogen relationships. The impetus here comes from an urgent requirement for the diagnosis and control of fish diseases in aquaculture, as well as from the intrinsic biological interest of these relationships themselves. In addition, because fish are aquatic animals of considerable ecological and economic importance, there is concern about the effects of environmental pollution on fish health. This has produced some interesting current studies on the effects of ecotoxicants on the fish immune system and disease susceptibility[191] and is likely to become an area attracting increasing attention.

3.6 REFERENCES

1. Romer, A.S. (1966). *Vertebrate Paleontology*, 3rd edn. University of Chicago Press, Chicago.
2. Manning, M.J. and Turner, R.J. (1976). *Comparative Immunobiology*. Blackie, Glasgow and London.
3. Young, J.Z. (1950). *The Life of Vertebrates*. Clarendon Press, Oxford.
4. Ellis, A.E. (ed.) (1988). *Fish Vaccination*. Academic Press, London.
5. Tatner, M.F. (1983). Fish Vaccines. In *Vaccines for Veterinary Applications*, Peters, A.R. (ed.). Butterworth-Heinemann, Oxford, pp. 199–224.
6. Secombes, C.J. (1981). *Comparative studies on the structure and function of teleost lymphoid organs*. PhD Thesis, University of Hull, Hull.
7. Tatner, M.F. and Manning, M.J. (1982). The morphology of the trout, *Salmo gairdneri* Richardson, thymus: some practical and theoretical considerations. *J. Fish Biol.* **21,** 27–32.
8. Chilmonczyk, S. (1985). Evolution of the thymus in rainbow trout. In *Fish Immunology*, Manning, M.J. and Tatner, M.F. (eds). Academic Press, London, pp. 285–292.
9. Fange, R. and Pulsford, A. (1985). The thymus of the angler fish (*Lophius piscatorius* [Pisces: Teleostei]): A light and electron microscopic study. In *Fish Immunology*, Manning, M.J. and Tatner, M.F. (eds). Academic Press, London, pp. 293–311.
10. Bigaj, J., Dulak, J. and Plytycz, B. (1987). Lymphoid organs of *Gasterosteus aculeatus. J. Fish Biol.* **31** (Suppl. A), 233–234.
11. Sailendri, K. and Muthukkaruppan, V.R. (1975). The immune response of the teleost, *Tilapia mossambica* to soluble and cellular antigens. *J. Exp. Zool.* **191,** 371–382.
12. Bly, J.E. (1985). The ontogeny of the immune system in the viviparous teleost *Zoarces viviparus* L. In *Fish Immunology*, Manning, M.J. and Tatner, M.F. (eds). Academic Press, London, pp. 327–341.
13. Manning, M.J. (1981). A comparative view of the thymus in vertebrates. In *The Thymus Gland*, Kendall, M.D. (ed.). Academic Press, London, pp.7–20.
14. Honma, Y. and Tamura, E. (1984). Histological changes in the lymphoid system of fish with respect to age, seasonal and endocrine changes. *Dev. Comp. Immunol.* (Suppl. 3), 239–244.
15. Tatner, M.F. and Manning, M.J. (1983). Growth of the lymphoid organs in rainbow trout from one to fifteen months of age. *J. Zool.* [*Lond.*] **199,** 503–520.
16. McCumber, L.J., Sigel, M.M., Trauger, R.J. and Cuchens, M.A. (1982). RES—Structure and function of the fishes. In *The Reticuloendothelial System, Vol. 3: Phylogeny and Ontogeny*, Cohen, N. and Sigel, M.M. (eds). Plenum Press, New York, pp. 393–422.
17. Hafter, E. (1952). Histological age changes in the thymus of the teleost, *Astyanex. J. Morphol.* **90,** 555–581.
18. Sailendri, K. (1973). Studies on the development of lymphoid organs and immune responses in the teleost, *Tilapia mossambica* (Peters). PhD Thesis, Madurai University, India.

19. Doggett, T.A. and Harris, J.E. (1987). The ontogeny of gut-associated lymphoid tissue in *Oreochromis mossambicus*. *J. Fish Biol.* **31,** (Suppl. A), 23–27.
20. Ellis, A.E. (1977). Ontogeny of the immune response in *Salmo salar*. Histogenesis of the lymphoid organs and appearance of membrane immunoglobulin and MLR. In *Developmental Immunology*, Solomon, J.B. and Horton, J.D. (eds). Elsevier/North Holland Biomedical Press, Amsterdam, pp.225–231.
21. Botham, J.W. and Manning, M.J. (1981). The histogenesis of the lymphoid organs in the carp *Cyprinus carpio* L. and the ontogenetic development of allograft reactivity. *J. Fish Biol.* **19,** 403–414.
22. Tatner, M.F. and Manning, M.J. (1983). The ontogeny of cellular immunity in the rainbow trout, *Salmo gairdneri* Richardson, in relation to the stage of development of the lymphoid organs. *Dev. Comp. Immunol.* **7,** 69–75.
23. Nakanishi, T. (1986). Ontogenetic development of the immune response in the marine teleost *Sebastiscus marmoratus*. *Bull. Jap. Soc. Sci. Fish.* **52,** 473–477.
24. O'Neill, J. (1989). Ontogeny of the lymphoid organs in an antarctic teleost, *Harpagifer antarcticus* (Notothenioidei: Perciformes). *Dev. Comp. Immunol.* **13,** 25–33.
25. Jósefsson, S. and Tatner, M.F. (1993). Histogenesis of the lymphoid organs in sea bream (*Sparus aurata* L.). *Fish Shellfish Immunol.* **3,** 35–49.
26. Hart, S., Wrathmell, A.B. and Harris, J.E. (1986). Ontogeny of gut-associated lymphoid tissue (GALT) in the dogfish *Scyliorhinus canicula* L. *Vet. Immunol. Immunopathol.* **12,** 107–116.
27. Secombes, C.J., Van Groningen, J.J.M. and Egberts, E. (1983). Separation of lymphocyte sub-populations in carp *Cyprinus carpio* L. by monoclonal antibodies: immunohistochemical studies. *Immunology* **48,** 165–175.
28. Tatner, M.F. (1985). The migration of labelled thymocytes to the peripheral lymphoid organs in the rainbow trout, *Salmo gairdneri* Richardson. *Dev. Comp. Immunol.* **9,** 85–91.
29. Ellsaesser, C.F., Bly, J.E. and Clem, L.W. (1988). Phylogeny of lymphocyte heterogeneity: the thymus of the channel catfish. *Dev. Comp. Immunol.* **12,** 787–799.
30. Manning, M.J., Grace, M.F. and Secombes, C.J. (1982). Ontogenetic aspects of tolerance and immunity in carp and rainbow trout; studies on the role of the thymus. *Dev. Comp. Immunol.* (Suppl. 2), 75–82.
31. Manning, M.J., Grace, M.F. and Secombes, C.J. (1982). Developmental aspects of immunity and tolerance in fish. In *Microbial Diseases of Fish*, Roberts, R.J. (ed.). Academic Press, London, pp.31–46.
32. Anderson, D.P. and Dixon, O.W. (1980). Immunological memory in rainbow trout to a fish disease bacterin administered by flush exposure. In *Phylogeny of Immunological Memory*, Manning, M.J. (ed.). Elsevier/North Holland Biomedical Press, Amsterdam, pp.103–111.
33. Manning, M.J. and Mughal, M.S. (1985). Factors affecting the immune responses of immature fish. In *Fish and Shellfish Pathology*, Ellis, A.E. (ed.). Academic Press, London, pp.27–40.
34. Tatner, M.F. (1986). The ontogeny of humoral immunity in rainbow trout, *Salmo gairdneri*. *Vet. Immunol. Immunopathol.* **12,** 93–105.
35. Jayaraman, S., Mohan, R. and Muthukkaruppan, V.R. (1979). Relationship between migration inhibition and plaque forming cell responses to sheep erythrocytes in the teleost *Tilapia mossambicus*. *Dev. Comp. Immunol.* **3,** 67–75.
36. Tatner, M.F., Adams, A. and Leschen, W. (1987). An analysis of the primary and secondary antibody responses in intact and thymectomized rainbow trout, *Salmo gairdneri* Richardson, to human gamma globulin and *Aeromonas salmonicida*. *J. Fish Biol.* **31,** 177–195.
37. Metcalf, D. (1965). Delayed effect of thymectomy in adult life on immunological competence. *Nature* [*Lond.*] **208,** 1336.
38. Zapata, A. (1991). Lymphoid organs of teleost fish. I. Ultrastructure of the thymus of *Rutilus rutilus*. *Dev. Comp. Immunol.* **5,** 427–436.
39. McKinney, E.C., Ortiz, G., Lee, J.C., Sigel, M.M., Lopez, D.M., Epstein, R.S. and McLeod, T.F. (1976). Lymphocytes of fish: multipotential or specialized? In *Phylogeny of Thymus and Bone Marrow-Bursa Cells*, Wright, R.K. and Cooper, E.L. (eds). Elsevier/North Holland Biomedical Press, Amsterdam, pp.73–82.
40. Riviere, H.B., Cooper, E.L., Reddy, A.L. and Hildemann, W.H. (1975). In search of the hagfish thymus. *Amer. Zool.* **15,** 38–49.
41. Page, M. and Rowley, A.F. (1982). A morphological study of pharyngeal lymphoid accumulations in larval lampreys. *Dev. Comp. Immunol.* (Suppl. 2), 35–40.
42. Fänge, R. (1982). A comparative study of lymphomyeloid tissue in fish. *Dev. Comp. Immunol.* (Suppl. 2), 23–33.
43. Zapata, A. (1979). Ultrastructural study of the teleost fish kidney. *Dev. Comp. Immunol.* **3,** 55–65.

44. Secombes, C.J., Manning, M.J. and Ellis, A.E. (1982). The effect of primary and secondary immunization on the lymphoid tissues of the carp, *Cyprinus carpio* L. *J. Exp. Zool.* **220,** 277–287.
45. Chiller, J.M., Hodgins, H.O. and Weiser, R.S. (1969). Antibody responses in rainbow trout (*Salmo gairdneri*). II. Studies on the kinetics and development of antibody-producing cells and on complement and natural haemolysin. *J. Immunol.* **102,** 1202–1207.
46. Anderson, D.P., Roberson, B.S. and Dixon, O.W. (1979). Cellular immune response in rainbow trout *Salmo gairdneri* Richardson to *Yersinia ruckeri* O-antigen monitored by the passive haemolytic plaque assay test. *J. Fish Dis.* **2,** 169–178.
47. Rijkers, G.T., Frederix-Wolters, E.M.H. and Van Muiswinkel, W.B. (1980). The effect of antigen dose and route of administration on the development of immunological memory in carp (*Cyprinus carpio*). In *Phylogeny of Immunological Memory*, Manning, M.J. (ed). Elsevier/North Holland Biomedical Press, Amsterdam, pp. 93–102.
48. Rijkers, G.T., Frederix-Wolters, E.M.H. and Van Muiswinkel, W.B. (1980). The immune system of cyprinid fish. Kinetics and temperature dependence of antibody-producing cells in carp (*Cyprinus carpio*). *Immunology* **41,** 91–97.
49. Miller, N.W. and Clem, L.W. (1984). Microsystem for *in vitro* primary and secondary immunization of channel catfish (*Ictalurus punctatus*) leucocytes with hapten-carrier conjugates. *J. Immunol. Methods* **72,** 367–379.
50. Secombes, C.J., White, A., Fletcher, T.C. and Houlihan, D.F. (1991). The development of an ELISPOT assay to quantify total and specific antibody-secreting cells in dab *Limanda limanda* (L.). *Fish Shellfish Immunol.* **1,** 87–97.
51. Roberts, R.J. (1975). Melanin-containing cells of teleost fish and their relation to disease. In *The Pathology of Fishes*, Ribelin, W.E. and Migaki, G. (eds). University of Wisconsin Press, Madison, pp. 399–428.
52. Ellis, A.E. and de Sousa, M.A.B. (1974). Phylogeny of the lymphoid system 1. A study of the fate of circulating lymphocytes in plaice. *Eur. J. Immunol.* **4,** 338–343.
53. Agius, C. (1985). The melano-macrophage centres of fish: a review. In *Fish Immunology*, Manning, M.J. and Tatner, M.F. (eds). Academic Press, London and San Diego, pp. 85–105.
54. Ellis, A.E. (1981). Non-specific defense mechanisms in fish and their role in disease processes. *Developments in Biological Standardization* **49,** 337–352.
55. Lamers, C.H.J. (1986). Histophysiology of a primary immune response against *Aeromonas hydrophila* in carp (*Cyprinus carpio* L.). *J. Exp. Zool.* **238,** 71–80.
56. Lamers, C.H.J. and De Haas, M.J.H. (1985). Antigen localization in the lymphoid organs of carp (*Cyprinus carpio* L.). *Cell Tissue Res.* **242,** 491–498.
57. Ellis, A.E. (1982). Differences between the immune mechanisms of fish and higher vertebrates. In *Microbial Diseases of Fish*, Roberts, R.J. (ed.). Academic Press, London, pp. 1–29.
58. Yoffey, J.M. (1929). A contribution to the study of the comparative histology and physiology of the spleen, with reference chiefly to its cellular constituents I. In fishes. *J. Anat.* **63,** 314–344.
59. Van Muiswinkel, W.B., Lamers, C.H.J. and Rombout, J.H.W.M. (1991). Structural and functional aspects of the spleen in bony fish. *Res. Immunol.* **142,** 362–366.
60. Secombes, C.J. and Manning, M.J. (1980). Comparative studies on the immune system of fishes and amphibians: antigen localization in the carp *Cyprinus carpio* L. *J. Fish Dis.* **3,** 399–412.
61. Ellis, A.E. (1980). Antigen-trapping in the spleen and kidney of the plaice, *Pleuronectes platessa* L. *J. Fish Dis.* **3,** 413–426.
62. Secombes, C.J., Manning, M.J. and Ellis, A.E. (1982). Localization of immune complexes and heat-aggregated immunoglobulin in the carp *Cyprinus carpio* L. *Immunology* **47,** 101–105.
63. Imagawa, T., Hashimoto, Y., Kon, Y. and Sugimura, M. (1991). Immunoglobulin containing cells in the head kidney of carp (*Cyprinus carpio* L.) after bovine serum albumin injection. *Fish Shellfish Immunol.* **1,** 173–185.
64. Good, R.A., Finstad, J., Pollara, B. and Gabrielsen, A.E. (1986). Morphologic studies on the evolution of the lymphoid tissues among the lower vertebrates. In *Phylogeny of Immunity*, Smith, R.T., Miescher, P.A. and Good, R.A. (eds). University of Florida Press, Gainesville, pp. 149–170.
65. Tomonaga, S., Kobayashi, K., Hagiwara, K., Sasaki, K. and Sezaki, K. (1985). Studies on immunoglobulin and immunoglobulin-forming cells in *Heterodontus japonicus*, a cartilaginous fish. *Dev. Comp. Immunol.* **9,** 617–626.
66. Tomonaga, S., Kobayashi, K., Kajii, T. and Awaya, K. (1984). Two populations of immunoglobulin-forming cells in the skate, *Raja kenojeii*: their distribution and characterization. *Dev. Comp. Immunol.* **8,** 803–812.

67. Fletcher, T.C. and Grant, P.T. (1969). Immunoglobulins in the serum and mucus of the plaice (*Pleuronectes platessa*). *Biochem. J.* **115,** 65–69.
68. Ourth, D.D. (1980). Secretory IgM, lysozyme and lymphocytes in the skin mucus of the channel catfish, *Ictalurus punctatus. Dev. Comp. Immunol.* **4,** 65–74.
69. St. Louise-Cormier, E.A., Osterland, C.K. and Anderson, P.D. (1984). Evidence for a cutaneous secretory immune system in rainbow trout (*Salmo gairdneri*). *Dev. Comp. Immunol.* **8,** 71–80.
70. Peleteiro, M.C. and Richards, R.H. (1985). Identification of lymphocytes in the epidermis of the rainbow trout, *Salmo gairdneri* Richardson. *J. Fish Dis.* **8,** 161–172.
71. Roberts, R.J. (ed.) (1989). *Fish Pathology*, 2nd edn. Baillière Tindall, London.
72. Hart, S., Wrathmell, A.B., Harris, J.E. and Grayson, T.H. (1988). Gut immunology in fish: a review. *Dev. Comp. Immunol.* **12,** 453–480.
73. Rowley, A.F., Hunt, T.C., Page, M. and Mainwaring, G. (1988). Fish. In *Vertebrate Blood Cells*, Rowley, A.F. and Ratcliffe, N.A. (eds). Cambridge University Press, Cambridge, pp. 19–27.
74. Tanaka, Y., Saito, Y. and Gotoh, H. (1981). Vascular architecture and intestinal haemopoietic nests of two cyclostomes *Eptatretus burgeri* and ammocoetes of *Entosphenus reissneri*: a comparative morphological study. *J. Morphol.* **170,** 71–93.
75. Davina, J.H.M., Rijkers, G.T., Rombout, J.H.W.M., Timmermans, L.P.M. and Van Muiswinkel, W.B. (1980). Lymphoid and non-lymphoid cells in the intestine of cyprinid fish. In *Development and Differentiation of Vertebrate Lymphocytes*, Horton, J.D. (ed.). Elsevier/North Holland Biomedical Press, Amsterdam, pp. 129–140.
76. Rombout, J.H.W.M. and Van den Berg, A.A. (1989). Immunological importance of the second gut segment of carp. I. Uptake and processing of antigens by epithelial cells and macrophages. *J. Fish Biol.* **35,** 13–22.
77. Jenkins, P.G., Harris, J.E. and Pulsford, A.L. (1991). Enhanced enteric uptake of human gamma globulin by Quil-A saponin in *Oreochromis mossambicus. Fish Shellfish Immunol.* **1,** 279–295.
78. Rombout, J.W.H.M., Blok, L.J., Lamers, C.H.J. and Egberts, E. (1986). Immunization of carp (*Cyprinus carpio*) with a *Vibrio anguillarum* bacterin: Indications for a common mucosal immune system. *Dev. Comp. Immunol.* **10,** 341–351.
79. Rombout, J.H.W.M., Bot, H.E. and Taverne-Thiele, J.J. (1989). Immunological importance of the second gut segment of carp. II. Characteristics of mucosal leucocytes. *J. Fish Biol.* **35,** 167–178.
80. Georgopoulou, U. and Vernier, J.M. (1986). Local immunological response in the posterior intestinal segment of the rainbow trout after oral administration of macromolecules. *Dev. Comp. Immunol.* **10,** 529–537.
81. Vigneulle, M. and Baudin-Laurencin, F. (1991). Uptake of *Vibrio anguillarum* bacterin in the posterior intestine of rainbow trout *Oncorhynchus mykiss*, sea bass *Dicentrarchus labrax* and turbot *Scophthalmus maximus* after oral administration or anal intubation. *Dis. Aquatic Organisms* **11,** 85–92.
82. Smith, P.D. (1988). Vaccination against vibriosis. In *Fish Vaccination*, Ellis, A.E. (ed.). Academic Press, London, pp. 67–84.
83. Kawai, K., Kusuda, R. and Itami, T. (1981). Mechanisms of protection in ayu orally vaccinated for vibriosis. *Fish Pathol.* **15,** 257–262.
84. Fletcher, T.C. and White, A. (1973). Antibody production in the plaice, *Pleuronectes platessa*, after oral and parenteral immunization with *Vibrio anguillarum* antigens. *Aquaculture* **1,** 417–428.
85. Lobb, C.J. (1987). Secretory immunity induced in catfish, *Ictalurus punctatus*, following bath immunization. *Dev. Comp. Immunol.* **11,** 727–738.
86. Rombout, J.H.W.M., Van den Berg, A.A., Van den Berg, C.T.G.A., Witte, P. and Egberts, E. (1989). Immunological importance of the second gut segment of carp. III. Systemic and/or mucosal immune responses after immunization with soluble or particulate antigen. *J. Fish Biol.* **35,** 179–186.
87. Lobb, C.J. and Clem, L.W. (1981). The metabolic relationships of the immunoglobulins of fish serum, cutaneous mucus, and bile. *J. Immunol.* **127,** 1525–1529.
88. Lobb, C.J. and Clem, L.W. (1981). Phylogeny of immunoglobulin structure and function. XII. Secretory immunoglobulins in the bile of the marine teleost *Archosargus probatocephalus. Molec. Immunol.* **18,** 615–619.
89. Doggett, T.A., Wrathmell, A.B. and Harris, J.E. (1993). Transport of ferritin and horseradish peroxidase into the systemic circulation of three species of teleost fish following oral and anal intubation. *Fish Shellfish Immunol.* **3,** 1–11.
90. Ingram, G. (1980). Substances involved in the natural resistance of fish to infection. A review.

J. Fish Biol. **16,** 23–60.
91. Fletcher, T.C. (1982). Non-specific defence mechanisms of fish. *Dev. Comp. Immunol.* (Suppl. 2), 123–132.
92. Ellis, A.E. (1989). The immunology of teleosts. In *Fish Pathology*, 2nd edn, Roberts, R.J. (ed.). Baillière Tindal, London, pp. 135–152.
93. De Kinkelin, P., Dorson, M. and Hattenberger-Baudouy, A.M. (1982). Interferon synthesis in trout and carp after viral infection. *Dev. Comp. Immunol.* (Suppl. 2), 167–174.
94. Nonaka, M. (1985). Evolution of complement system. *Dev. Comp. Immunol.* **9,** 377.
95. Courtney-Smith, L. and Davidson, E.H. (1992). The echinoid immune system in the phylogenetic occurrence of immune mechanisms in deuterostomes. *Immunol. Today* **13,** 356–361.
96. Nonaka, M., Fujii, T., Kaidoh, T., Natsuume-Sakai, S., Nonaka, M., Yamaguchi, N. and Takahashi, M. (1984). Purification of a lamprey complement protein homologous to the third component of the mammalian complement system. *J. Immunol.* **133,** 3242–3249.
97. Jensen, J.A. and Festa, E. (1981). The 6-component complement system of the nurse shark *Ginglymostoma cirratum*. In *Aspects of Developmental and Comparative Immunology 1*, Solomon, J.B. (ed.). Pergamon Press, Oxford, pp. 485–486.
98. Koppenheffer, T.L. (1987). Serum complement systems of ectothermic vertebrates. *Dev. Comp. Immunol.* **11,** 279–286.
99. Nonaka, M., Yamaguchi, N., Natsuume-Sakai, S. and Takahashi, M. (1981). The complement system of the rainbow trout (*Salmo gairdneri*). 1. Identification of the serum lytic system homologous to mammalian complement. *J. Immunol.* **126,** 1489–1494.
100. Nonaka, M., Iwaki, M., Nakai, C., Nozaki, M., Kaidoh, T., Nonaka, M., Natsuume-Sakai, S. and Takahashi, M. (1984). Purification of a major serum protein of rainbow trout (*Salmo gairdneri*), homologous to the third component of mammalian complement. *J. Biol. Chem.* **259,** 6327–6333.
101. Matsuyama, H., Yano, T., Yamakawa, T. and Nakao, M. (1992). Opsonic effect of the third complement (C3) of carp (*Cyprinus carpio*) on phagocytosis by neutrophils. *Fish Shellfish Immunol.* **2,** 69–78.
102. Litman, G.W. (1976). Physical properties of immunoglobulins of lower species: A comparison with immunoglobulins of mammals. In *Comparative Immunology*, Marchalonis, J.J. (ed.). Blackwell Scientific Publications, Oxford, pp. 239–275.
103. Warr, G.W. and Marchalonis, J.J. (1982). Molecular basis of self/non-self discrimination in the ectothermic vertebrates. In *The Reticuloendothelial System, Vol. 3: Phylogeny and Ontogeny*, Cohen, N. and Sigel, M.M. (eds). Plenum Press, New York, pp. 541–567.
104. Lobb, C.J. (1986). Structural diversity of channel catfish immunoglobulins. *Vet. Immunol. Immunopathol.* **12,** 7–12.
105. Atwell, J.L. and Marchalonis, J.J. (1982). Immunoglobulin classes of lower vertebrates distinct from IgM immunoglobulin. In *Comparative Immunology*, Marchalonis, J.J. (ed.). Blackwell Scientific Publications, Oxford, pp. 276–279.
106. Gitlin, D., Pericelli, A. and Gitlin, J.D. (1973). Multiple immunoglobulin classes among sharks and their evolution. *Comp. Biochem. Physiol.* **66B,** 225.
107. Litman, G.W., Shamblot, M.J., Haire, R., Amemiya, C., Nichikata, H., Hinds, K., Harding, F., Litman, R. and Varner, J. (1989). Evolution of immunoglobulin gene complexity. In *Progress in Immunology VII*, Melchers, F. *et al.* (eds). Springer-Verlag, Berlin, pp. 361–368.
108. Du Pasquier, L. and Schwager, J. (1989). Evolution of the immune system. In *Progress in Immunology VII*, Melchers, F. *et al.* (eds). Springer-Verlag, Berlin, pp. 1246–1255.
109. Hohman, V.S., Schluter, S.F. and Marchalonis, J.J. (1992). Complete sequence of a cDNA clone specifying sandbar shark immunoglobulin light chain: Gene organization and implications for the evolution of light chains. *Proc. Natl. Acad. Sci. US* **89,** 276–280.
110. DuPasquier, L. (1982). Antibody diversity in lower vertebrates—why is it so restricted? *Nature* [*Lond.*] **296,** 311–313.
111. Fiebig, H. and Ambrosius, H. (1977). Studies on the regulation of the IgM immune response I. Changes of the affinity of anti-dinitrophenyl antibodies of carp during immunization, dependence on the antigen dose. *Acta Biol. Med. Germ.* **36,** 79–86.
112. Etlinger, H.M., Hodgins, H.O. and Chiller, J.M. (1976). Evolution of the lymphoid system 1. Evidence for lymphocyte heterogeneity in rainbow trout revealed by the organ distribution of mitogenic responses. *J. Immunol.* **116,** 1547–1553.
113. DeLuca, D., Wilson, M. and Warr, G. (1983). Lymphocyte heterogeneity in the trout, *Salmo gairdneri*, defined with monoclonal antibodies to IgM. *Eur. J. Immunol.* **13,** 546–551.
114. Thuvander, A., Fossum, C. and Lorenzen, N. (1990). Monoclonal antibodies to salmonid

immunoglobulin: characterization and applicability in immunoassays. *Dev. Comp. Immunol.* **14,** 415–423.
115. Secombes, C.J., Van Groningen, J.J.M. and Egberts, E. (1983). Separation of lymphocyte subpopulations in carp, *Cyprinus carpio* L., by monoclonal antibodies; Immunohistochemical studies. *Immunology* **48,** 165–175.
116. Sizemore, R.G., Miller, N.W., Cuchens, M.A., Lobb, C.J. and Clem, L.W. (1984). Phylogeny of lymphocyte heterogeneity: the cellular requirements for *in vitro* mitogenic responses of channel catfish leukocytes. *J. Immunol.* **133,** 2920–2924.
117. Clem, L.W., Sizemore, R.C., Ellsaesser, C.F. and Miller, N.W. (1985). Monocytes as accessory cells in fish immune responses. *Dev. Comp. Immunol.* **9,** 803–809.
118. Clem, L.W., Miller, N.W. and Bly, J.E. (1991). Evolution of lymphocyte subpopulations, their interactions and temperature sensitivities. In *Phylogenesis of Immune Function*, Warr, G.W. and Cohen, N. (eds). CRC Press, Boca Raton, pp. 191–213.
119. Vallejo, A.N., Ellsaesser, C.F., Miller, N.W. and Clem, L.W. (1991). Spontaneous development of functionally active long term monocyte-like cell lines from channel catfish. *In Vitro (Cell Dev. Biol.)* **27A,** 279–286.
120. Vallejo, A.N., Miller, N.W. and Clem, L.W. (1991). Phylogeny of immune recognition: Processing and presentation of structurally-defined proteins in channel catfish immune responses. *Dev. Immunol.* **1,** 137–148.
121. Vallejo, A.N., Miller, N.W. and Clem, L.W. (1991). Phylogeny of immune recognition: role of alloantigens in antigen presentation in channel catfish immune responses. *Immunology* **74,** 165–168.
122. Rijkers, G.T., Wiegerink, J.A.M., Van Oosterom, R. and Van Muiswinkel, W.B. (1980). The immune system of cyprinid fish. Kinetics and temperature dependence of antibody-producing cells in carp (*Cyprinus carpio*). *Immunology* **41,** 91–97.
123. Lamers, C.H.J. (1985). *The reaction of the immune system of fish to vaccination.* Doctoral Thesis, Agricultural University, Wageningen, The Netherlands.
124. Lamers, C.H.J., De Haas, M.J.H. and Van Muiswinkel, W.B. (1985). Humoral response and memory formation in carp after injection of *Aeromonas hydrophila* bacterin. *Dev. Comp. Immunol.* **9,** 65–75.
125. Dunier, M. (1985). Absence of anamnestic antibody response to DNP-haemocyanin and DNP-ficoll in rainbow trout. In *Fish Immunology*, Manning, M.J. and Tatner, M.F. (eds). Academic Press, London, pp. 171–184.
126. Yui, M.A. and Kaattari, S.L. (1987). *Vibrio anguillarum* antigen stimulates mitogenesis and polyclonal activation of salmonid lymphocytes. *Dev. Comp. Immunol.* **11,** 539–549.
127. Secombes, C.J. and Resnik, J.W. (1984). The immune response of carp, *Cyprinus carpio* L., following injection of antigen-antibody complexes. *J. Fish Biol.* **24,** 193–200.
128. Arkoosh, M.R. and Kaattari, S.L. (1991). Development of immunological memory in rainbow trout (*Oncorhynchus mykiss*). 1. An immunological and cellular analysis of the B cell response. *Dev. Comp. Immunol.* **15,** 279–293.
129. Etlinger, H.M., Chiller, J.M. and Hodgins, H.O. (1979). Evolution of the lymphoid system IV. Murine T-independent but not T-dependent antigens are very immunogenic in rainbow trout *Salmo gairdneri. Cell. Immunol.* **47,** 400–406.
130. Van Loon, J.J.A., Van Oosterom, R. and Van Muiswinkel, W.B. (1981). Development of the immune system in carp (*Cyprinus carpio*). In *Aspects of Developmental and Comparative Immunology 1*, Solomon, J.B. (ed.). Pergamon Press, Oxford, pp. 469–470.
131. Mughal, M.S., Farley-Ewens, E.K. and Manning, M.J. (1986). Effects of direct immersion in antigen on immunological memory in young carp, *Cyprinus carpio. Vet. Immunol. Immunopathol.* **12,** 181–192.
132. Serero, M. and Avtalion, R.R. (1978). Regulatory effect of temperature and antigen upon immunity in ectothermic vertebrates III. Establishment of immunological suppression in fish. *Dev. Comp. Immunol.* **2,** 87–94.
133. Avtalion, R.R., Wishkovsky, A. and Katz, D. (1980). Regulatory effect of temperature on specific suppression and enhancement of the humoral response in fish. In *Phylogeny of Immunological Memory*, Manning, M.J. (ed.). Elsevier/North Holland Biomedical Press, Amsterdam, pp. 113–121.
134. Avtalion, R.R. (1981). Environmental control of the immune response in fish. *CRC Critical Review of Environmental Control* **11,** 163–188.
135. Bly, J.E. and Clem, L.W. (1991). Temperature-mediated processes in teleost immunity: *in vivo* low temperature immunisation does not induce tolerance in channel catfish. *Fish Shellfish Immunol.* **1,** 229–231.
136. Bly, J.E. and Clem, L.W. (1992). Temperature and teleost immune functions. *Fish Shellfish Immunol.* **2,** 159–171.
137. Ward, P.D., Tatner, M.F. and Horne, M.T.

(1985). Factors influencing the efficiency of vaccines against vibrosis caused by *Vibrio anguillarum*. In *Fish Immunology*, Manning, M.J. and Tatner, M.F. (eds). Academic Press, London, pp. 221–229.
138. Bly, J.E., Grimm, A.S. and Morris, I.G. (1986). Transfer of passive immunity from mother to young in a teleost fish: Haemagglutinating activity in the serum and eggs of plaice, *Pleuronectes platessa* L. *Comp. Biochem. Physiol.* **84A,** 309–313.
139. Mor, A. and Avtalion, R.R. (1989). Transfer of antibody activity from immunized mother to embryo in tilapias. *J. Fish Biol.* **37,** 249–254.
140. Hayman, J.R. and Lobb, C.J. (1993). Immunoglobulin in the eggs of the channel catfish (*Ictalurus punctatus*). *Dev. Comp. Immunol.* **17,** 241–248.
141. Takahashi, Y. and Kawahara, E. (1987). Maternal immunity in newborn fry of the ovoviviparous guppy. *Nippon Suisan Gakkaishi* **53,** 721–735.
142. Hine, P.M. (1992). The granulocytes of fish. *Fish Shellfish Immunol.* **2,** 79–98.
143. Griffin, B.R. (1984). Random and directed migration of trout (*Salmo gairdneri*) leukocytes: activation by antibody, complement, and normal serum components. *Dev. Comp. Immunol.* **8,** 589–598.
144. MacArthur, J.I., Thomson, A.W. and Fletcher, T.C. (1985). Aspects of leucocyte migration in the plaice, *Pleuronectes platessa* L. *J. Fish Biol.* **27,** 667–676.
145. Howell, C.J.G. (1987). A chemokinetic factor in the carp, *Cyprinus carpio. Dev. Comp. Immunol.* **11,** 139–146.
146. Pettitt, T.R., Rowley, A.F. and Secombes, C.J. (1989). Lipoxins are major lipoxygenase products of rainbow trout macrophages. *FEBS Letters* **259,** 168–170.
147. Rowley, A.F., Pettitt, T.R., Secombes, C.J., Sharp, G.J.E., Barrow, S.E. and Mallet, A.I. (1991). Generation and biological activities of lipoxins in the rainbow trout—an overview. In *Advances in Prostaglandin, Thromboxane and Leukotriene Research: Vol. 21B*, Samuelsson, B. *et al.* (eds). Raven Press, New York, pp. 557–560.
148. Rowley, A.F. (1991). Eicosanoids: aspects of their structure, function and evolution. In *Phylogenesis of Immune Function*, Warr, G.W. and Cohen, N. (eds). CRC Press, Boca Raton, pp. 269–294.
149. Nash, K.A., Fletcher, T.C. and Thomson, A.W. (1986). Migration of fish leucocytes *in vitro*: The effect of factors which may be involved in mediating inflammation. *Vet. Immunol. Immunopathol.* **12,** 83–92.
150. Sharp, G.J.E., Pike, A.W. and Secombes, C.J. (1991). Leucocyte migration in rainbow trout (*Oncorhynchus mykiss* [Walbaum]): Optimization of migration conditions and responses to host and pathogen (*Diphyllobothrium dendriticum* [Nitzsch]) derived chemoattractants. *Dev. Comp. Immunol.* **15,** 295–305.
151. MacArthur, J.I. and Fletcher, T.C. (1985). Phagocytosis in fish. In *Fish Immunology*, Manning, M.J. and Tatner, M.F. (eds). Academic Press, London, pp. 29–46.
152. O'Neill, J.C. (1985). An *in vitro* study of polymorphonuclear phagocytosis and the effect of temperature. In *Fish Immunology*, Manning, M.J. and Tatner, M.F. (eds). Academic Press, London, pp. 47–55.
153. Fujii, T. (1981). Antibody-enhanced phagocytosis of lamprey polymorphonuclear leucocytes against sheep erythrocytes. *Cell Tissue Res.* **219,** 41–51.
154. Haynes, L., Fuller, L. and McKinney, E.C. (1988). Fc receptors for shark IgM. *Dev. Comp. Immunol.* **12,** 561–571.
155. Griffin, B.R. (1983). Opsonic effect of rainbow trout (*Salmo gairdneri*) antibody on phagocytosis of *Yersinia ruckeri* by trout leukocytes. *Dev. Comp. Immunol.* **7,** 253–259.
156. Sakai, D.K. (1984). Opsonization by fish antibody and complement in the immune phagocytosis by peritoneal exudate cells isolated from salmonid fishes. *J. Fish Dis.* **7,** 29–38.
157. Scott, A.L. and Klesius, P.H. (1981). Chemiluminescence: A novel analysis of phagocytosis in fish. *Developments in Biological Standardization* **49,** 243–254.
158. Nash, K.A., Fletcher, T.C. and Thomson, A.W. (1987). Effect of opsonization on oxidative metabolism of plaice (*Pleuronectes platessa* L.) neutrophils. *Comp. Biochem. Physiol.* **86B,** 31–36.
159. Plytycz, B., Flory, C.M., Galvan, I. and Bayne, C.J. (1989). Leukocytes of rainbow trout (*Oncorhynchus mykiss*) pronephros. Cell types producing superoxide anion. *Dev. Comp. Immunol.* **13,** 217–224.
160. Chung, S. and Secombes, C.J. (1988). Analysis of events occurring within teleost macrophages during the respiratory burst. *Comp. Biochem. Physiol.* **89B,** 539–544.
161. Secombes, C.J., Chung, S. and Jeffries, A.H. (1988). Superoxide anion production by rainbow trout macrophages detected by the reduction of ferricytochrome c. *Dev. Comp. Immunol.* **12,** 201–206.

162. Graham, S., Jeffries, A.H. and Secombes, C.J. (1988). A novel assay to detect macrophage bactericidal activity in fish: factors influencing the killing of *Aeromonas salmonicida*. *J. Fish Dis.* **11,** 389–396.
163. Whyte, S.K., Chappell, L.H. and Secombes, C.J. (1989). Cytotoxic reactions of rainbow trout, *Salmo gairdneri* Richardson, macrophages for larvae of the eye fluke *Diplostomum spathaceum* (Digenea). *J. Fish Biol.* **35,** 333–345.
164. Hinuma, S., Abo, T., Kumagai, K. and Hata, M. (1980). The potent activity of freshwater fish kidney cells in cell killing. Characterization and species distribution of cytotoxicity. *Dev. Comp. Immunol.* **4,** 653–666.
165. Carlson, R.L., Evans, D.L. and Graves, S.S. (1985). Non-specific cytotoxic cells in fish (*Ictalurus punctatus*) V. Metabolic requirements of lysis. *Dev. Comp. Immunol.* **9,** 271–280.
166. Evans, D.L., Jaso-Friedmann, L., Smith, E.E., St. John, A., Koren, H.S. and Harris, D.T. (1988). Identification of a putative antigen receptor on fish non-specific cytotoxic cells with monoclonal antibodies. *J. Immunol.* **141,** 324–332.
167. Graves, S.S., Evans, D.L. and Dawe, D.L. (1985). Antiprotozoan activity of non-specific cytotoxic cells (NCC) from the channel catfish (*Ictalurus punctatus*). *J. Immunol.* **134,** 78–85.
168. Haynes, L. and McKinney, E.C. (1991). Shark spontaneous cytotoxicity: characterization of the regulatory cell. *Dev. Comp. Immunol.* **15,** 123–134.
169. Ellis, A.E. (1985). Eosinophilic granular cells (EGC) and histamine responses to *Aeromonas salmonicida* toxins in rainbow trout. *Dev. Comp. Immunol.* **9,** 251–260.
170. Jurd, R.D. (1987). Hypersensitivity in fishes: a review. *J. Fish Biol.* **31** (Suppl. A), 1–7.
171. Sigel, M.M., Hamby, B.A. and Huggins, E.M. (1986). Phylogenetic studies on lymphokines. Fish lymphocytes respond to human IL-1 and epithelial cells produce an IL-1 like factor. *Vet. Immunol. Immunopathol.* **12,** 47–58.
172. Caspi, R.R. and Avtalion, R.R. (1984). Evidence for the existence of an IL-2 like lymphocyte growth promoting factor in a bony fish (*Cyprinus carpio*). *Dev. Comp. Immunol.* **8,** 51–60.
173. Grondel, J.L. and Harmsen, E.G.M. (1984). Phylogeny of interleukins: growth factors produced by leucocytes of the cyprinid fish, *Cyprinus carpio* L. *Immunology* **52,** 477–482.
174. Graham, S. and Secombes, C.J. (1988). The production of a macrophage activating factor for rainbow trout, *Salmo gairdneri* leucocytes. *Immunology* **65,** 293–297.
175. Graham, S. and Secombes, C.J. (1990). Do fish lymphocytes secrete interferon-γ? *J. Fish Biol.* **36,** 563–573.
176. Bridges, A.F. and Manning, M.J. (1991). The effects of priming immersions in various human gamma globulin (HGG) vaccines on humoral and cell-mediated immune responses after intraperitoneal HGG challenge in the carp, *Cyprinus carpio* L. *Fish Shellfish Immunol.* **1,** 119–129.
177. Miller, N.W., Bly, J.E., Van Ginkel, F., Ellsaesser, C.F. and Clem, L.W. (1987). Identification and separation of functionally distinct subpopulations of channel catfish lymphocytes with monoclonal antibodies. *Dev. Comp. Immunol.* **11,** 739–747.
178. Miller, N.W., Deuter, A. and Clem, L.W. (1986). Phylogeny of lymphocyte heterogeneity: the cellular requirements for the mixed leucocyte reaction with channel catfish. *Immunology* **59,** 123–128.
179. Graham, S. and Secombes, C.J. (1990). Cellular requirements for lymphokine secretion by rainbow trout, *Salmo gairdneri*, leucocytes. *Dev. Comp. Immunol.* **14,** 59–68.
180. Finstad, J. and Good, R.A. (1964). The evolution of the immune response III. Immunological responses in the lamprey. *J. Exp. Med.* **120,** 1151–1167.
181. Bartos, J.M. and Sommer, C.V. (1981). *In vivo* cell-mediated immune response to *M. tuberculosis* and *M. salmoniphilium* in rainbow trout (*Salmo gairdneri*). *Dev. Comp. Immunol.* **5,** 75–83.
182. Stevenson, R.M.W. and Raymond, B. (1990). Delayed-type hypersensitivity skin reactions. *Techniques in Fish Immunology I*, Stolen, J.S. *et al.* (eds). SOS Publications, Fair Haven, New Jersey, pp.173–178.
183. Hildemann, W.H. (1970). Transplantation immunity in fishes: Agnatha, Chondrichthyes and Osteichthyes. *Transplant. Proc.* **11,** 253–259.
184. Stet, R.J.M. and Egberts, E. (1991). The histocompatibility system in teleostean fishes: from multiple histocompatibility loci to a major histocompatibility complex. *Fish Shellfish Immunol.* **1,** 1–16.
185. Matsuzaki, T. and Shima, A. (1989). Number of major histocompatibility loci in inbred strains of the fish *Oryzias latipes. Immunogenetics* **30,** 226–228.
186. Kaastrup, P., Stet, R.J.M., Tigchelaar, A.J., Egberts, E. and Van Muiswinkel, W.B. (1989). A major histocompatibility locus in fish: serological identification and segregation of transplantation antigens in common carp *Cyprinus carpio*

L. *Immunogenetics* **30,** 284–290.

187. Stet, R.J.M., Kaastrup, P., Egberts, E. and Van Muiswinkel, W.B. (1990). Characterization of new immunogenetic markers using carp alloantisera: evidence for the presence of major histocompatibility complex (MHC) molecules. *Aquaculture* **85,** 119–124.
188. Hashimoto, K., Nakanishi, T. and Kurosawa, Y. (1990). Isolation of carp genes encoding major histocompatibility complex antigens. *Proc. Nat. Acad. Sci. USA* **87,** 6863–6868.
189. Hashimoto, K., Nakanishi, T. and Kurosawa, Y. (1992). Identification of a shark sequence resembling the major histocompatibility complex class Iα3 domain. *Proc. Nat. Acad. Sci. USA* **89,** 2209–2212.
190. Tatner, M.F. and Findlay, C. (1991). Lymphocyte migration and localisation patterns in rainbow trout *Oncorhynchus mykiss*, studies using the tracer sample method. *Fish Shellfish Immunol.* **1,** 107–117.
191. Bucke, D. (1991). Current approaches to the study of pollution-related diseases in fish. *Bull. Eur. Ass. Fish Pathol.* **11,** 46–53.

4

AMPHIBIANS

J.D. Horton

Department of Biological Sciences, University of Durham, UK

4.1 INTRODUCTION

Amphibians are ectothermic tetrapods and were the first vertebrate group to emerge onto land about 350 million years ago. They are mainly represented today by two orders, the Urodela, which have retained a long-bodied, fish-like form (e.g. newts and salamanders), and the tailless Anura (frogs and toads). A third order, the Apoda, are an aberrant group of legless amphibians of which there are only a few genera, and about which relatively little is known immunologically. Although the term amphibian implies a double mode of life, most amphibians live near water to which they must return for breeding. Embryos develop into aquatic tadpoles which possess gills; later, at metamorphosis, lungs replace gills and limbs mature, making terrestrial life possible. Some urodeles (e.g. axolotls) retain external gills and reproduce in what is essentially a larval condition, although there is evidence that these neotenous animals in fact undergo cryptic 'metamorphosis', involving haematological changes[1].

This chapter concentrates on the immune system of anurans, because today they are providing especially useful model systems for exploring ontogenetic aspects of immunity, which are highlighted here. Although pertinent studies on urodele immunology are included, the reader is referred to a number of previous reviews[2–12], which provide a comprehensive background to the diversity of immune systems found at the amphibian level of evolution. Overall the anuran immune system appears more sophisticated than that found in urodeles, which may well be related to physiological and morphological adaptations (e.g. improved circulation of body fluids) necessary for active life of frogs and toads on land.

Today the most-studied amphibian immunological model is the clawed toad, *Xenopus* (actually a pipid frog)[13]. A variety of techniques for exploring the *Xenopus* immune system are available[14]. These entirely aquatic animals are easy to breed and maintain (ideally at 23°C) in the laboratory and, in recent years, several isogeneic and inbred families of *Xenopus* have become available for immunological research. The technique for producing isogeneic *Xenopus* involves the use of hybrid *Xenopus*[13,14]. For example, *X. laevis/X. gilli* and *X. laevis/X. muelleri* hybrids lay some

Immunology: A Comparative Approach. Edited by R.J. Turner

diploid eggs (due to endoreduplication of chromosomes during oocyte development), and gynogenetic development of these eggs results in a clone. Members of the clone are identical when tested by a variety of criteria, including mixed leucocyte reactivity and allograft responsiveness. Different *Xenopus* clones that are either major histocompatibility complex (MHC) compatible (but express minor histocompatibility differences) or possess one or two MHC haplotype differences, are proving invaluable for a whole range of immunological investigations.

Section 4.2 illustrates the cellular basis of the amphibian immune system, by describing the diversity and development of lymphomyeloid tissues and immune cell types. Section 4.3 reviews ongoing studies which attempt to characterize the structure and genetics of the immune molecules involved in antigen-specific responses (e.g. MHC proteins, immunoglobulins and T-cell receptor molecules) and those that are not antigen-specific (e.g. complement proteins and cytokines). The overall functioning of the amphibian immune system *in vivo* is briefly considered in Section 4.4. Finally, Section 4.5 focuses attention on the particular advantages of using free-living amphibian embryonic and larval model systems to explore important issues in developmental immunology. Topics covered include origins of lymphocytes, ontogeny of cellular immunity and antibody responses, role of the thymus in T-cell education, development of transplantation tolerance and the immunology of metamorphosis.

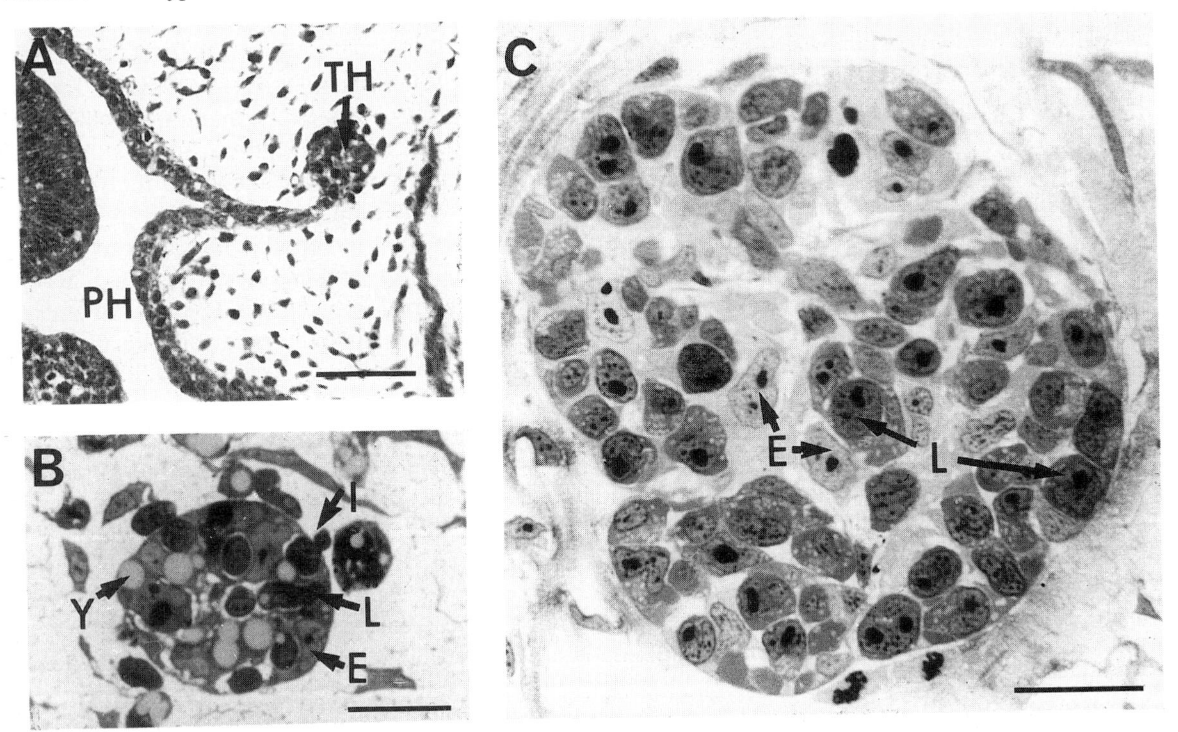

Figure 4.1. Early stages of thymic development in *Xenopus*. (A) Thymic bud (TH) is still in continuity with the pharyngeal epithelium in this 2.5–3-day-old larva. PH = pharynx in the region of the second branchial pouch. 8-μm section: H & E stain. Scale bar = 50 μm. (B) Basophilic lymphoid 'stem cells' (L) can be seen within the thymic bud at 3 days, alongside paler-staining epithelial stromal cells (E). Similar basophilic cells are often seen around the thymus (now detached from the pharyngeal epithelium) and one such cell is seen immigrating (I) through the thymic capsule. Y = yolk platelets. 1-μm section: toluidine blue stain. Scale bar = 20 μm. (C) At 7 days the thymus comprises approximately 1000 cells of two major types. The lymphoid cells (L) are readily distinguishable from the paler-staining epithelial cells (E). 1-μm section: toluidine blue stain. Scale bar = 20 μm.

4.2 CELLULAR BASIS OF THE AMPHIBIAN IMMUNE SYSTEM: LYMPHOID TISSUES AND IMMUNE CELL TYPES

Lymphoid and myeloid components are more intermingled in amphibians (and fishes) than in mammals. The organ distribution of lymphoid tissues in the three amphibian orders is reviewed elsewhere[8,10].

4.2.1 The Thymus, Thymectomy and T-cell Development

Thymic Histogenesis

The embryological origin of the thymus from the pharyngeal epithelium in apodans, urodeles and anurans has been described elsewhere in detail[8]. In *Xenopus*, the thymus arises dorsally from the second pharyngeal pouch around day 3 after fertilization (Figure 4.1). (Immunocyte development of *Xenopus* in relation to age is shown in Figure 4.2). By 7 days, two major cell types are found in the thymus, the epithelial cells and the immigrant lymphoid cells (Figure 4.1), the latter previously having colonized the thymus at 3–4 days (Figure 4.1)[15], on arrival from the lateral plate mesoderm[16].

The differentiation of surface MHC-encoded molecules on thymic cell types has been studied in *Xenopus*[17], since these molecules play a crucial role in T-cell education in mammals. MHC class II molecules are not expressed on *Xenopus* thymic epithelium until 7 days, indicating that class II is not required to attract stem cells to the thymus. The question whether class I MHC proteins are expressed on thymic epithelial cells is dealt with in Section 4.3.1. Larval thymic lymphoid cells do not express class II MHC proteins.

At 7 days, a T-cell differentiation antigen—the XT-1 antigen (120kDa) recognized by an anti-thymocyte mouse monoclonal antibody and expressed on a major T-cell subset[18]—begins to appear on thymic lymphoid cells. This occurs

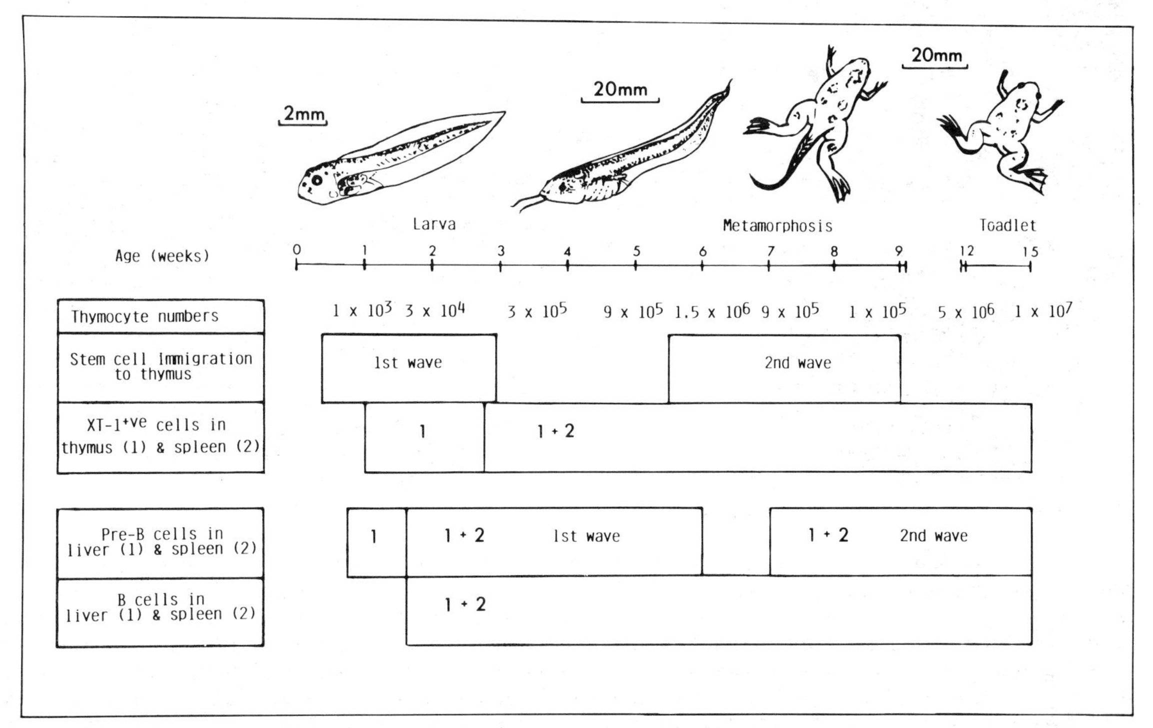

Figure 4.2. Ontogeny of T- and B-cells in *Xenopus*. Based on data from references 13, 48, 106 and 137. Animal drawings reproduced by courtesy of Dr L. DuPasquier.

before the emergence of XT-1^{+ve} cells in the periphery (Figure 4.2). Evidence for migration of lymphocytes from thymus to periphery has been obtained by *in situ* labelling of axolotl (*Ambystoma mexicanum*) thymus with fluorescein isothiocyanate and observing migrant cells[19].

During metamorphosis the *Xenopus* thymus is translocated to the ear region. At the end of

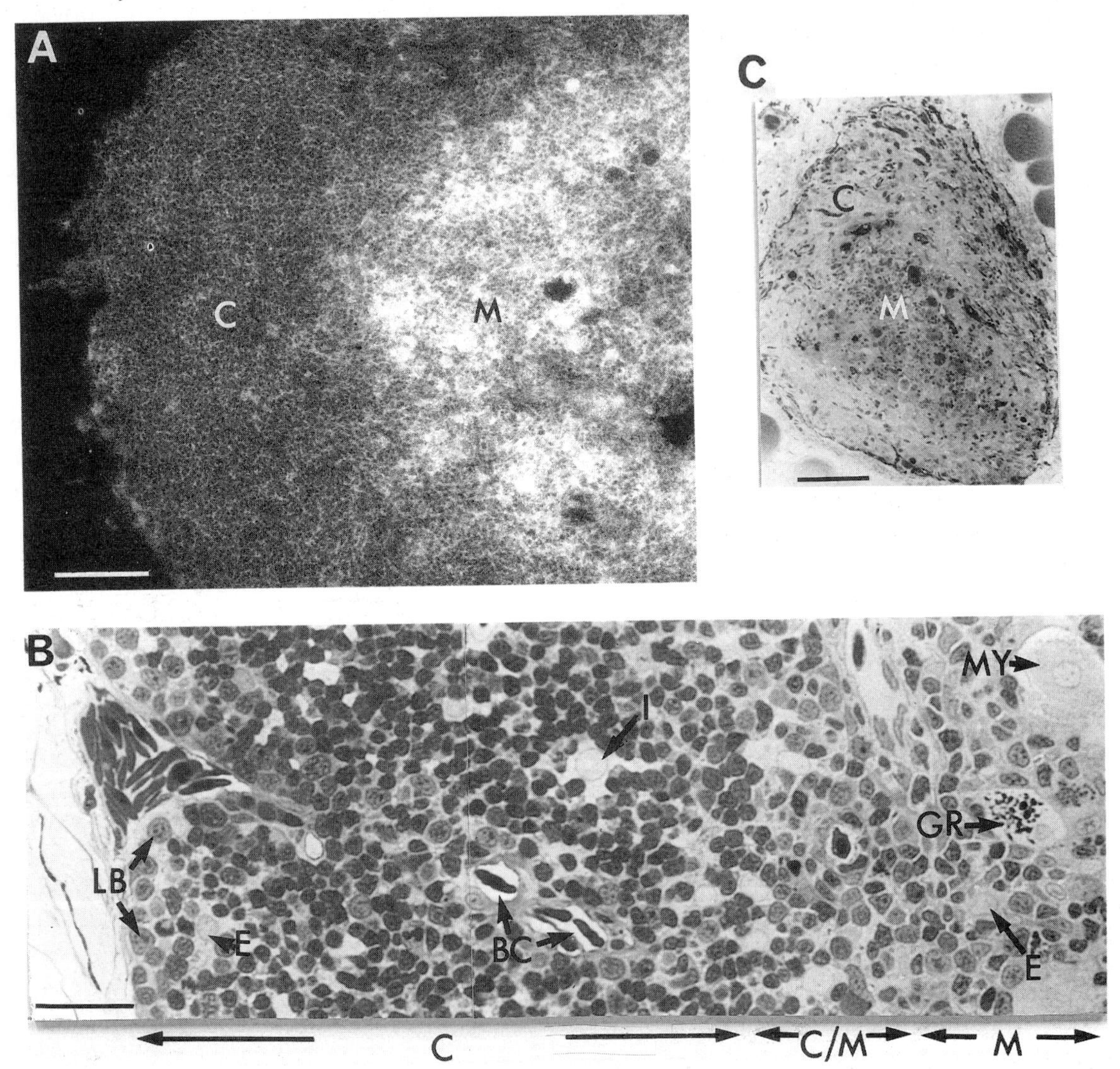

Figure 4.3. Characterization of post-metamorphic *Xenopus* thymus. (A) Thymus cryostat (6-μm) section of 6-month-old animal stained with an anti-MHC class II monoclonal antibody (AM20)[17,32], using indirect immunofluorescence. The epithelial-rich medulla (M) stains brightly, compared with the more delicately-stained cortex (C). (Thymic lymphocytes prior to metamorphosis are MHC class II^{-ve}.) Scale bar = 100 μm.
(B, C) Toluidine blue-stained, 1-μm plastic-embedded sections of control and *in vivo* gamma-irradiated thymus (9 days after 3000 rads) of 8-month-old froglets. (B) Montage through segment of cortex (C), corticomedullary junction (C/M) and part of medulla (M) of control thymus. The cortex is richly populated with lymphocytes, with lymphoblasts (LB) seen just under the capsule. BC = blood capillaries with large, nucleated erythrocytes; E = epithelial cells; GR = granulocyte; I = putative 'interdigitating cell'; MY = myoid cell. Scale bar = 30 μm. (C) Low-power view of irradiated thymus. Note dramatic reduction in size and loss of lymphocytes from cortex (C). Some lymphocytes are retained in medulla (M). Scale bar = 100 μm.

metamorphosis it is temporarily but extensively depleted of its lymphocytes[9] (Figure 4.2). It then undergoes a second histogenesis (see Section 4.5.1) with newly-formed adult T-lymphocytes now expressing readily-detectable surface class II MHC, although the epithelial-rich medulla remains more heavily stained with anti-class II monoclonal antibodies (Figure 4.3). The rapidly-proliferating cortical lymphocytes are particularly sensitive to γ-irradiation (Figure 4.3) whereas a variety of stromal cell types are radiation-resistant[20]. The thymus undergoes regression at sexual maturity.

A variety of cell types, in addition to lymphocytes and epithelial cells, are found within the amphibian thymus[20,21]; these include large dendritic (interdigitating) cells, macrophages, cysts (which increase in number after irradiation) and myoid cells (Figure 4.3). It has been suggested that the latter are involved in promoting circulation of tissue fluids within the thymus, or that they may provide a source of self antigen. These cells, along with thymic epithelial cells and interdigitating cells must be included as candidates for educating T-lineage cells (Section 4.5.2).

Diverse granulated cells and secretory cells are also found in the thymus, especially in the adult medulla[21]. The production of thymic hormones in amphibians is suggested from studies on urodeles[22]. IgM-secreting plasma cells have also been identified in the *Xenopus* thymus, although this organ appears not to be involved in the production of these cells[13]. Thymic nurse cell-like complexes are found in frogs. In endotherms, such thymic nurse cells are thought to be complexes of epithelial cells with enclosed thymocytes, and are considered as possible sites where self-MHC restriction of developing T-cells is learned. In *Rana pipiens* larvae, these complexes have been investigated through *in vitro* observations and appear to be comprised of several stromal cell types, which enclose thymocytes[23].

Thymectomy

Early removal of the thymus has a profound effect on immunological development in both anurans and urodeles (see review in reference 8). The free-living larva of *Xenopus* is ideally suited for probing the effects of early thymectomy on T-cell development, since the thymic buds can be visualized through the transparent skin (Figure 4.4). Thymectomy can be achieved early in life at

Figure 4.4. *Xenopus* thymectomy model system. In the control larva (right) the pigmented paired thymus lies posterior to the eyes; its absence is readily apparent in the sibling larva (38 days old) thymectomized at 7 days (left). Scale bar = 1 cm.

a rudimentary stage of thymic histogenesis[24,25] without the animal runting, although early-thymectomized *Xenopus* adults appear more susceptible to subcutaneous hydrops, of unknown aetiology.

Thymectomy of *Xenopus* from 4–8 days of age has clearly demonstrated the existence of T-dependent and T-independent components of immunity in anuran amphibians (Figure 4.5). Such lymphoid cell heterogeneity is less well defined in urodeles[10,26], partly due to the rather poor response of their lymphocytes to T-cell mitogens and alloantigens and also because it is more difficult to remove the multi-lobed thymuses at early developmental stages. It should be noted, however, that monoclonal and polyclonal antibodies that can differentiate between T-like and B-like cells of axolotls have now been produced[27]. A monomeric 38-kDa glycoprotein can be recognized on axolotl T-cell populations, which are surface Ig^{-ve}; this antigen is, at best, poorly expressed after thymectomy[139]. It has also been demonstrated that axolotl lymphocytes can be made to respond reproducibly to phytohaemagglutinin (PHA), provided that culture medium is supplemented with bovine serum albumin[138]. In anurans, early thymectomy abrogates acute allograft rejection and mixed leucocyte reactivity; responsiveness to the T-cell mitogens concanavalin A (Con A) and PHA is severely diminished[24,25]. Cellular and humoral responses to T-cell-dependent erythrocyte antigens are abolished, as is the low molecular weight (IgY) response to DNP-KLH[28] (see also Section 4.3.2). *In vivo* IgM antibody and *in vitro* proliferative responses to T-cell independent antigens/mitogens—e.g. *E. coli* lipopolysaccharide (LPS), polyvinylpyrrolidone (PVP) and anti-IgM—are not affected by early thymectomy[8,25]. Various categories of T-independent antigens

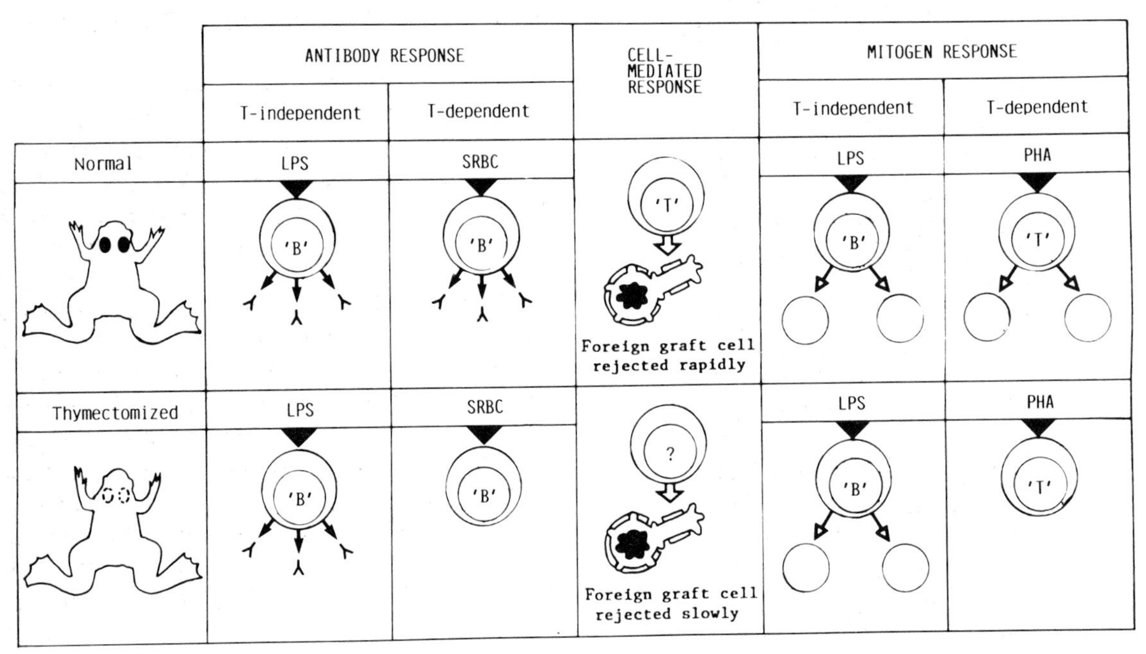

Figure 4.5. T-dependent and T-independent components of immunity demonstrated by early thymectomy. *Xenopus* thymectomized at 4–8 days of age display impaired antibody responses to thymus-dependent antigens (e.g. sheep red blood cells—SRBC, but not to thymus-independent antigens (e.g. lipopolysaccharide from *E. coli*—LPS. Similarly, *in vitro* polyclonal mitogenesis can be induced in lymphocytes from thymectomized animals by LPS (a T-independent mitogen), but not by PHA (phytohaemagglutinin), a T-dependent mitogen. Allograft rejection is dramatically impaired in thymectomized animals, although chronic graft destruction can still take place.

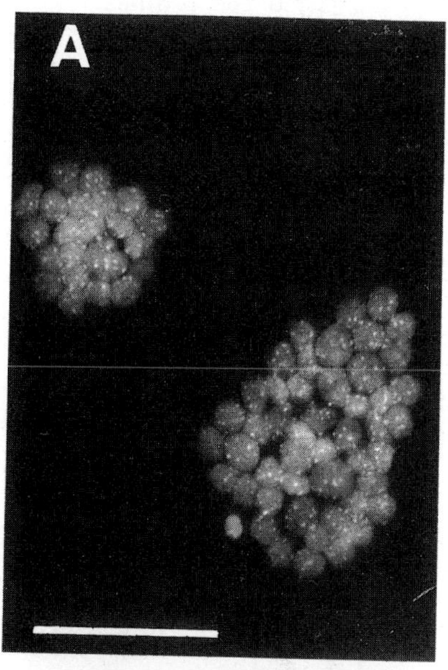

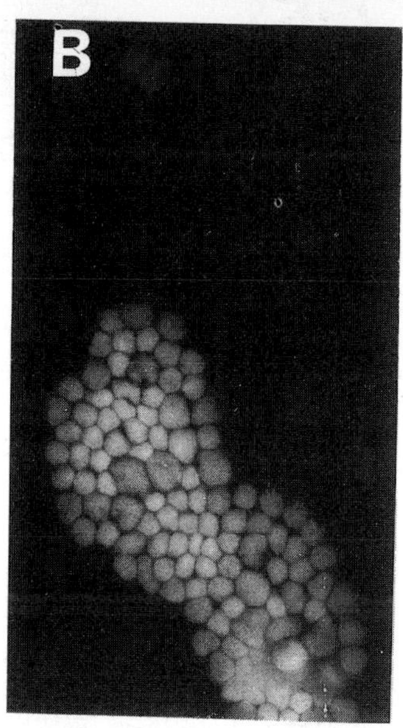

Figure 4.6. The *Xenopus borealis* quinacrine fluorescence cell marker. Quinacrine-stained thymocytes from 4-month-old *Xenopus*, viewed by fluorescence microscopy. (A) *X. borealis* thymocytes, with spotted nuclear fluorescence. (B) *X. laevis* thymocytes, with unstained nuclei. Scale bar = 50 μm.

have now been characterized in *Xenopus*[29].

Immune responses abrogated by early thymectomy can be restored by grafting of thymuses or injection of dissociated thymocytes into thymectomized *Xenopus*[24,25]. Reconstitution of allograft rejection is effected through direct involvement of donor cells, whereas the latter have been shown to simply act as helper cells in restoring antibody responses to T-dependent antigens[25]. The use of cytogenetically distinct donor cells has been employed in the above restoration experiments. Triploid embryos can readily be produced, for example, by 'cold shock' treatment of recently-fertilized eggs, which prevents extrusion of the second polar body[30]. Species-specific markers, such as the quinacrine fluorescence uniquely displayed by *X. borealis*, have also been recently used in this respect (Figure 4.6)[31].

Extrathymic Development of 'T-like' Cells?

Use of monoclonal antibodies specific for *Xenopus* T-cell epitopes—e.g. the aforementioned XT-1 and the AM22 antibody that recognizes a CD8-like molecule (35kDa) on a T-cell subset[32]—are currently proving useful for exploring whether early thymectomy totally precludes T-cell maturation. The spleen of larval and young (3–4-month-old) post-metamorphic thymectomized *Xenopus* no longer contain detectable XT-1^{+ve} or AM22^{+ve} cells by flow cytometric analysis, whereas surface immunoglobulin positive (sIg^{+ve}) B-lymphocytes are plentiful[33] (Figure 4.7). However, the possibility of an extrathymic maturation pathway of 'T-like' cells in thymectomized amphibians (comparable to the development of T-like cells in nude mice and rats[34]) is hotly debated. Such a pathway of T-cell development in anurans was, in part, suggested because of the finding that early thymectomized *Xenopus*, even those thymectomized at 4 days of age, can sometimes chronically reject MHC-disparate skin grafts. Furthermore, following graft rejection, splenocytes of thymectomized animals become able to display a non-donor-specific mixed lymphocyte reactivity, but are still unable to respond

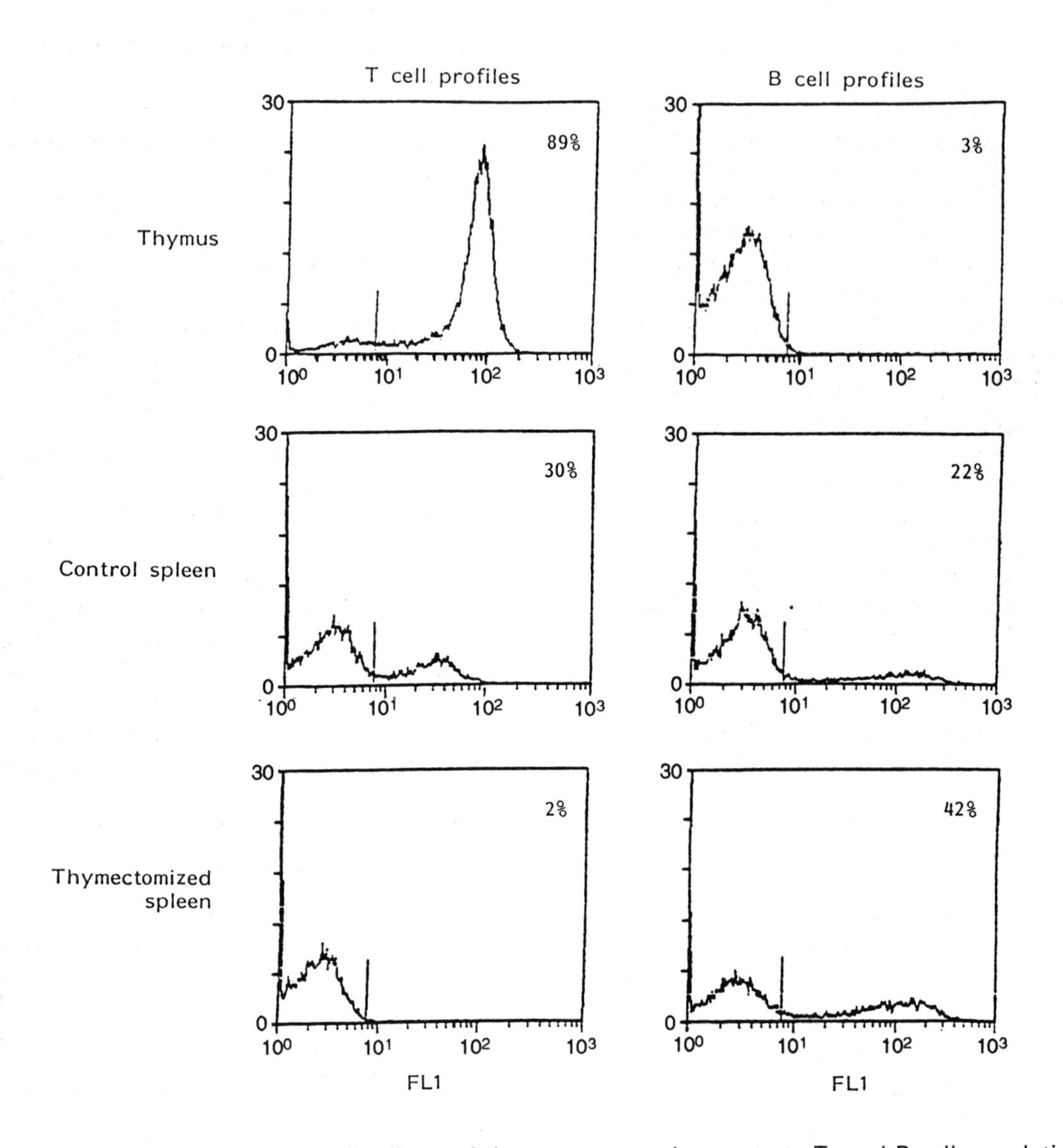

Figure 4.7. Use of monoclonal antibodies and thymectomy to demonstrate T- and B-cell populations in *Xenopus*. Fluorescence-activated cell sorter (FACS) profiles of cells from thymus and spleen of a control *Xenopus* (4 months old) and from spleen of a 7-day thymectomized sibling. Cells were stained first with either mouse monoclonal antibody XT-1, which identifies a 120 kDa protein on most thymocytes and on a proportion of peripheral T-cells, or with monoclonal antibody 8E4:57 (anti-*Xenopus* IgM), and were then stained with a FITC-labelled anti-mouse Ig secondary antibody. Early larval thymectomy deletes XT-1 positive T-cells from the spleen of young post-metamorphic animals, which now show a proportional increase in B-cell numbers. The flow cytometric profiles show fluorescence intensity (log scale) on *x* axis and relative cell number on *y* axis. The percentage represents the proportion of positive cells displaying more intense fluorescence than the marker set to exclude approximately 98% of those cells stained with secondary antibody alone.

normally to the T-cell mitogen PHA[35].

The possibility that thymectomy activates alternative pathways of lymphopoiesis was indicated in studies on diploid *Rana pipiens* froglets that, as embryos, received triploid ventral mesoderm (ventral blood island transplants). Control froglets did not develop with significant numbers of triploid splenic lymphocytes, whereas early-thymectomized froglets possessed approximately 60% splenocytes derived from the ventral blood island haemopoietic source[36]. (Since it is now known that the island contributes to both T- *and* B-cell populations [see Section 4.5.1], the significance of this early transplantation experiment is rather unclear.)

Work in my laboratory (unpublished) has recently shown that the residual splenocyte response to T-cell mitogens seen in thymectomized *Xenopus* by some authors[9] is uniformly absent in larvae and also 1 month post-metamorphosis. However, such responsiveness frequently emerges in 6-month-old thymectomized animals, although stimulation indices are <10% of those found with control cells. Interestingly this emergence of functional 'T-like' cells is coincident with the appearance, as demonstrated by flow cytometry, of a small population of XT-1^{+ve} and AM22^{+ve} splenocytes in thymectomized animals aged 6 months. We have also recently demonstrated XT-1^{+ve} cells immunohistologically in skin of both control *and* thymectomized *Xenopus*[37]. Work is in progress to determine if these various 'T-like' cells in thymectomized animals represent non-thymus-derived lymphocytes. The use of monoclonal antibodies that are being developed against $\alpha\beta$ and $\gamma\delta$-like T-cell receptor molecules of *Xenopus* (Section 4.3.3) should be useful in this respect. Thus the T-like cells frequently found in aged nude mammals are often $\gamma\delta$-expressing[34].

Sequential T-cell Development

Thymectomy of anuran larvae at sequential times during development suggests that different T-cell functions require the presence of the thymus for varying periods in order to become established in the periphery (Figure 4.8)[38]. Studies on the ontogeny of T-cell-dependent immunity in intact *Xenopus* reveal that alloimmune reactivity *in vivo* and *in vitro*, together with the ability of splenocytes to respond to T-cell mitogens, develops early in life, whereas good IgY primary antibody responses are only seen in the froglet (Figure 4.8).

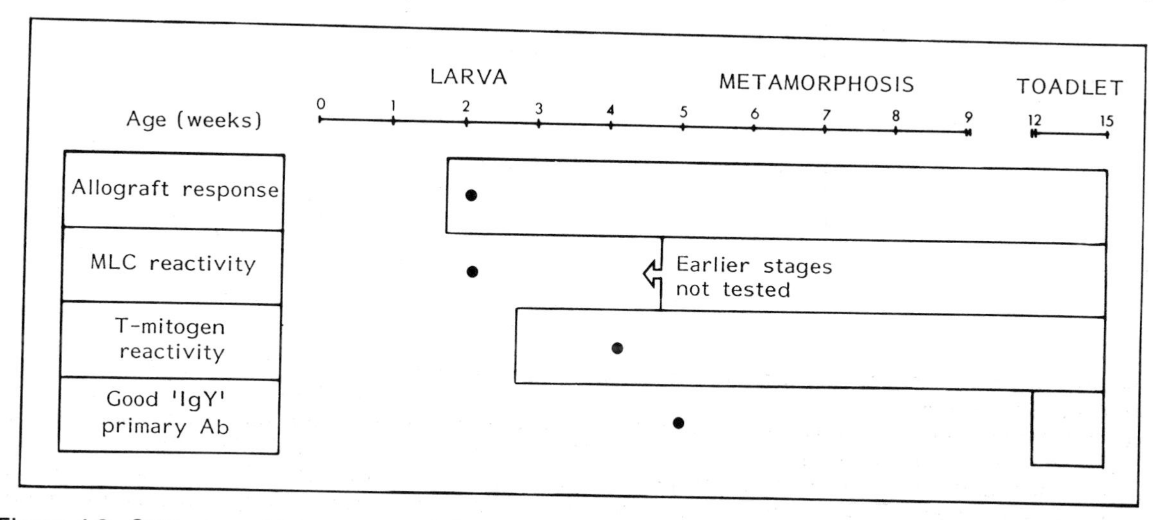

Figure 4.8. Ontogeny of immune reactivity and the effect of sequential thymectomy in *Xenopus*. Open boxes show that the allograft response, mixed lymphocyte culture (MLC) reactivity and T-mitogen reactivity appear early in ontogeny. These are followed by T-helper cells, e.g. those involved in the IgY primary antibody response, which emerge after metamorphosis. ● Indicates the age until which thymectomy still impairs function.

T-cell Subpopulations

Functional assays indicate the presence of specific helper, cytotoxic and suppressor T-cell activities in *Xenopus*. Whether these various functions reside in separate subpopulations is not yet known, but may soon be answered in view of the increasing array of anti-*Xenopus* T-cell antibodies[32]. *In vitro* collaboration between splenic or peripheral blood-derived T-cells (identified by their nylon wool non-adherence, lack of surface immunoglobulin (sIg) staining and their thymic dependence) and B-cells in secondary IgY antibody production has been shown to be MHC-restricted in *Xenopus*[13,39]. Carrier-dependent helper function is involved in generating anti-hapten responses in newts and this function may well be associated with T-like cells[7,29]. The generation of cytotoxic T-lymphocytes in *Xenopus* requires *in vivo* priming and restimulation of responders in mixed lymphocyte culture[40]. There is evidence indicating that cytotoxic T-cells can be generated against targets expressing class I or class II MHC, and also against minor histocompatibility antigens[41,42]. Thymic T-cells are able to suppress *in vitro* antibody production by *Xenopus*[43] and this suppressive ability has been shown to be MHC unrestricted[44]. T-suppressor activity is dependent upon the thymic medullary area, since it is insensitive to *N*-methyl-*N*-nitrosourea, a lymphotoxic agent which results in loss of thymic cortex in *Xenopus* and abrogation of helper T-cell function and allograft rejection capacity[29]. Suppressor function in *Xenopus* shows the cyclophosphamide and radiation sensitivity also found in mammals[44]. Thymus-related suppression of *in vivo* antibody production has been demonstrated in urodeles; thus in the axolotl, thymectomy increases antibody levels elicited by horse erythrocytes[45]. In *Xenopus*, thymic T-cells are poor helpers and graft-versus-host inducers, whereas splenic T-cells are much stronger in these respects[13].

4.2.2 The Spleen

This is a major peripheral lymphoid tissue in all amphibians, involved in trapping antigen and housing the proliferating lymphoid cells after stimulation by antigen. The spleen also provides for appropriate release of activated cells and their products[8], and is intimately involved in production of erythrocytes, thrombocytes and granulocytes in amphibians. Its role in B-cell development is discussed in Section 4.2.3.

The spleen appears at about 2 weeks of age in *Xenopus*. The mature spleen is clearly demarcated into T-dependent and T-independent lymphoid zones separated by a boundary layer. The white pulp follicles are rich in B-cells[33,46] as shown by the selective staining of this region with anti-Ig monoclonal antibodies (Figure 4.9). Splenic

Figure 4.9 (opposite). Immunohistological characterization of adult *Xenopus* spleen. (A) Immunofluorescence to show antigen trapping. A soluble protein antigen (human IgG) had been injected 3 weeks before preparation of frozen sections (10 μm) and their incubation with fluorescein-labelled anti-human IgG. The bright fluorescence indicates presence of antigen within white pulp (WP) follicles. The antigen is trapped in a dendritic pattern, and appears to be held on the surfaces of primitive follicular dendritic cells. Arrows point to boundary layer that surrounds the white pulp follicles. RP = red pulp. Scale bar = 100 μm. (B) Silver-stained preparation showing reticulin network in a white pulp (WP) follicle. CA = central arteriole, with capillary branches (BR) passing to the periphery of white pulp. Arrows point to boundary layer. RP = red pulp. Scale bar = 50 μm.
(C, D) Immunofluorescence cryostat sections (6-μm) stained with either anti-IgM monoclonal antibody (C) or anti-T-cell monoclonal antibody, XT-1 (D), followed by fluorescein-labelled secondary antibody, to show location of B- and T-cell zones respectively. (C) This section (of spleen from an early-thymectomized *Xenopus*) shows that the white pulp (WP), here cut longitudinally, is heavily stained for B-cells and that the marginal zone, just outside of the boundary layer (arrowed) is also rich in these IgM^{+ve} lymphocytes. A few intensely-staining plasma cells can be seen. CA = central arteriole; RP = red pulp, relatively unstained. Scale bar = 100 μm. (D) Section from a control *Xenopus* stained with XT-1, reveals that T-cells expressing this surface marker (arrowheads) are found in the marginal zone and also in the red pulp (RP). The white pulp (WP) follicles tend to be XT-1^{-ve}. Scale bar = 100 μm.

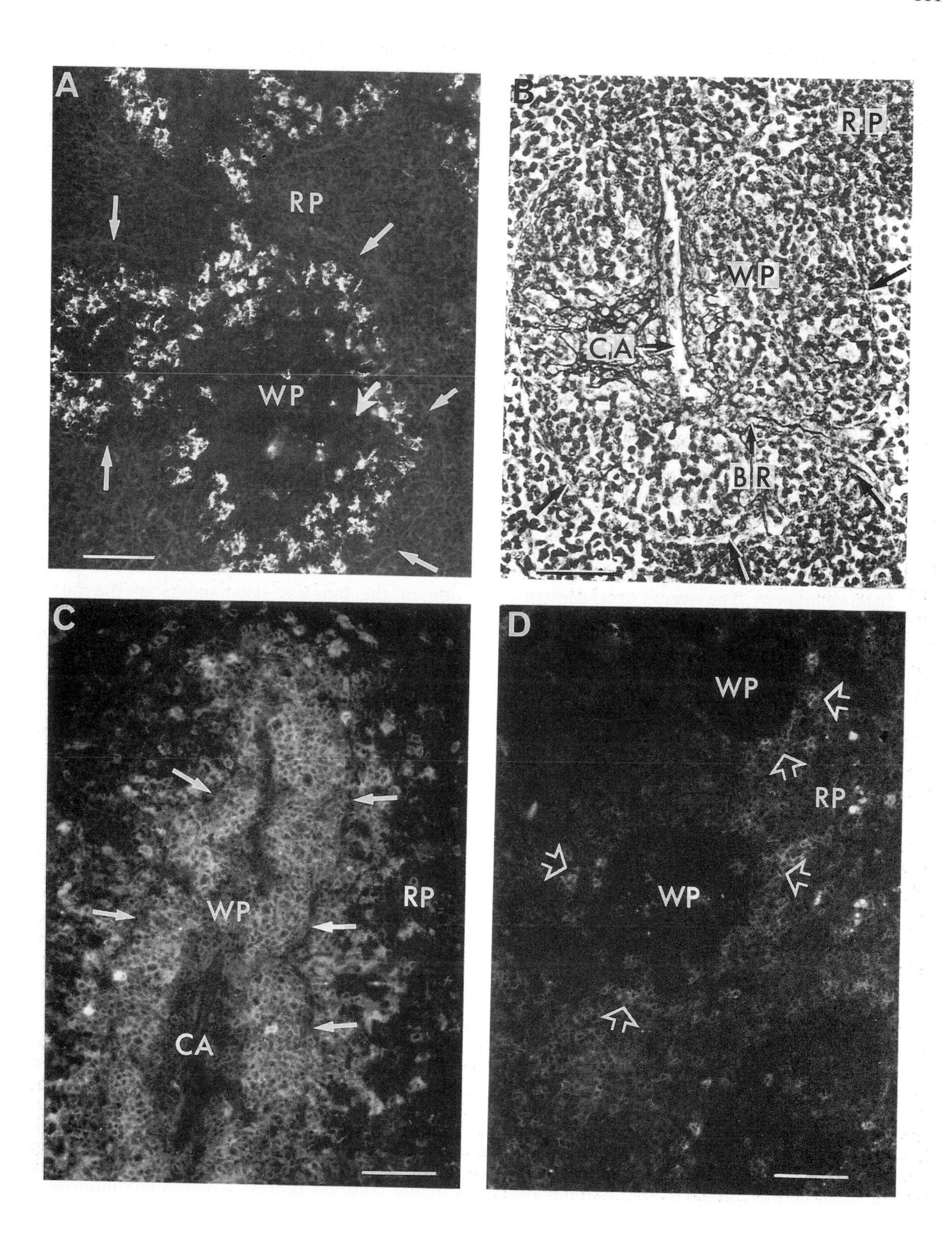
A
RP
WP
B
RP
WP
CA
BR
C
WP
RP
CA
D
WP
RP
WP

T-cells found especially in the perifollicular (marginal) zone, are surface Ig^{-ve} but a population will bind to the anti-T-cell antibodies XT-1 and AM22 (Figure 4.9). Adult splenic T-cells, but not larval T-cells, constitutively express surface class II MHC[17].

Blood enters the spleen through the white pulp central arteriole, from where capillaries leave and empty into the surrounding red pulp marginal zone (Figure 4.9), capillary walls contributing to the boundary layer[8]. Experimental studies with Indian ink and fluoresceinated antigens reveal that it is the red pulp that initially receives foreign material circulating in the blood. Circulating antigens are later trapped within the white pulp follicles (Figure 4.9), that is they are closely associated with potential antibody-producing cells[8]. Antigen is held on the surfaces of large dendritic cells, whose cytoplasmic processes extend pseudopods through the boundary layer and into the T-lymphocyte-rich marginal zone. These cells may be primitive follicular dendritic cell types[47]. The overall arrangement of the amphibian spleen is similar to that of the mammalian spleen.

4.2.3 Organs Involved in B-cell Development: Spleen, Liver, Kidney and Bone Marrow

The *Xenopus* spleen has been shown to be a major site of B-cell differentiation, in both larva and (more especially) adult. Three ontogenetic stages of B-cell differentiation—pre-B, which are cytoplasmic IgM^{+ve} and surface IgM^{-ve} ($C\mu^+$, $S\mu^-$), B ($C\mu^-$, $S\mu^+$), and plasma cells ($C\mu^+$, $S\mu^+$)—are found in this organ[48]. Most splenic B-cells produce IgM, rather than the other Ig isotypes[13]. In view of the importance of the spleen as a major lymphoid organ in both anurans and urodeles and as a site of B-cell differentiation, it is surprising that splenectomy has relatively little effect on cell-mediated and humoral immune reactivity in amphibians[49].

The first organ in which pre-B-cells can be found in *Xenopus* is the larval liver, sIg^{+ve} B-cells being identified several days later, at about 10–12 days of age in the subcapsular region. Two waves of B-cell development can be recognized in liver and spleen, one before and one after metamorphosis[48] (Figure 4.2). The liver is an important source of various haemopoietic cells in both anurans and urodeles: the perihepatic layer being a site of B-cell differentiation[8] in urodeles.

In *Xenopus* the bone marrow is rudimentary and is not a lymphopoietic tissue. In adults it is involved in the differentiation of neutrophilic granulocytes[48]. Absence of B-cell differentiation in *Xenopus* bone marrow contrasts with the active lymphopoiesis seen in ranid frogs. Indeed experimental studies suggest that *Rana* bone marrow, like mammalian bone marrow, is a source of antibody-producing cells[50]. Perhaps there has been a reduction of bone marrow space in *Xenopus* in adaptation to a totally aquatic existence. Alternatively in pipid frogs, the bone marrow might not have evolved as an appropriate site of B-cell differentiation. Bone marrow is absent in apodans and urodeles, with the exception of the lungless salamanders of the family Plethodontidae, where the marrow cavity contains large numbers of developing neutrophils, eosinophils, lymphocytes and plasma cells[8].

The kidney (rather than liver) appears to be the initial source of B-cells in *R. pipiens* larvae[51]. In *Xenopus* this organ accumulates B-cells between the renal tubules, and retains injected antigen[9].

4.2.4 Lymphomyeloid Nodes

In the ranid and bufid anurans a number of lymphoid structures are found in the neck and upper thoracic regions[8]. These lymphomyeloid nodes are mainly blood-filtering organs (contrast with mammalian lymph nodes), although trapping of material from surrounding lymph is also believed to occur. Such nodes are absent from 'lower' anurans and also from urodeles and apodans. The lymphomyeloid nodes are involved in long-term retention of antigen and are a major site of antibody-producing plasma cells. The larvae of many 'higher' anurans possess a paired lymphomyeloid

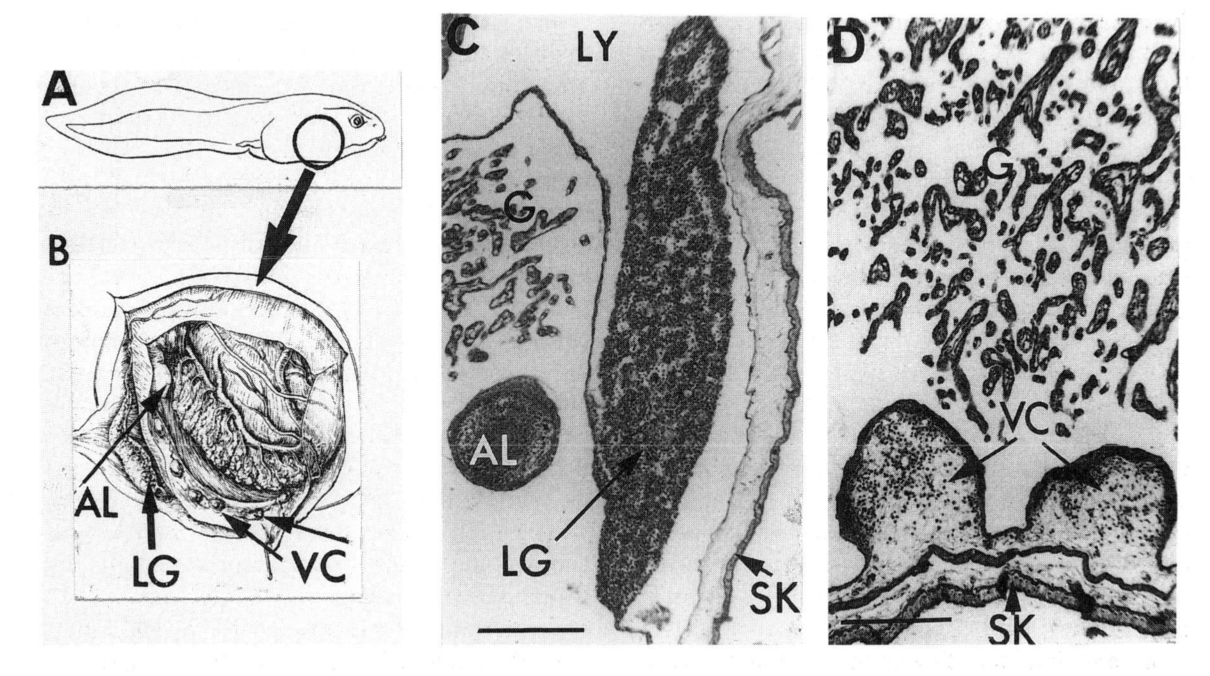

Figure 4.10. Lymph gland and ventral cavity bodies in larval *Rana*. (A, B) The branchial region of the tadpole outlined in A is dissected in B to display positions of the lymph gland (LG) and the (series of) ventral cavity bodies (VC). AL = anterior limb. Reproduced by courtesy of Dr E.L. Cooper.
(C, D) Haematoxylin and eosin-stained 8-μm sections showing histology of lymph gland (LG) and ventral cavity bodies (VC). The lymph gland (Figure 4.10C) consists of an extensive lymphoid parenchyma (stains darkly) and a network of intervening sinusoids (pale-staining regions). The organ is dorso-ventrally elongated, is attached ventrally to the epithelium lining the gill chamber (G) and projects into the anterior lymphatic channel (LY). AL = digit of anterior limb; SK = lateral skin. Two ventral cavity bodies (VC; Figure 4.10D) are shown lying below the gills (G). Lymphocytes are found within the epithelium and connective tissue region of the VC. SK = skin of ventro-lateral surface. Scale bars = 200 μm.

node called the lymph gland, in the branchial region (Figure 4.10)[5]. The lymph gland is active in phagocytosis, filtering both blood and lymph, and plays a key role in larval humoral immunity[8].

4.2.5 Gut-associated Lymphoid Tissue (GALT)

Nodular GALT is seen in adults of diverse anuran species but is apparently lacking in urodeles. This tissue, which is rich in IgM and IgX-secreting plasma cells[13], is conveniently located to form a first line of defence against antigens in the gut (see also Section 4.3.2). In the larval branchial region, associated with the pharyngeal epithelium, are found dorsal- and ventral-cavity bodies (Figure 4.10), which occupy a sentinel position in relation to the water current passing over the gills and opercular openings from the pharyngeal cavity. These lymphoid accumulations are rich in XT-1^{+ve} T-cells (our unpublished observations) and are thymus-dependent[8]; they disappear at metamorphosis.

It has been shown that MHC class II expression is associated with the epithelial surfaces of the gills, suggesting the possibility that class II MHC proteins may be used by diverse cell types to present conventional antigens to T-cells. Such epithelial-associated MHC class II may also be important in binding native bacterial toxins[17]. The widespread distribution of class II in tadpoles compared with adults suggests that this pattern might represent the way in which antigen was presented in a more primitive immune system[17].

In the adult anurans and urodeles 'tonsils' are found in the oropharyngeal cavity. These lymphoepithelial structures are phagocytic and, together with more posteriorly-situated accumulations of lymphocytes in the lamina propria, may be sites of lymphopoiesis[8].

4.2.6 Skin

The epidermis of the adult *Xenopus* skin is rich in MHC class II^{+ve} dendritic cells (Figure 4.11)[17]. These may be homologous to mammalian Langerhans cells which (in association with MHC class II) present antigens entering the skin to T-cells in the vicinity. The inner epidermal layer also contains XT-1^{+ve} cells (Figure 4.11), and such 'T-like' cells are still found following early thymectomy.

4.2.7 Accessory Cells

The ability of amphibian monocytes and macrophages to engulf antigens and to elaborate lymphokine-like factors has been studied[10]. However, the roles that accessory cells—such as dendritic cells of the skin, thymus and spleen and the macrophages found throughout the lymphomyeloid system—play in the induction of amphibian immune responses await elucidation. Although peritoneal exudate cells (macrophages) of *Xenopus* have been shown to secrete an interleukin-1 (IL-1)-like cytokine[52] and can take up and degrade antigens to small peptides[53], there is no conclusive evidence that antigen processing is required before presentation to T-cells at this phylogenetic level[26]. It has been suggested that

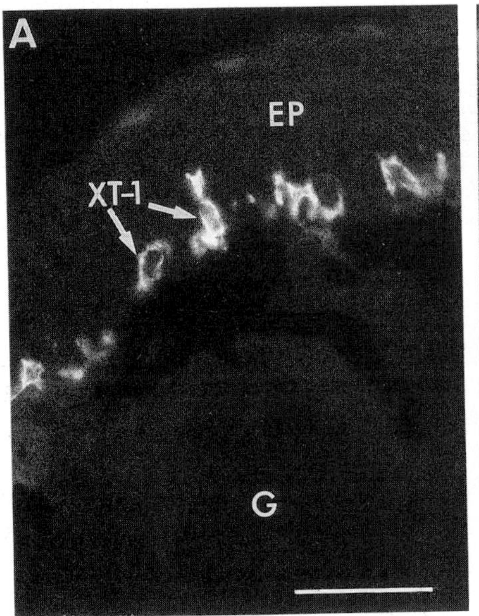

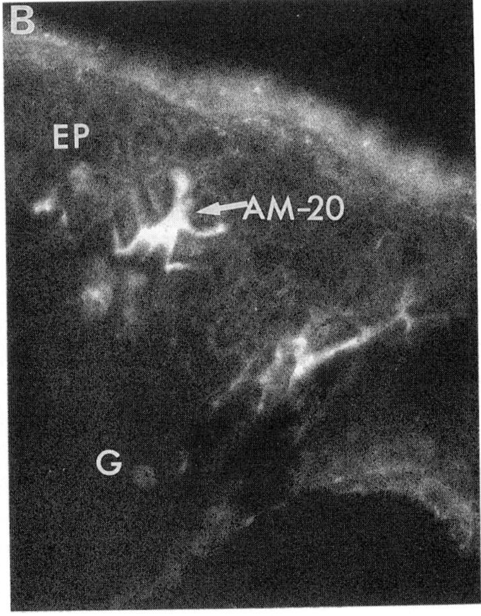

Figure 4.11. Immunostaining of adult *Xenopus* skin. Immunofluorescence. Cryostat sections (6 μm) of skin from a 6-month-old animal have been stained with either XT-1 monoclonal antibody (stains a subset of T-cells) (A), or AM20 (an anti-class II MHC monoclonal antibody) (B), followed by secondary, fluorescein-labelled anti-mouse Ig. The basal epidermal (EP) region in A is replete with XT-1^{+ve} cells. Large dendritic, class II MHC-positive cells are also found in the epidermal layer (B); these may be homologous to mammalian Langerhans cells. G = skin glands. (See Figure 4.18C for phase contrast view through entire skin layers.) Scale bar = 50 μm.

the failure of newts to respond to soluble protein antigens (that are immunogenic to anurans) may relate to an inability of newt macrophages, but not anuran macrophages, to process soluble immunogens[53].

Membrane Fc receptors for antigen-complexed or heat-aggregated *Xenopus* immunoglobulins (IgY and IgM), but not human Ig, have been demonstrated on *Xenopus* splenocytes (including most B-lymphocytes) and there is also evidence for similar Ig receptors and complement (C3) receptors on peritoneal macrophages[54,55]. Such receptors may well be important in phagocytosis, antigen presentation, regulation of antibody production and antibody-dependent cell cytotoxicity. A useful review of the evolution of macrophage-lymphocyte immune regulation which highlights differences between anurans and urodeles is available[56].

4.2.8 Non-specific Cytotoxic Cells

Non-antigen-specific natural killer (NK) cell activity against xenogeneic tumour targets and viruses has been described in several anuran species and the role of these cells in tumour immunity discussed[57]. Frog and toad spleen, peripheral blood and bone marrow are rich sources of these NK cells. Antibody-dependent cell-mediated cytotoxicity (ADCC) has also been reported in urodeles and anurans[58]. An interesting cytotoxic reaction has been described in viviparous salamanders: splenocytes from pregnant or post-partum females exhibit cytotoxicity towards paternal antigens expressed on embryonic epidermal cells. Protection of embryos from the maternal immune system may not have developed adequately in these first viviparous tetrapods[59].

4.2.9 Granulocytes

The morphology and immunological roles of neutrophils, eosinophils and basophils/mast cells have been reviewed in depth[10]. The role of eosinophils in anti-parasitic immunity and basophils/mast cells in the immediate-type hypersensitivity reaction described in *R. pipiens* await clarification at the amphibian level of evolution[10]. There is also scant information on the role that leukotrienes and other lipid mediators (collectively called eicosanoids) play in inflammatory responses in amphibians.

4.3 MOLECULAR BASIS OF THE AMPHIBIAN IMMUNE SYSTEM

4.3.1 The Major Histocompatibility Complex (MHC)

In mammals, the MHC plays a vital role in the immune system. This genetic complex codes for polymorphic histocompatibility antigens that are associated with acute graft rejection and which are now known to play a crucial role in presentation of antigenic peptides to T-cell receptor molecules. The MHC is intimately involved with 'education' of T-cells within the thymus and also codes for certain complement components. The existence of an MHC in *Rana* and in *Xenopus* was first suggested in the 1970s and has now been fully documented. Most work has been carried out on *Xenopus*[12,13,60]. The *Xenopus* MHC has been defined by functional criteria (cosegregation of acute graft rejection, mixed lymphocyte reactivity and certain surface antigens) and, more recently, by molecular and genetic characterization[61]. In *Xenopus*, phenomena such as T-B-cell collaboration, the generation of antigen-specific cytotoxic responses (e.g. to allogeneic cells) and thymic education of T-lineage cells are all under control of its MHC, called the XLA (by analogy with human HLA). The MHC remains functionally diploid in *X. laevis*, which is a tetraploid species, displaying many duplicated gene loci[140]. It has been argued[140] that too many copies of the MHC gene complex would be detrimental to T-cell functioning.

The genetic organization of XLA is shown in Figure 4.12, where it is compared with B of the chicken and H-2 of the mouse. The basic structure must have appeared by the time frogs and mammals diverged from a common ancestor.

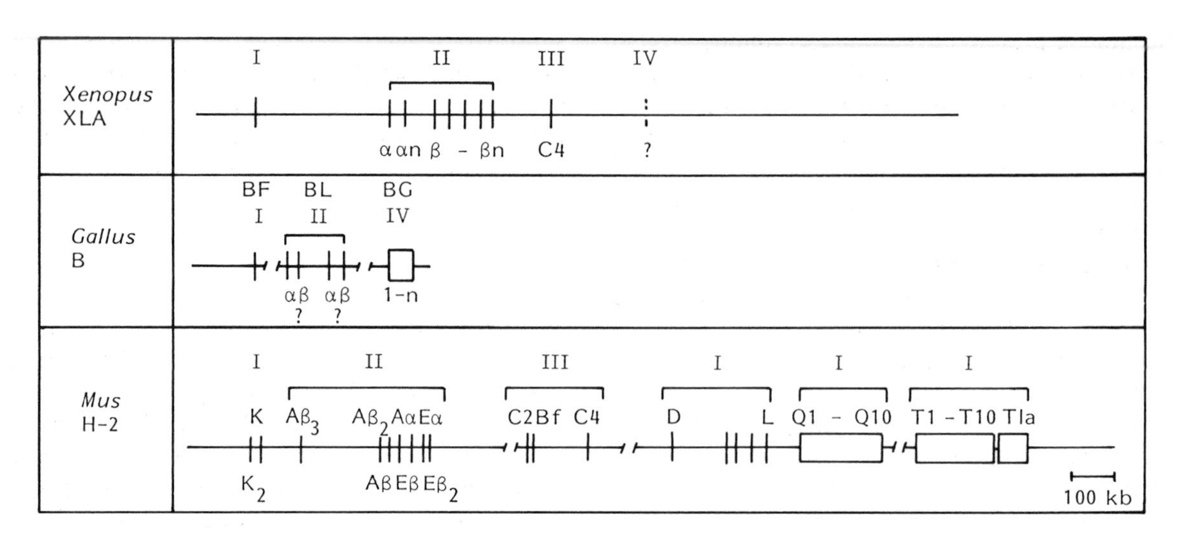

Figure 4.12. MHC of different species. Speculative organization of MHC gene loci in the clawed toad, *Xenopus*, and the chicken, *Gallus*. These loci are compared with known architecture of the mouse H-2 complex. Distances for *Xenopus* and *Gallus* are arbitrary. Reproduced by courtesy of Dr L. DuPasquier.

Molecular cloning of *Xenopus* class I MHC[141] predicts amino acid sequences similar to MHC class I molecules of higher vertebrates. (It should be pointed out that the original *Xenopus* class I cDNA clone isolated is now known to relate to a large family of non-MHC-linked class I molecules [M.F. Flajnik, personal communication].) However, the *Xenopus* α_3 domain is more similar to the immunoglobulin-like domains of mammalian class IIβ chains than to those of mammalian class I molecules. *Xenopus* class Iα chains are 40–44kDa molecules which are non-covalently bound to non-MHC-coded β_2-microglobulin[62]. The predicted amino acid sequences of class I peptide-binding domains (α_1 and α_2) indicate structural similarities to certain peptide-binding heat shock proteins[142]. Class I molecules are expressed on the surfaces of all adult cells, highest expression being on haemopoietic cells, with very low expression in the brain. Class I molecules expressed by erythrocytes and leucocytes appear to be distinct (the α chain of the former is not associated with β_2-microglobulin, thus class IV rather than I?[60]). Several reagents specific for class I antigens of *Xenopus* are now available[143].

Xenopus class II molecules are composed of MHC-encoded α and β chains, both of which are 30–35kDa transmembrane glycoproteins[63]. cDNA clones for *Xenopus laevis* MHC class IIβ chain genes have recently been isolated[140]. Amino acid sequence identity with representative mammalian MHC class IIβ chains is predicted to be nearly 50%. As in mammals, genes encoding *Xenopus* β_1 domains are subject to positive Darwinian selection. An invariant chain is found transiently associated with class II during biosynthesis[61]. Class II proteins are expressed constitutively on only a limited range of adult cells, including thymocytes, B-cells *and* T-cells (contrast some but not all mammals—see Chapter 6) and various antigen-presenting cells that include putative 'Langerhans-like' cells of skin epidermis (Section 4.2 and reference 17). *Xenopus* MHC antigens are polymorphic: about 20 class I and 30 class II alleles are suspected. Complement component C4 is a class III region gene product in *Xenopus*, as it is in mammals.

One particularly interesting feature of MHC expression in *Xenopus* is that, in contrast to the early larval appearance of class II molecules (on B-cells and several tadpole epithelia—see

Section 4.2), class I proteins have not been found on the surface of any cell surface prior to metamorphosis. Thus class I expression is not essential for early development or for functioning of the larval immune system. Lack of surface class I expression on larval thymus stromal cells would appear to preclude the selection of class I-restricted cytotoxic T-cells prior to metamorphosis. This may well be advantageous to an animal which must become immunocompetent when still possessing few lymphocytes. The larval T-cell repertoire can therefore be dedicated to interact with class II-expressing B-cells (and other antigen-presenting cells) and effecting class II MHC-restricted killing[42]. It should be noted, however, that class I MHC α chains are found intracytoplasmically in thymic epithelial cells of young tadpoles[60,64] and that the putative CD8 (cytotoxic T-cell-'specific'?) molecule is expressed on larval T-cells[64]. The possibility that the emergence of surface class I MHC at metamorphosis plays a crucial role during this important period is discussed in Section 4.5. Experiments involving manipulation of metamorphosis with agents that inhibit or promote thyroid function have shown that class I and II gene loci are independently regulated, as is the situation in mammals[60,65].

Axolotls (*Ambystoma mexicanum*), which display relatively poor T-cell reactivity to alloantigens, possess α and β MHC class II molecules, which are not very polymorphic[61]. This urodele expresses MHC-encoded erythrocyte antigens, which show similarities to class I α chains (44 kDa) and also to polymorphic class IV molecules found on nucleated erythrocytes of chickens (the B-G MHC gene product), which may also exist in *Xenopus*. Axolotls normally fail to undergo the metamorphosis typical of most urodeles, namely from an aquatic juvenile with broad-finned tail and external gills to a terrestrial adult with thin tail and no gills. It does, however, undergo 'cryptic metamorphosis' involving several biochemical changes[1], including changes in class II tissue distribution. Thus post-cryptic metamorphosed axolotls display, for the first time, class II^{+ve} T-cells (in addition to B-cells) and regions of the thymic stroma now stain brightly for class II[61]. The latter immunohistologically-recognized regions may represent the urodele thymic medulla, which cannot be distinguished from cortex by conventional histology.

4.3.2 Immunoglobulins

As in other vertebrates, amphibian immunoglobulins are comprised of multi-domain, heavy and light polypeptide chains. All amphibians produce IgM. *Xenopus* IgM (heavy chain 61 kDa) is hexameric and is associated with a J chain[13]. Non-μ heavy chain isotypes (with four constant heavy chain domains—contrast three for IgG) have also been identified in *Xenopus* (IgY and IgX), *Rana* and the axolotl (IgY)[12]. Although the precise roles of these different isotypes are not known, *Xenopus* IgY (66 kDa heavy chain) is thymus-dependent, whereas IgM and IgX (64 kDa H chain) are not. *Xenopus* IgX is polymeric and possibly acts as the equivalent of mammalian IgA, since this isotype is most frequently found in the gut[12]. The IgY of axolotls may also be a secretory immunoglobulin, as in the gut this immunoglobulin isotype becomes associated with 'secretory component-like' molecules. Three types of light chain, two of which are antigenically distinct, have been found in *Xenopus* (25–29 kDa)[13].

Antibody diversity (V region differences) in *Xenopus* is quite low, there being only some 5×10^5 different antibody molecules in adults. This restricted diversity is not due to a special gene organization in *Xenopus*. Thus the heavy chain gene locus resembles that seen in mammals, with the V_H region being coded for by numerous (80–100)V, about 15 Da and at least seven J gene segments[66,152]. Both framework and complementarity determining regions are found[12]. Multiple rearrangements of *Xenopus* Ig genes occur during B-cell development and allelic exclusion exists[67], resulting in immunospecific B-lymphocytes. Possible reasons for restricted antibody diversity in *Xenopus* are considered in Section 4.5.3.

4.3.3 T-cell Receptors for Antigen

Monoclonal antibodies against *Xenopus* $\alpha\beta$ and $\gamma\delta$ T-cell receptor candidates have been produced[145]. Identification of these two antigen receptor types, together with additional molecules specific for T-cell subsets[32], will undoubtedly promote a far greater understanding of T-cell differentiation events at this level of evolution. Characterization of TCR genes in anuran and urodele amphibians is now eagerly awaited. Recently, cDNA clones from Mexican axolotl thymocytes and splenocytes have been sequenced; these have considerable homology with avian and mammalian TCR β chains[146].

4.3.4 Cytokines

Cytokines are peptides which play a crucial role in regulating immune responses in a non-antigen-specific, paracrine or autocrine fashion. In amphibians, molecules with functional similarities to mammalian interferons (IFNs), interleukin-1 (IL-1) and IL-2 have been described[68]. The biochemistry of some of these molecules is under study, but molecular and genetic characterization of amphibian cytokines and their cellular receptors is still awaited.

Cytokines with antiviral activity have not been demonstrated in amphibians, but culture supernatants of antigen-treated splenocytes from antigen-stimulated frogs, toads and salamanders display factors that inhibit the migration of macrophages, as do supernatants from Con A-stimulated splenocytes[69].

Factors derived from 24-hour LPS-stimulated peritoneal exudate macrophages (of *Xenopus*), which can enhance T-cell proliferative responses of thymocytes from young adults to submitogenic doses of T-cell mitogen, appear to represent IL-1-like molecules[52]. Cross-reactivity between mammalian IL-1 and amphibian IL-1-like factors (human to *Xenopus* or vice versa) was not revealed in *in vitro* co-stimulation studies on *Xenopus* lymphocytes, but has been suggested in *in vivo* studies involving immune regulation in this anuran and also in newts[68,70].

Xenopus T-cell growth factor (TCGF), in contrast to the above IL-1-like cytokine, promotes proliferation and growth of T-cell lymphoblasts. This 'IL-2-like' material has been identified from culture supernatants of T-cell mitogen-stimulated or MLC-activated, control (but not thymectomized) *Xenopus* splenocytes (Figure 4.13)[71,72]. Purification of *Xenopus* TCGF indicates a protein of molecular mass 16 kDa[73,153]. As yet there is no evidence for TCGF production in

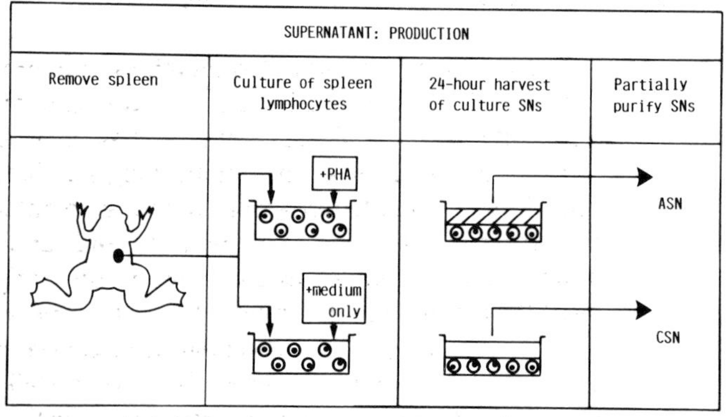

SUPERNATANT: ASSAYS		
Induces ^{3}H-thymidine incorporation in:		Supports growth of T cell lines
Resting splenocytes	T lymphoblasts	
+	+ + +	+ + +
–	–	–

Figure 4.13. T cell growth factors in *Xenopus*. Active culture supernatant (ASN) is harvested from PHA-stimulated splenocytes and compared with supernatants from control cultures (CSN). The supernatants are partially purified by ammonium sulphate precipitation, dialysis and removal of PHA by incubation with chicken erythrocytes. The ASNs induce considerable proliferation of T-lymphoblasts (but are less stimulatory for resting splenocytes) and support the growth of alloreactive T-cell lines; CSNs have no comparable effects. A similar but reduced level of activity can be generated in MLC supernatants and may be attributable to molecules functionally homologous to mammalian T-cell growth factor (i.e. IL-2).

urodeles[68]. It is pertinent to point out that crude supernatants from (lectin-free) mitogen or alloantigen-activated splenocytes undoubtedly contain a mixture of cytokines. It is therefore not surprising that these supernatants can also induce proliferation of surface Ig^{+ve} (B) cells (enriched by panning with anti-Ig monoclonal antibodies) from control *Xenopus*[68] and surface Ig^{-ve} splenocytes (enriched by cell sorting) from early-thymectomized animals[72].

Larval *Xenopus* thymocytes (contrast larval splenocytes) are unresponsive to T-cell mitogens; moreover, supernatant from such cultures does not contain TCGF activity. Larval thymocytes can be co-stimulated by addition of T-cell mitogen together with either TCGF or IL-1-rich supernatant[74]. This indicates that larval thymic cell populations are unable to produce, but can respond to, IL-1 and IL-2-like material.

Xenopus TCGF and mammalian IL-2 are not functionally cross-reactive *in vitro*[71]. In contrast, there is evidence that human recombinant IL-2 (rIL-2) can modulate a variety of *in vivo* immune reactivities in *Xenopus* (e.g. substitution of carrier priming[75]). *Xenopus* splenocytes can bind an anti-human IL-2 receptor antibody (against the P55 Tac peptide), an interaction blocked by preincubation of cells with the rIL-2[76]. However, at present the physiological significance of these findings remains controversial[77,78]. Molecular characterization of the receptor for homologous IL-2 in *Xenopus* is now in progress[68] and will be aided by the development of novel monoclonal antibodies that may bind to this receptor (M. Flajnik, personal communication).

4.3.5 Complement

Both classical and alternative pathways of complement activation occur in amphibians. Biochemical properties of mammalian complement (thermolability, requirements for Ca^{++} and Mg^{++}) are shared by amphibian complement, although the temperature range over which amphibian complement remains active is greater (activity remains at 4°C) and heat-inactivation can be achieved at a lower temperature, for example serum from *Xenopus* and *Triturus* (newt) is completely decomplemented by treatment at 45°C for 40minutes. Guinea pig complement may be used successfully in haemolytic antibody assays *in vitro* for adult amphibians, whereas in larval *Xenopus*, homologous complement must be used[79]. A *Xenopus* serum protein involved in haemolysis shares functional, structural and physicochemical properties with mammalian C3[80]. Molecular and genetic characterization of C3 from *Xenopus* and axolotls is in progress[81,82]. Characterization of other complement components in anurans includes C1q, C4, C5, and the membrane attack complex[83]. Demonstration of factor B-like activity (involved in C3 convertase production in the alternative complement system) in *Xenopus* serum has also been described[84].

4.4 IN VIVO IMMUNE RESPONSES

The cellular and molecular components of the amphibian immune system described above interplay to provide cellular and antibody-mediated protection at the organismal level. An outline of the spectrum of protective mechanisms afforded by amphibians is given here.

4.4.1 Transplantation Immunity

There has been a wealth of research on amphibian transplantation reactions. Tissue transplantation between amphibian embryos has proved rewarding for experimental embryologists since the 1920s. A detailed description of transplantation reactions that can occur in adult and larval anurans and urodeles was available by the early 1970s[85]. By that time it was clear that (at temperatures of 20–25°C) allo- and xenograft rejection in urodeles and apodans is generally chronic (>30 days to indefinite survival), whereas rejection times of allografts in anurans were either acute (<14 days) in ranid frogs, or subacute (*c.* 20 days) in more 'primitive' forms such as *Xenopus* and *Alytes*, the midwife toad. Skin grafts were and continue to be the tissue most frequently used in amphibian transplantation work, but other organs, including heart, gonads, pituitaries, eyes and lymphoid organs have all been grafted. The chronicity of graft rejection in urodeles led to the suggestion that these animals, in contrast to anurans, lacked a major histocompatibility complex[86]. We now know from molecular studies described in Section 4.3.1 that the chronicity of urodele graft rejection is connected with lack of MHC (class II) polymorphism between individuals and limited expression (on erythrocytes) of class I α chains, rather than to absence of an MHC at this level of evolution. Minor histocompatibility antigenic differences are undoubtedly involved in the chronic graft rejection seen in urodeles, and frogs can also mount chronic reactions to multiple minor H antigens[87].

The earlier work on amphibian transplantation immunity revealed either accelerated second-set responses or even delayed second-set responses (e.g. in some urodeles, where enhancement is suspected), described the histology of allograft rejection (vascular disturbances, lymphocyte infiltration and subsequent tissue destruction), demonstrated thymus-dependency of allograft rejection, and probed the ontogeny of alloimmunity. Graft-versus-host reactivity, often inducing recipient death, of implanted allogeneic spleens has been described in *Xenopus*[88,89], but appears not to occur in urodeles[86].

The relative contribution made by T- and B-cells to skin allo- and xenograft destruction in amphibians is currently being explored. Thymus-dependence of rapid graft rejection appears less evident when xenograft rejection is studied, especially when donor and host are distantly related, e.g. *X. tropicalis* grafts on *X. laevis* recipients[37,90]. Skin xenograft (*X. tropicalis*) destruction in control *X. laevis* has recently been studied immunohistologically[37] and is associated with B-cell accumulation under the foreign skin, rather than to a heavy T-cell infiltration that occurs within allografts. Since T-cells are selected to respond preferentially with antigens presented by self MHC proteins (during their development in the thymus—see Section 4.5.2), it seems quite likely that graft antigens presented by donor cells expressing *xenogeneic* MHC appear 'too different' to be directly recognized by the host's cytotoxic T-cells. Perhaps donor xenoantigens can only be successfully presented (to T-cells) by host antigen-presenting cells, i.e. in the context of the host's MHC proteins, which would tend to lead to antibody responses, rather than to T-cell cytotoxicity or delayed-type hypersensitivity reactions.

4.4.2 Delayed-type Hypersensitivity

Delayed-type hypersensitivity responses have been described in the axolotl (*Ambystoma*) and in *Xenopus*, for example when the latter is sensitized to *Mycobacterium tuberculosis* and subsequently challenged with a purified protein derivative of tuberculin[91]. These responses are similar to mammals and involve the elaboration of cytokines that can inhibit macrophage migration.

4.4.3 Cancer

Malignant neoplasia (cancer) occurs infrequently in natural amphibian populations, whereas lesions or granulomas probably caused by bacterial or fungal infections (i.e. not tumours) are frequently described[92]. Rare examples of amphibian tumours include the virally-induced renal adenocarcinomas found in leopard frogs, various kinds of tumours 'caused' by organic chemical pollution found in neotenic, but not adult, salamanders and a lymphoblastic lymphoma described in *Xenopus*[93]. Experimental chemical carcinogenesis resulting in proven malignant neoplasia is also difficult[92], but not impossible[57,94], to achieve in amphibians. The immunological basis for the efficient control of aberrant cells in amphibians awaits clarification, although it is possible that the 'immunological' events (e.g. destruction of larval tissues) occurring at metamorphosis might play an important role[92]. There is evidence that T-cells may not play a major role in prevention of tumour development in anurans, since thymectomized leopard frogs, injected as embryos with tumour herpes virus, displayed no greater incidence of renal adenocarcinoma than did controls[95].

A spontaneous lymphoid thymus tumour has recently been discovered in *Xenopus*[147], from which stable cell lines have been derived. These cell lines will be extremely useful for identifying T-cell-specific molecules, especially the T-cell receptor for antigen.

4.4.4 Cellular Responses to Viruses

Relatively few viral infections of amphibians have been documented. One example is the herpes-like virus, frog virus-3, which inserts virus-specific proteins in the infected cell membrane. The cellular immune response to this virus appears to involve lymphokine-activated, NK-like cells rather than cytotoxic T-lymphocytes[96].

4.4.5 Antibody-mediated Immunity

Amphibians are able to generate specific antibodies to a range of antigens. Different antigens elicit different antibody isotypes—for example antibacterial agglutinins involve IgM exclusively, whereas viruses and serum proteins initially elicit IgM production, which is later accompanied (but not replaced) by a switch to IgY production. A moderate affinity maturation of IgM is seen during an immune response. T-dependent and T-independent antigens have been categorized in *Xenopus*[29,97]. Urodeles respond relatively poorly to T-dependent antigens[45].

The heterogeneity of the antibody response in amphibians is low compared with mammals, the number of different anti-DNP antibodies not exceeding 40 in *Xenopus*, compared to as many as 500 in mammals[98]. However, the apparently small antibody repertoire appears not to restrict the ability of this anuran to respond to diverse antigens. It has recently been suggested that the poor affinity maturation in *Xenopus* is not due to lack of somatic mutations of Ig genes, but reflects the lack of germinal centres necessary to select these mutants[152].

Interestingly, antibody release in amphibians can occur at low temperatures, which may have a significance in relation to disease resistance during cold spells.

4.4.6 Immunopathology

An excellent review on this topic is available[99]. A variety of infectious diseases have been described in amphibians, these often being promoted by damage to the skin, allowing a foothold for pathogenic bacteria such as pseudomonads (cause of 'red-leg') and, more rarely, mycobacteria. Fungal, protozoan and helminth parasitic infections can also cause problems. For example, nematodes of the genus *Capillaria*, which can live in the skin of *Xenopus*, cause 'flaky skin' disease, commonly seen in this anuran. The disease is especially likely to occur in early-thymectomized *Xenopus*[100], suggesting that *Xenopus* can mount a protective T-dependent response to this parasite. On the other hand, the relative good health of early-thymectomized *Xenopus* points to the effectiveness of thymus-independent mechanisms in dealing with naturally occurring pathogenic microorganisms. One is reminded here of the fact that the skin of *Xenopus* contains a group of active peptides, called magainins, that are powerful antibiotics sharing a common general structure with the defensins found in invertebrates (e.g. the cecropins of moths) and in mammalian epithelial cells[148].

4.5 AMPHIBIAN MODEL SYSTEMS FOR EXPLORING ONTOGENETIC ISSUES

The free-living embryo and larva, exempt from direct maternal influence and amenable to bold surgical manipulations[101], together with the phenomenon of metamorphosis, have made amphibian model systems of considerable value in exploring ontogenetic aspects of the immune system.

4.5.1 Embryological Origins of Lymphocytes

Despite initially controversial experiments on anurans, suggesting an intrinsic origin of thymic lymphocytes from elements of the thymic rudiment, it became clear by the end of the 1970s that thymocytes develop from extrinsic precursor cells which colonize the epithelial thymic buds. These conclusions, which concur with avian and mammalian experimental findings, came from transplantation of gill buds (containing the thymic rudiments), or tissues containing presumptive stem cell precursors, between ploidy-marked anuran embryos and from the establishment of anterior-posterior, embryonic chimaeras. Two of these experimental approaches are illustrated in Figure 4.14. Confirmation of the extrinsic origin of thymic lymphocytes in amphibians has come

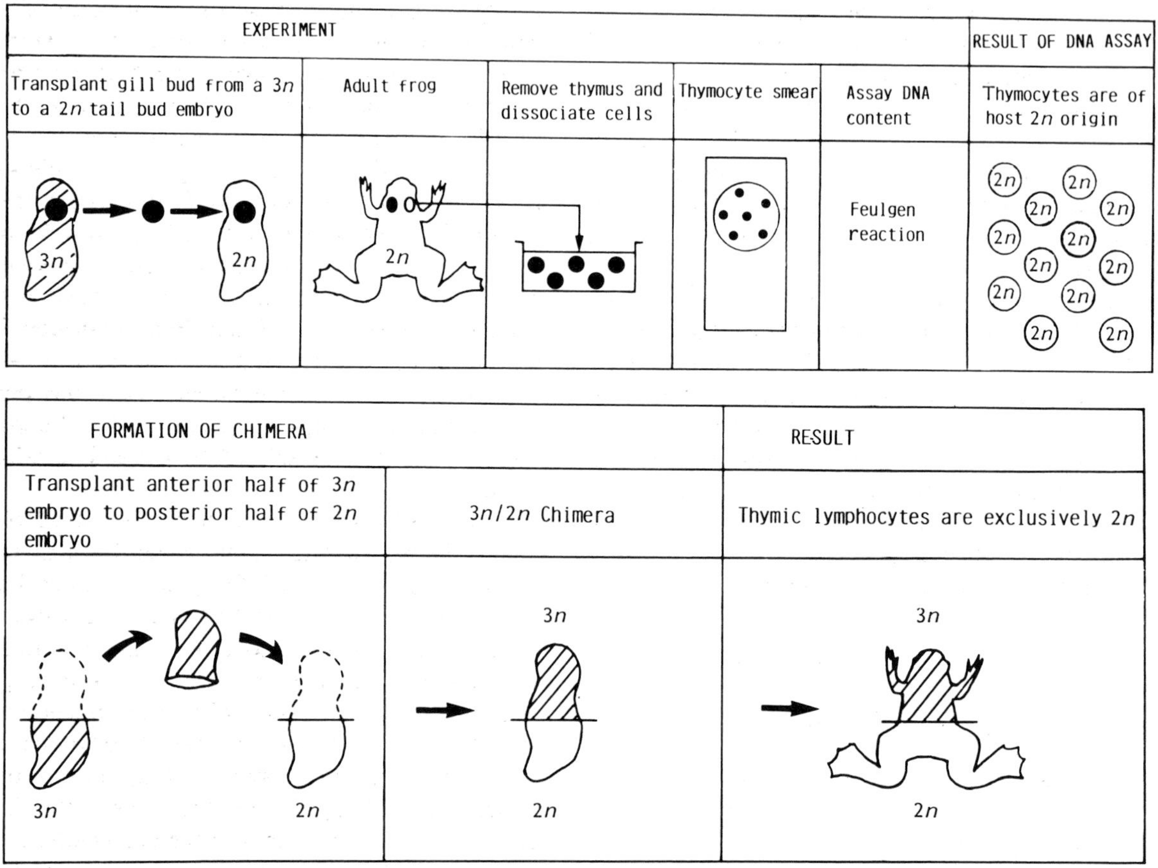

EXPERIMENT					RESULT OF DNA ASSAY
Transplant gill bud from a $3n$ to a $2n$ tail bud embryo	Adult frog	Remove thymus and dissociate cells	Thymocyte smear	Assay DNA content	Thymocytes are of host $2n$ origin
$3n$ $2n$	$2n$			Feulgen reaction	$2n$ $2n$ $2n$ $2n$ $2n$ $2n$ $2n$ $2n$ $2n$ $2n$ $2n$ $2n$ $2n$

FORMATION OF CHIMERA		RESULT
Transplant anterior half of $3n$ embryo to posterior half of $2n$ embryo	$3n/2n$ Chimera	Thymic lymphocytes are exclusively $2n$
$3n$ $2n$	$3n$ $2n$	$3n$ $2n$

Figure 4.14. Experimental models demonstrating extrinsic (stem cell) origin of thymic lymphocytes in amphibians. The experiment shown on the first line involves transplantation of (triploid) gill bud (containing the thymus precursor tissue) to a ploidy-distinct ($2n$) tail-bud embryo. When the embryo host developed into a normal adult frog, the transplanted thymus was removed, and the thymus cells were dissociated. A cell smear was prepared and the cell's DNA content was determined by the Feulgen reaction (more recently this is achieved by flow cytometry of propidium iodide-stained nuclei). Thymocytes were shown to be host-derived (i.e. here of $2n$ origin).

The second experiment (shown below) involves creation of a $3n$:$2n$ chimeric animal by exchanging the anterior and posterior regions of two embryos 24 h after fertilization (see also Figure 4.17). Although the thymus epithelial bud differentiates from the anterior, $3n$, part of the animal, thymic lymphocytes can be shown to arise extrinsically from $2n$ cells that immigrate to the thymus from posteriorly-derived sources (see Figure 4.15).

from work on *X. laevis*, *R. pipiens* and the newt *Pleurodeles waltlii*[16,102].

The embryonic location of stem cells that migrate to the thymus (and of those that become B-lymphocytes, erythrocytes, etc.) has been probed in considerable depth in *Xenopus*. Flow cytometry of propidium iodide-stained nuclei has recently[103] taken over from microdensitometry of Feulgen-stained cells and microscopic analysis of cells in metaphase as the more sensitive and reproducible assay for identifying the ploidy of the differentiated cell populations. The *X. borealis* fluorescent cell marker has also been employed by some workers[104]. It has emerged that two areas within the embryonic lateral plate mesoderm, one located dorsally and the other ventrally, give rise to haemopoietic cells in *Xenopus* (Figure 4.15).

The relative contribution of dorsal lateral plate mesoderm and ventral mesoderm (the ventral blood island comparable to the avian and mammalian yolk sac) to lymphopoiesis has begun to emerge. It appears that the ventral region contributes predominantly to erythropoiesis, and to early thymocyte populations, thymic accessory cells, peritoneal macrophages and B-lymphocyte precursors[103]. The dorsal lateral plate also contributes to thymocyte precursors in later larval life

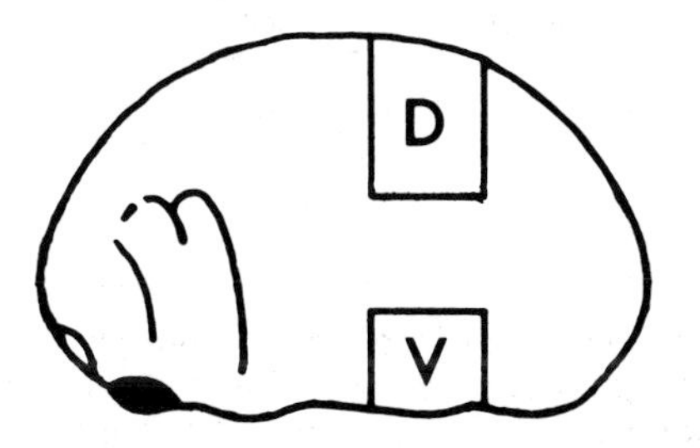

Figure 4.15. Location of embryonic regions that give rise to haemopoietic cells in the amphibian embryo. Two areas within the lateral plate mesoderm of the 24-h *Xenopus* embryo, namely the dorsolateral plate and the ventral blood islands, have been shown to be the major sources of haemopoietic cells.

and is a primary haemopoietic centre in the anuran, but apparently not in the newt[103]. Transplantation experiments using varying numbers of ventral mesoderm cells have revealed that thymocyte and B-lymphocyte lineage-specific precursors differentiate in this embryonic layer by 20 h of development in *Xenopus*[82]. Thymocytes and thymic accessory cells in this species arise from a bipotential (ventral mesoderm) precursor, that diverges into the two separate lineages after colonization of the epithelial rudiment[105]. Transplantation experiments have shown that there are at least two periods of stem cell entry into the thymus in *Xenopus*, one during early larval life (at 3–4 days), and a second during metamorphosis, between 38 and 57 days post-fertilization[106] (see Figure 4.2), which involves both ventral blood island and dorsal stem cell compartment-derived components[149]. Presumably this second wave of T-cell development allows T-cells to be 'educated' in an environment where adult-specific antigens are expressed. Although some larval T-cells may persist beyond metamorphosis, this early T-cell population is significantly diluted by expanding adult populations[150].

A monoclonal antibody has been prepared which identifies a determinant called XL-1 that is expressed on all types of *Xenopus* leucocytes, but not on erythrocytes, and which therefore appears to be similar to the leucocyte-common antigen reported in birds and mammals[107]. Fairly large numbers of XL-1^{+ve} cells with macrophage-like morphology are found in several locations in mesenchyme in the 2-day-old larva, prior to the appearance of lymphocytes. This draws attention to the possibility that these cells are involved in non-specific defence mechanisms required by the young free-living tadpole, before its antigen-specific lymphoid system matures[3,10].

4.5.2 Thymic 'Education' of T-cells

Implantation of a thymus, which may be either lymphoid or lymphocyte-depleted following γ-irradiation, to early-thymectomized (larval or adult) *Xenopus* reveals that foreign thymus grafts

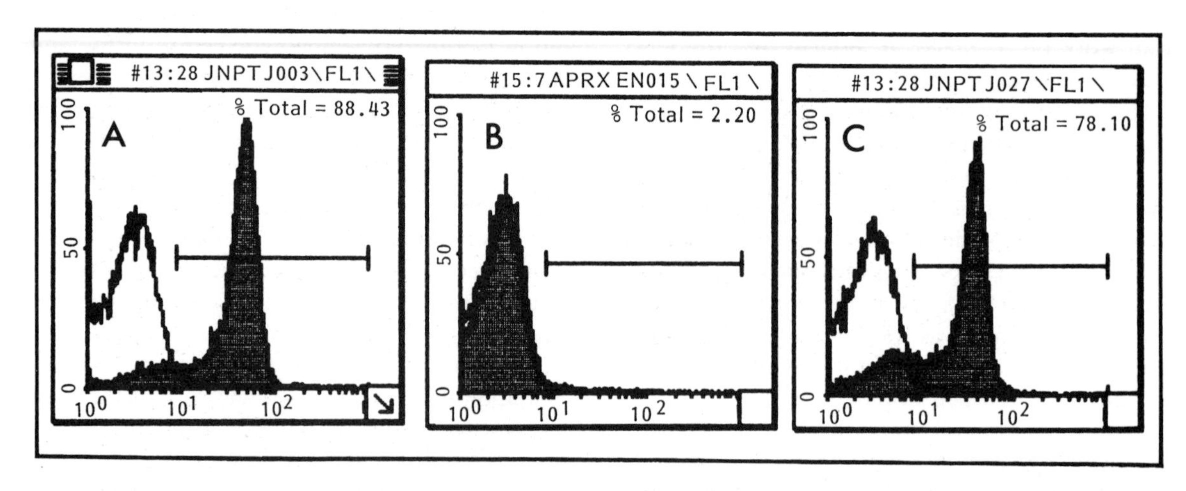

Figure 4.16. Use of a species-specific T-cell marker to demonstrate host derivation of thymocytes in thymus-implanted, thymectomized *Xenopus*. Thymocytes were obtained from 9-month-old LG15 *Xenopus* (A) and *X. tropicalis* (B); additionally (C) thymocytes were also prepared from an *X. tropicalis* thymus implant 8 months after grafting to an LG15, early-thymectomized host. Thymocytes were incubated with the T-lineage-specific monoclonal antibody XT-1, and then with fluorescein-labelled secondary antibody (anti-mouse Ig). Figure shows FACS profiles of the three thymocyte populations, with fluorescence intensity (log scale) shown on *x* axis and relative cell number on *y* axis. The percentages represent the proportion of XT-1-positive cells displaying more intense fluorescence than the marker set to exclude approximately 98% of those cells stained with secondary antibody alone (unshaded profiles). Experimental profiles (primary + secondary antibodies) are shown in grey shading. Note that *X. tropicalis* cells (B) do not express the antigen recognized by XT-1, whereas nearly 90% of LG15 thymocytes (A) are positive. The *X. tropicalis* thymus implant (C) is by now replete with host-derived (XT-1^{+ve}) thymocytes.

can promote the differentiation of host precursor cells along a T-cell pathway[108]. Use of xenogeneic donor/host combinations has shown that various donor thymic stromal cell types persist in the implant[31], whereas the thymic lymphocyte population becomes host-derived (Figure 4.16). The *in vivo* construction of thymuses with MHC-disparate epithelial and lymphoid compartments can readily be achieved by a different surgical approach that is feasible with the amphibian embryo. This approach involves joining the anterior part of one 24-h *Xenopus* embryo, containing the thymic epithelial buds, to the posterior portion of an MHC-incompatible embryo, from which the haemopoietic stem cells, including lymphocytes, arise[109] (Figures 4.14 and 4.17).

These two experimental approaches have been used to explore the role which is played by thymic stromal cells in establishing tolerance (negative selection) of T-lineage lymphocytes to

Figure 4.17. Chimaeric *Xenopus*. Chimaeric *Xenopus*, here produced by exchanging anterior and posterior regions between a normal *Xenopus* embryo and an albino variant with white skin (see Figure 4.14), are proving useful for studying thymic education. Reproduced by courtesy of Dr Martin Flajnik and Dr Louis DuPasquier. Scale bar = 2 cm.

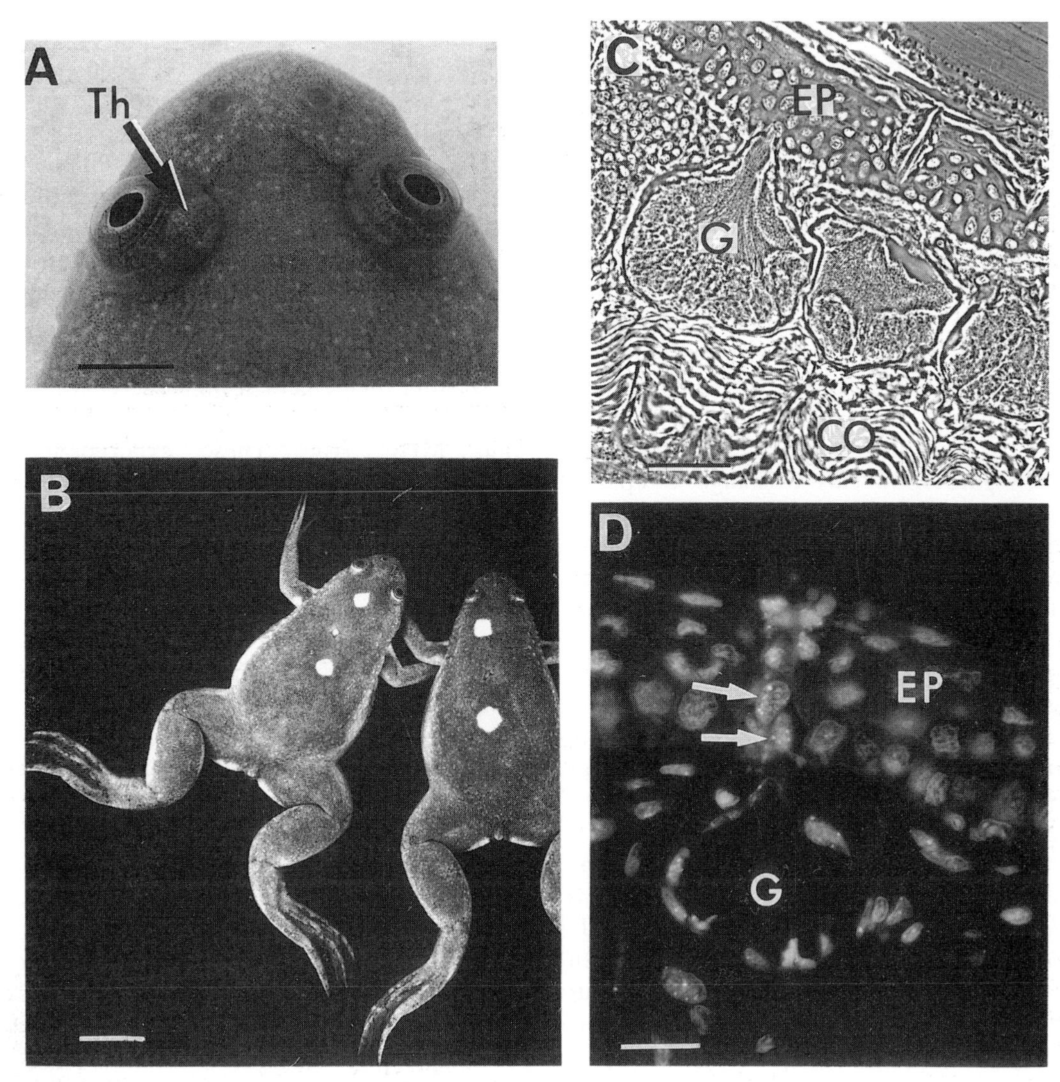

Figure 4.18. Studies exploring the nature of skin transplant tolerance induced in thymus-grafted, thymectomized *Xenopus*. (A) A thymus implant (Th) placed subcutaneously, medial to the larval eye, can be readily visualized following metamorphosis of the thymectomized recipient. Scale bar = 2 mm. (B) Appearance of skin grafts (of white belly skin) on thymectomized clonal (LM) *Xenopus*, each reconstituted with thymus from an MHC-disparate (two MHC haplotype different) LM donor when 5 weeks old. Photograph taken 50 days after test skin was grafted to 6-month-old recipients. Anterior grafts from the host clone and the posterior ones from the allogeneic thymus donor clone are permanently tolerated. In contrast, the middle skin graft, from a third-party donor (expressing additional histocompatibility antigens) has been rejected. The experiment demonstrates alloimmune restoration and specificity of allotolerance[28]. Scale bar = 1 cm.
(C, D) Histological observations investigating origin of cells in tolerated skin xenografts applied to xenothymus-implanted, thymectomized animals. Hosts = thymectomized LG clonal *Xenopus*, implanted with *X. borealis* thymus in late larval life and subsequently grafted (at 5 months of age) with *X. borealis* thymus donor skin. Skin grafts embedded in historesin and sections cut at 4 μm. (C) Phase contrast view of graft tolerated for 50 days reveals that the xenogeneic skin graft remains histologically healthy, with epidermal (EP), glandular (G) and collagen (CO) layers all intact. Scale bar = 40 μm. (D) Quinacrine fluorescence reveals that the epidermis of the graft has been replaced by host cells (latter do not stain with quinacrine). The glands (G), on the other hand, remain donor-derived and quinacrine-positive. Gland neck cells (arrowed) can be seen passing through the epidermal layer. Scale bar = 25 μm.

self- and alloantigens, and in restricting the MHC antigen specificities (positive selection) with which helper and effector T-cell populations preferentially interact during recognition and/or destruction of non-MHC antigens.

It appears that foreign thymic epithelium is involved in inducing tolerance of T-cells towards subsequent skin grafts of thymus MHC type[25,28,108–111]. Use of the *X. borealis* quinacrine marker reveals that the glandular and collagen layers of tolerated skin grafts remain donor-derived, whereas the graft epidermis is soon replaced by host epidermis (our unpublished findings; Figure 4.18). This 'exchange' of epidermal components (also noted to occur in skin allografting experiments with *Xenopus*[112] and with newts[113]) could be immunologically significant, since this region is rich in dendritic 'Langerhans-like' cells (Section 4.2.6). Loss of a substantial population of donor-derived dendritic cells, believed to be important in stimulating skin allograft rejection in larval *Xenopus*[114], might aid the establishment of 'tolerance' to thymus donor skin in 'immunologically-compromised' thymus-implanted animals.

Although one of the reports[111] on allothymus-induced tolerance indicates that this tolerance is effected by selective deletion of anti-donor T-lymphocytes in the thymus, it has generally been found that tolerance induced to skin of thymus donor type does *not* appear to prevent a proliferative response when host splenic lymphocytes (and even host thymocytes in the foreign thymus implant) are cultured with donor stimulators in one-way mixed lymphocyte culture[108–110]. Presumably, peripheral suppressive or anergic mechanisms are at play in these 'split-tolerance' situations. Similar findings with respect to thymus epithelium-induced 'split-tolerance' have emanated from experiments on avian and mammalian embryos[115]. In mammals there is considerable evidence to indicate that thymic interdigitating cells (dendritic, antigen-presenting cells), which are a stromal population of extrinsic origin, play a crucial role in inducing self-tolerance of developing thymic lymphocytes. This type of tolerance is effected within the thymus by clonal deletion of T-cells which possess T-cell receptors with high affinity for self antigen/MHC complexes delivered by the interdigitating cells[116].

The thymectomy/allothymus implantation approach has not revealed a clear-cut involvement of the thymus in MHC restriction; for example, T-B collaboration appears relatively normal in such *Xenopus*[28]. In contrast, studies on MHC-mismatched chimeric *Xenopus* have clearly shown that IgY responses are impaired, thereby suggesting that T-cells from these chimeras cannot interact effectively with B-cells of non-thymus MHC type. These experiments also suggested the thymus is involved in selection of T-cells that preferentially visualize minor histocompatibility antigens in the context of thymus MHC type[109].

4.5.3 Ontogeny of Transplantation Immunity and Antibody Production

Onset of alloimmunity (to MHC antigens) in the anuran larva correlates with the lymphoid maturation of the thymus and, presumably, the appearance of the necessary T-cell population in the periphery[9,87,112]. Graft rejection capacity begins when the lymphoid system contains only 0.5×10^6 lymphocytes. Early embryonic and larval grafting can lead to sensitization[117]. For example, a single neural fold grafted in embryonic life is rejected (Figure 4.19); this can lead to second-set rejection of donor skin in the adult[101].

Anuran larvae synthesize mostly IgM antibodies, but can be made to produce IgY if challenged repeatedly. IgX is also expressed in larvae[152]. Their antibodies are antigen-specific; IgM (but not IgY) antibodies increase in affinity during an immune response. Larval antibody titres are considerably lower than in the adult[118]. Secondary antibody responses can occur in adults, following primary injection of larvae with mammalian red cells[119] or hapten-carrier conjugates[13]. It is not yet known whether transfer of immunological memory over metamorphosis reflects transfer of memory cells or is due to antigen persistence.

In view of the diminished library of antibodies

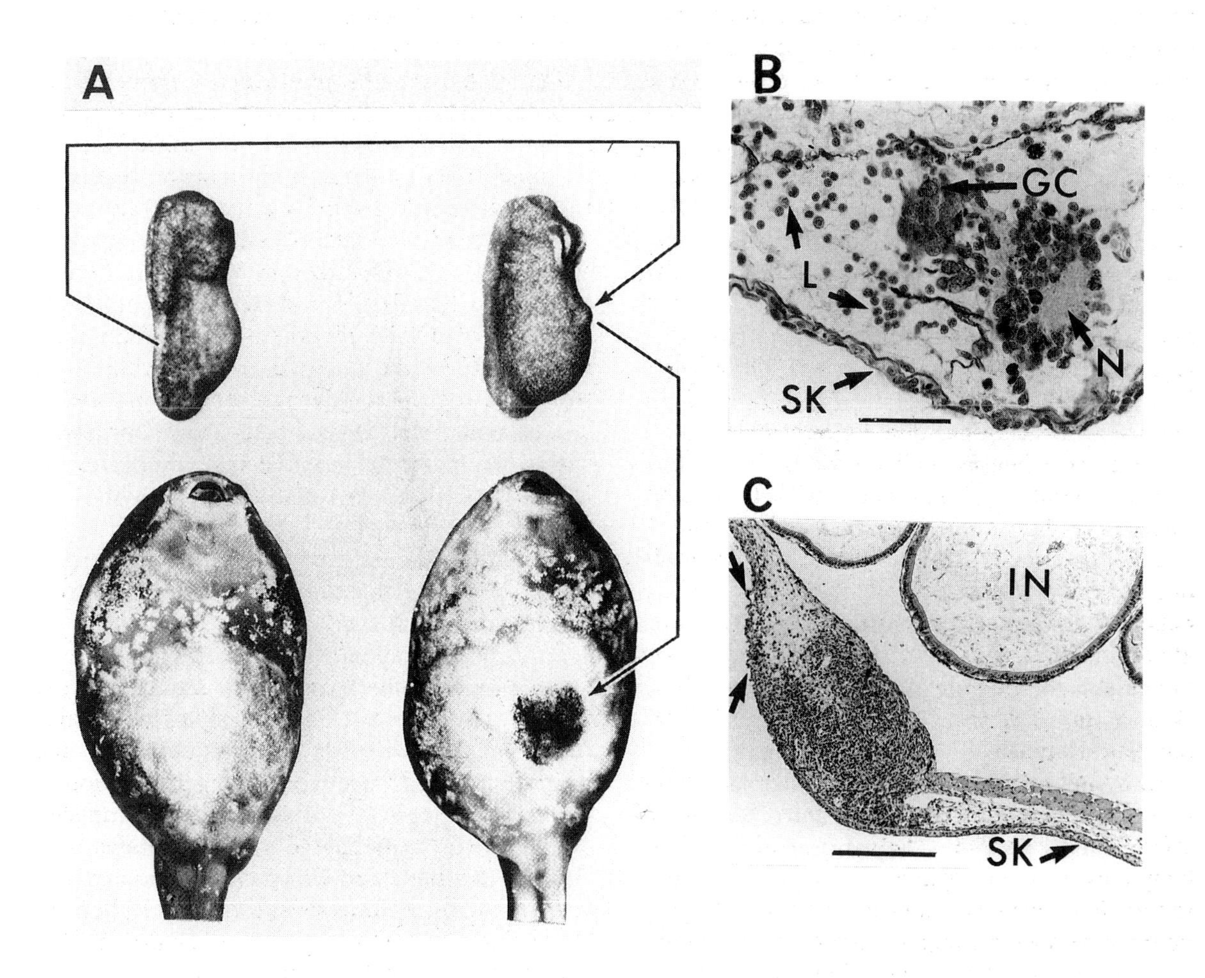

Figure 4.19. Transplantation of embryonic tissue in *Rana*—ontogeny of alloimmunity. (A) A piece of neural fold removed from one embryo (tail-bud stage, top left) is transplanted to the mid-ventral surface of another host embryo (top right). Intimately associated with the neural folds are the neural crest elements which are precursors of diverse cell types, including pigment cells. The pigment cells that differentiate provide an externally visible means of following the progress of the embryonic transplant. The host larva (bottom right) has developed with a distinctive mass of graft-derived pigment cells. Illustration by courtesy of Dr E.P. Volpe.

(B, C) 8-μm sections through neural fold allografts stained in H & E. (B) Differentiated graft elements (large ganglion cells (GC) with prominent nucleoli, other nervous tissue (N) and melanin) 15 days after transplantation. Despite the earliness of the transplantation, leucocytes (L)—both lymphocytes and granulocytes—are seen invading the graft. Scale bar = 100 μm. (C) By 45 days post-transplantation, the neural fold graft is undergoing extensive destruction, the differentiated graft tissues now being virtually unrecognizable in the milieu of host invasive cells. A portion of skin overlying the graft is necrotic (arrows). IN = intestine; SK = healthy skin of mid-ventral surface. Scale bar = 300 μm.

found in amphibians compared with mammals[98], it is interesting that amphibian tadpoles, whose antibodies are even less heterogeneous than in the adult, are able to mount specific antibody responses to a range of antigens, despite having less than 1×10^6 lymphocytes[118,120]. Cell economy is essential in these larvae and since cell wastage would be a threat, B-cells appear to generate different immunoglobulin specificities through Ig V gene rearrangements, but possibly not through somatic mutational mechanisms. In fact it is thought that two waves of such gene rearrangements occur in *Xenopus* ontogeny[13]. The first is in the larva and involves relatively few B-cells over a limited period. The somewhat different adult antibody repertoire is effected by gene rearrangements occurring in a new wave of B-cell development after metamorphosis[66,154] (Figure 4.2).

4.5.4 Ontogeny of Transplantation Tolerance

Although the thymus plays a crucial role in self tolerance induction, it is clearly not the exclusive site where T-cell tolerance occurs. Transplantation of tissues to embryonic, larval and metamorphic stages of amphibians is readily feasible and is providing valuable information on the cellular and genetic basis of such 'peripheral' tolerance. Most of this work has been done with allogeneic tissues, which by extrapolation provide information on tolerance induction to self. Some experiments, however, have directly probed *self* tolerance. For example, the demonstration that *Xenopus* grown without eyes or pituitaries are tolerant of subsequently-grafted, appropriate organs from either isogeneic or MHC-compatible *Xenopus*[121,122] refuted the earlier, much quoted work of Triplett[123], and illustrated that self-tolerance is not restricted to an early phase of ontogeny.

Immunocompetent (thymus intact) anuran larvae, but not adults, can be rendered tolerant by transplantation of allogeneic (e.g. one MHC-haplotype disparate) skin from adult donors[124]. Some authors have found that allotolerance induction is particularly easy to achieve at metamorphosis[125]. Others have found that the size of skin grafts applied to the larva and degree of histoincompatibility appear to be critical to the outcome. For example, minor histoincompatibility disparate skin grafts are routinely tolerated by larval and peri-metamorphic *Xenopus*[87,126], whereas a high degree of rejection occurs to two MHC-haplotype disparate skin grafts[126]. We have recently found[151] that the transplantation of *larval* allogeneic skin, thymus or spleen (1 or 2 MHC-disparate) to larval *Xenopus* (LG clonal animals) can lead to donor-specific skin tolerance in the adult. Larval skin proves less effective in this respect than the thymus and spleen, which could well relate to the 'lack' of class II MHC-positive cells in larval skin compared with the class II-rich cells found in larval thymus and spleen (Section 4.2).

The tolerance induced in the foregoing experiments is judged on the basis of skin tolerance, specific to the donor, induced to secondary skin grafts applied to young adults. In general this tolerance is incomplete (or 'split'—see also Section 4.5.2) as it does not cause deletion of lymphocytes reactive with the tolerizing MHC[124]. Such anti-donor MLC-reactive spleen cells can also be visualized *in vivo*[127], and tolerated skin grafts are subjected to some degree of T-lymphocyte infiltration and increased MHC class II expression[150] (Figure 4.20). Failure to reject test skin grafts in these allotolerant animals appears to be maintained by active suppression (of cytotoxic T-cells?[127]), as evidenced by breakdown of tolerance following cyclophosphamide treatment[128], and the finding that tolerance can be transferred with splenic lymphoid cells[129].

It would seem that complete tolerance (*in vivo* and *in vitro*) can only be achieved when embryonic stem cells are transplanted, and when animals become chimaeric, as in embryonic flank transplantation studies[130]. It has been suggested[60] that such tolerance may be achieved by haemopoietically-derived antigen-presenting cells. The latter may effect tolerance peripherally, but could conceivably gain access to the thymus and achieve 'central' deletion of appropriate T-cell specificities. 'Split-tolerance', on the other hand,

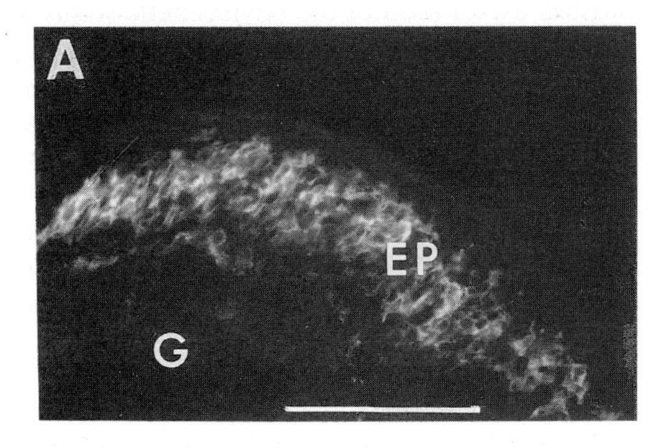

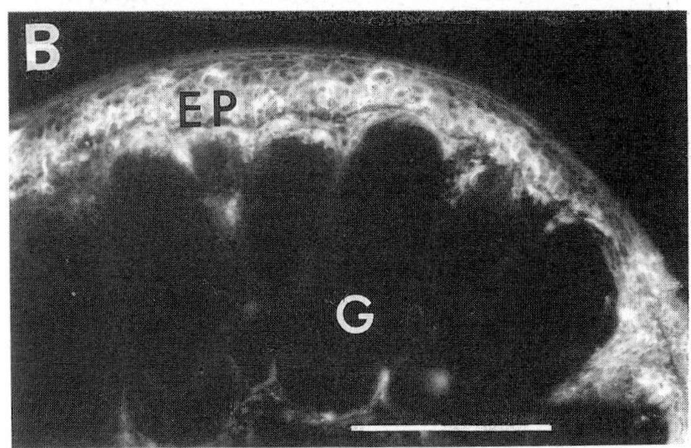

Figure 4.20. Immunohistology of 'tolerated' skin allografts. Immunofluorescence cryostat sections (6 μm) through LG5 (MHC = bc) skin allografts 3 weeks after grafting to an LG15 (ac) 5-month-old *Xenopus*. Recipient had previously been implanted in larval life with a spleen from a larval LG5 donor, a procedure that is effective at inducing donor-specific, skin 'tolerance'. The section in (A) has been stained with XT-1 mAb (stains T-cell subset); section (B) with AM20 (anti-class II MHC mAb), prior to incubation with secondary, FITC-labelled anti-mouse Ig antibody. Note the bright staining of graft epidermis (EP), and superficial glandular layer (G) with both mAbs, indicating T-lymphocyte infiltration and increased MHC class II antigen expression (compare with normal skin—see Figure 4.11). These phenomena, not seen in skin isografts, indicate that the otherwise healthy graft is subject to some degree of immunological reactivity. Scale bars = 100 μm (A), 200 μm (B).

can be established by various tissues that are introduced during early ontogeny, and is maintained through suppression or anergy.

4.5.5 Metamorphosis

Immunologists are intrigued to know how amphibians escape the risk of dying from an autoimmune disease at metamorphosis, since there is a need for the immunocompetent animal to become tolerant to adult-specific determinants that are first expressed at this time. Autoimmune-like phenomena ('red disease') can occur in *Xenopus* at metamorphosis following application of adult skin in larval life, possibly as a result of immunization to adult antigens during larval stages[9].

The question of how self-tolerance is generated at metamorphosis, and in larval life, continues to be explored in some depth in *Xenopus*. The involvement of suppressor cell populations at metamorphosis has been suggested, since lymphocytes (e.g. thymocytes) from transforming animals (but not adults) can suppress the response of young isogeneic toadlets to minor histocompatibility-disparate grafts[131]. This work was also important in showing memory cells were not as sensitive to suppression as cells responding in a first-set reaction; this would allow immune experiences of the larva to be remembered after metamorphosis. Thymectomy of 5-week-old *Xenopus* larvae impairs the ease with which transplantation tolerance can be achieved at metamorphosis, possibly by lowering T-suppressor cell numbers in the periphery[132].

Hormonal control of altered immunoregulation at metamorphosis also appears likely[133]. Thyroid hormones provide the driving force for both the regressive (removal of larval structures) and subsequent progressive (appearance of adult structures) developmental alterations at metamorphosis[65]. High titres of thyroid hormones at metamorphosis are accompanied by a comparable dramatic transitory increase in both corticosteroid serum titres and corticosteroid receptors (especially on thymic and splenic lymphocytes). Corticosteroids appear to inhibit the ability of thymocytes to suppress antibody production at metamorphosis[134]. This finding contrasts with suggestions above that enhanced suppressor function is intimately involved with generation of allotolerance at this time. It has been suggested that corticosteroids directly inhibit T-cell clonal expansion at metamorphosis through interference with regulatory cytokines[133]. The possibility that injected human rIL-2 can induce a higher percentage of metamorphosing *Xenopus* to reject skin allografts has been reported[133], although our own follow-up experiments have as yet failed to confirm these preliminary observations.

A possible role for the late emergence (at metamorphosis) of MHC class I surface expression in targeting the destruction of certain larval tissues such as gills and tail at this time has recently been suggested[64,135]. Thus CD8^{+ve} T-cells (stained with monoclonal antibody AM22—Section 4.2.1) are found in larvae and it is postulated that these cells are responsible for the tadpole anti-adult mixed lymphocyte reactivity that has been noticed in *Xenopus*[64]. These T-cells would have the capacity to destroy newly emerging class I^{+ve} cells that co-express (present?) class II—i.e. tadpole tissues such as pharynx, gut and skin, and the class II-rich B-cells. However, such 'autoimmune' T-cells would, at the same time, have to be prevented from killing newly-arising adult cells—through the suppressive and corticosteroid-induced mechanisms discussed above?! If this model were true, the normal metamorphosis of thymectomized animals might depend upon extra-thymic 'T-cell-like' maturation in *larval* life, which has not been demonstrated to date.

4.6 CONCLUDING REMARKS: FUTURE DIRECTIONS FOR AMPHIBIAN IMMUNOLOGY

This chapter has attempted to highlight features of the amphibian immune system that are of interest phylogenetically and those that aid in our understanding of immunological ontogeny. Further exploration of the phylogenetic status of the amphibian immune system will undoubtedly centre on obtaining a greater understanding of the molecular and genetic basis of immunoglobulin superfamily members (MHC proteins, Ig molecules, T-cell receptors, etc.) and cytokines. Ontogenetic studies are, in my opinion, the *raison d'être* for employing amphibian models. Investigations on lymphocyte differentiation and tolerance will benefit hugely from Ig superfamily characterization and from the development of a much wider array of monoclonal antibodies to mark larval and adult cell populations. Finally, interplay between endocrine and immune systems in the regulation of cellular and humoral immunity should be explored in more detail. Amphibian metamorphosis is already proving to be a fascinating period in this respect. A functional association between the peripheral nervous system (in the form of the neurotransmitter noradrenaline) and the adult *Xenopus* immune system has been identified[136]. Such investigations will undoubtedly have significance outside purely phylogenetic considerations.

Acknowledgements

I am indebted to my wife, Trudy, for skilfully transforming my handwritten scribble into manuscript form and for unfailing encouragement throughout the preparation of this chapter. Special thanks also go to Pamela Ritchie for help with preparation of the illustrations, especially the line drawings. The photographic assistance of David Hutchinson is gratefully acknowledged.

4.7 REFERENCES

1. Jurd, R.D. (1985). Haematological and immunological 'metamorphosis' in neotenous urodeles. In *Metamorphosis*, Balls, M. and Bownes, M.E. (eds). Clarendon Press, Oxford, pp.313–331.
2. Volpe, E.P. (ed.) (1971). Biology of immunity in amphibians. *Amer. Zool.* **11,** 169–237.
3. Manning, M.J. and Turner, R.J. (1976). *Comparative Immunobiology*. Blackie, Glasgow.
4. Marchalonis, J.J. (ed.) (1976). *Comparative Immunology*. Blackwell Scientific Publications, Oxford.

5. Cooper, E.L. (1976). Immunity mechanisms. In *Physiology of the Amphibia, Vol. 3*, Lofts, B. (ed.). Academic Press, New York, pp. 164–272.
6. Cohen, N. (1977). Phylogenetic emergence of lymphoid tissues and cells. In *The Lymphocyte: Structure and Function*, Marchalonis, J.J. (ed.). Marcel Dekker, New York, pp. 149–202.
7. Edwards, B.F. and Ruben, L.N. (1982). Aspects of amphibian immunity. In *Animal Models of Immunological Processes*, Hay, J.B. (ed.). Academic Press, London, pp. 255–286.
8. Manning, M.J. and Horton, J.D. (1982). RES structure and function of the Amphibia. In *The Reticuloendothelial System: Phylogeny and Ontogeny*, Cohen, N. and Sigel, M.M. (eds). Plenum Press, New York, pp. 423–459.
9. Du Pasquier, L. (1982). Ontogeny of immunological functions in amphibians. In *The Reticuloendothelial System: Phylogeny and Ontogeny*, Cohen, N. and Sigel, M.M. (eds). Plenum Press, New York, pp. 633–657.
10. Turner, R.J. (1988). Amphibians. In *Vertebrate Blood Cells,* Rowley, A.F. and Ratcliffe, N.A. (eds). Cambridge University Press, pp. 129–209.
11. Horton, J.D. and Ratcliffe, N.A. (1993). Evolution of immunity. In *Immunology*, 3rd edn, Roitt, I.M., Brostoff, J. and Male, D.K. (eds). Mosby, London, pp. 14.1–14.22.
12. Du Pasquier, L. (1989). Evolution of the immune system. In *Fundamental Immunology*, Paul, W.E. (ed.). Raven Press, New York, pp. 139–165.
13. Du Pasquier, L., Schwager, J. and Flajnik, M.F. (1989). The immune system of *Xenopus. Ann. Rev. Immunol.* **7,** 251–275.
14. Du Pasquier, L. *et al.* (1985). Methods used to study the immune system of *Xenopus* (Amphibia, Anura). *Immunol. Methods* **3,** 425–465.
15. Tochinai, S. (1978). Thymocyte stem cell inflow in *Xenopus laevis* after grafting diploid thymic rudiments into triploid tadpoles. *Dev. Comp. Immunol.* **2,** 627–635.
16. Turpen, J.B. *et al.* (1982). Ontogeny of amphibian haemopoietic cells. In *The Reticuloendothelial System: Phylogeny and Ontogeny*, Cohen, N. and Sigel, M.M. (eds). Plenum Press, New York, pp. 569–588.
17. Du Pasquier, L. and Flajnik, M.F. (1990). Expression of MHC class II antigens during *Xenopus* development. *Dev. Immunol.* **1,** 85–95.
18. Nagata, S. (1988). T cell-specific antigens in *Xenopus* identified with a mouse monoclonal antibody: biochemical characterization and species distribution. *Zool. Sci.* **5,** 77–83.
19. Charlemagne, J. and Mottet, A. (1986). Quantitative analysis of lymphocyte migration from the thymus to periphery in the mexican axolotl. *Dev. Comp. Immunol.* **10,** 603–608.
20. Russ, J.H. and Horton, J.D. (1987). Cytoarchitecture of the *Xenopus* thymus following γ-irradiation. *Development* **100,** 95–105.
21. Clothier, R.H. and Balls, M. (1985). Structural changes in the thymus glands of *Xenopus laevis* during development. In *Metamorphosis*, Balls, M. and Bownes, M.E. (eds). Clarendon Press, Oxford, pp. 332–359.
22. Dardenne, M. *et al.* (1973). Studies on thymus products VII. Presence of thymic hormone in urodele serum. *Ann. Immunol. Inst. Pasteur* **124C,** 465–469.
23. Holtfreter, H.B. and Cohen, N. (1987). *In vitro* behaviour of thymic nurse cell-like complexes from mechanically and enzymatically dissociated frog tadpole thymuses. *Amer. J. Anat.* **179,** 342–355.
24. Horton, J.D. and Horton, T.L. (1975). Development of transplantation immunity and restoration experiments in the thymectomized amphibian. *Amer. Zool.* **15,** 73–84.
25. Katagiri, C. and Tochinai, S. (1987). Ontogeny of thymus-dependent immune responses and lymphoid cell differentiation in *Xenopus laevis. Dev. Growth Differ.* **29,** 297–305.
26. Clem, L.W., Miller, N.W. and Bly, J.E. (1991). Evolution of lymphocyte subpopulations, their interactions and temperature sensitivities. In *Phylogenesis of Immune Function*, Warr, G. and Cohen, N. (eds). CRC Press, Boca Raton, pp. 191–213.
27. Tournefier, A. *et al.* (1988). Surface markers of axolotl lymphocytes as defined by monoclonal antibodies. *Immunology* **63,** 269–276.
28. Du Pasquier, L. and Horton, J.D. (1982). Restoration of antibody responsiveness in early thymectomized *Xenopus* by implantation of major histocompatibility complex-mismatched larval thymus. *Eur. J. Immunol.* **12,** 546–551.
29. Ruben, L.N. *et al.* (1985). T-lymphocyte regulation of humoral immunity in *Xenopus laevis*, the South African clawed toad. *Dev. Comp. Immunol.* **9,** 811–818.
30. Kawahara, H. (1978). Production of triploid and gynogenetic diploid *Xenopus* by cold treatment. *Dev. Growth Differ.* **20,** 227–236.
31. Horton, J.D. *et al.* (1987). Thymocyte/stromal cell chimaerism in allothymus-grafted *Xenopus*: developmental studies using the *X. borealis* fluorescence marker. *Development* **100,** 107–117.
32. Flajnik, M.F. *et al.* (1990). Evolution of the MHC: antigenicity and unusual tissue distribution of *Xenopus* (frog) class II molecules. *Molec.*

Immunol. **27,** 451–462.
33. Bleicher, P.A. and Cohen, N. (1981). Monoclonal anti-IgM can separate T cell from B cell proliferative responses in the frog, *Xenopus laevis*. *J. Immunol.* **127,** 1549–1555.
34. Bell, E.B. (1989). Thymus-derived and non-thymus-derived T-like cells: the origin and function of cells bearing $\gamma\delta$ receptors. *Thymus* **14,** 3–17.
35. Nagata, S. and Cohen, N. (1983). Specific *in vivo* and nonspecific *in vitro* alloreactivities of adult frogs (*Xenopus laevis*) that were thymectomized during early larval life. *Eur. J. Immunol.* **13,** 541–545.
36. Turpen, J.B. and Cohen, N. (1976). Alternate sites of lymphopoiesis in the amphibian embryo. *Ann. Immunol. Inst. Pasteur* **127C,** 841–848.
37. Horton, J.D. *et al.* (1992). Skin xenograft rejection in *Xenopus*: immunohistology and effect of thymectomy. *Transplantation* **53,** 473–476.
38. Solomon, J.B. and Horton, J.D. (eds) (1977). *Developmental Immunobiology*. Elsevier, Amsterdam.
39. Bernard, C.C.A. *et al.* (1981). Genetic control of T helper cell function in the clawed toad *Xenopus laevis*. *Eur. J. Immunol.* **11,** 151–155.
40. Bernard, C.C.A. *et al.* (1979). Immunogenetic studies on the cell-mediated cytotoxicity in the clawed toad *Xenopus laevis*. *Immunogenetics* **9,** 443–454.
41. Watkins, D., Harding, F. and Cohen, N. (1988). *In vitro* proliferative and cytotoxic responses against *Xenopus* minor histocompatibility antigens. *Transplantation* **45,** 499–501.
42. Horton, T.L., Horton, J.D. and Varley, C.A. (1988). *In vitro* cytotoxicity in adult *Xenopus* generated against larval targets and minor histocompatibility antigens. *Transplantation* **47,** 880–882.
43. Hsu, E., Julius, M.H. and Du Pasquier, L. (1983). Effector and regulator functions of splenic and thymic lymphocytes in the clawed toad *Xenopus*. *Ann. Immunol. Inst. Pasteur* **134D,** 277–292.
44. Clothier, R.H. *et al.* (1989). Differential cyclophosphamide sensitivity of suppressor function in *Xenopus*, the clawed toad. *Dev. Comp. Immunol.* **13,** 159–166.
45. Charlemagne, J. and Tournefier, A. (1977). Anti-horse red blood cell antibody synthesis in the Mexican axolotl (*Ambystoma mexicanum*). Effect of thymectomy. In *Developmental Immunobiology*, Solomon, J.B. and Horton, J.D. (eds). Elsevier, Amsterdam, pp.267–275.
46. Obara, N., Tochinai, S. and Katagiri, C. (1982). Splenic white pulp as a thymus-independent area in the African clawed toad, *Xenopus laevis*. *Cell Tissue Res.* **226,** 327–335.
47. Baldwin, W.M. and Cohen, N. (1981). A giant cell with dendritic cell properties in spleens of the anuran amphibian *Xenopus laevis*. *Dev. Comp. Immunol.* **5,** 461–473.
48. Hadji-Azimi, I., Coosemans, V. and Canicatti, C. (1990). B-lymphocyte populations in *Xenopus laevis*. *Dev. Comp. Immunol.* **14,** 69–84.
49. Turner, R.J. (1973). Response of the toad, *Xenopus laevis*, to circulating antigens. II. Responses after splenectomy. *J. Exp. Zool.* **183,** 35–45.
50. Zettergren, L.D., Boldt, B.W. and Schmid, L.S. (1988). Cellular and serum immune response to dinitrophenol in adult *Rana pipiens*. *Dev. Comp. Immunol.* **11,** 99–107.
51. Zettergren, L.D., Kubagawa, H. and Cooper, M.D. (1980). Development of B cells in *Rana pipiens*. In *Phylogeny of Immunological Memory*, Manning, M.J. (ed.). Elsevier, Amsterdam, pp.177–185.
52. Watkins, D., Parsons, S.C. and Cohen, N. (1987). A factor with interleukin-1-like activity is produced by peritoneal cells from the frog, *Xenopus laevis*. *Immunology* **62,** 669–673.
53. Gammie, A.E. and Ruben, L.N. (1986). The phylogeny of macrophage function: antigen uptake and degradation by peritoneal exudate cells of two amphibian species and CAF_1 mice. *Cell. Immunol.* **100,** 577–583.
54. Coosemans, V. and Hadji-Azimi, I. (1988). Immunoglobulin Fc receptor molecules on *Xenopus laevis* splenocytes. *Immunology* **65,** 641–645.
55. Sekizawa, A., Fujii, T. and Tochinai, S. (1984). Membrane receptors on *Xenopus* macrophages for two classes of immunoglobulins (IgM and IgY) and the third complement component (C3). *J. Immunol.* **133,** 1431–1435.
56. Ruben, L.N. (1984). Some aspects of the phylogeny of macrophage-lymphocyte immune regulation. *Dev. Comp. Immunol.* **8,** 247–256.
57. Ghoneum, M., Cooper, E.L. and Sadek, I. (1990). Variability of natural killer cell activity in anuran amphibians. *Dev. Comp. Immunol.* **14,** 359–365.
58. Jurd, R.D. and Doritis, A. (1977). Antibody-dependent cellular cytotoxicity in poikilotherms. *Dev. Comp. Immunol.* **1,** 341–352.
59. Evans, D.L. and McKinney, E.C. (1991). Phylogeny of cytotoxic cells. In *Phylogenesis of Immune Functions*, Warr, G. and Cohen, N. (eds). CRC Press, Boca Raton, pp.215–239.
60. Flajnik, M.F. and Du Pasquier, L. (1990). The major histocompatibility complex of frogs. *Immunol. Rev.* **113,** 47–63.

61. Kaufman, J., Flajnik, M.F. and Du Pasquier, L. (1991). The MHC molecules of ectothermic vertebrates. In *Phylogenesis of Immune Functions*, Warr, G. and Cohen, N. (eds). CRC Press, Boca Raton, pp. 125–149.
62. Flajnik, M.F. *et al.* (1984). Identification of class I major histocompatibility complex encoded molecules in the amphibian *Xenopus*. *Immunogenetics* **20,** 433–442.
63. Kaufman, J.F. *et al.* (1985). *Xenopus* MHC class II molecules. I. Identification and structural characterization. *J. Immunol.* **134,** 3248–3257.
64. Flajnik, M.F. and Du Pasquier, L. (1990). Changes in the expression of the major histocompatibility complex during the ontogeny of *Xenopus*. In *Developmental Biology. UCLA Symposium on Molecular and Cellular Biology*, Davidson, E., Rudeman, J. and Posakony, J. (eds.) Alan R. Liss, New York, pp. 215–224.
65. Rollins-Smith, L.A., Parsons, S.C.V. and Cohen, N. (1988). Effects of thyroxine-driven precocious metamorphosis on maturation of adult-type allograft rejection responses in early thyroidectomized frogs. *Differentiation* **37,** 180–185.
66. Schwager, J. *et al.* (1989). Genetic basis of the antibody repertoire in *Xenopus*: analysis of the V_H diversity. *EMBO J.* **8,** 2989–3001.
67. Du Pasquier, L. and Hsu, E. (1983). Immunoglobulin expression in diploid and polyploid interspecies hybrids of *Xenopus*: evidence for allelic exclusion. *Eur. J. Immunol.* **13,** 585–590.
68. Cohen, N. and Haynes, L. (1991). The phylogenetic conservation of cytokines. In *Phylogenesis of Immune Functions*, Warr, G. and Cohen, N. (eds). CRC Press, Boca Raton, 231–268.
69. Gearing, A. and Rimmer, J.J. (1985). Amphibian lymphokines. 2. Migration inhibition factor produced by antigenic and mitogenic stimulation of amphibian leucocytes. *Dev. Comp. Immunol.* **9,** 291–300.
70. Ruben, L.N. *et al.* (1988). The substitution of carrier priming of helper function in the common American newt, *Notophthalmus viridescens*, by lectins and human lymphokines. *Thymus* **11,** 77–87.
71. Watkins, D. and Cohen, N. (1987). Mitogen-activated *Xenopus laevis* lymphocytes produce a T-cell growth factor. *Immunology* **62,** 119–125.
72. Turner, S.L. *et al.* (1991). Splenocyte response to T cell-derived cytokines in thymectomized *Xenopus. Dev. Comp. Immunol.* **15,** 319–328.
73. Watkins, D. and Cohen, N. (1987). Description and partial characterization of a T-cell growth factor from the frog, *Xenopus laevis*. In *Immune Regulation by Characterized Polypeptides*, Goldstein, G., Bach, J.F. and Wigzell, H. (eds). A.R. Liss, New York, pp. 495–508.
74. Watkins, D., Parsons, S.V. and Cohen, N. (1988). The ontogeny of interleukin production and responsivity in the frog, *Xenopus. Thymus* **11,** 113–122.
73. Ruben, L.N. (1986). Recombinant DNA produced human IL-2, injected *in vivo*, will substitute for carrier priming of helper function in the South African clawed toad, *Xenopus laevis. Immunol. Letters* **13,** 227–230.
76. Langeberg, L. *et al.* (1987). Toad splenocytes bind human IL-2 and anti-human IL-2 receptor antibody specifically. *Immunol. Letters* **14,** 103–110.
77. Ruben, L.N. *et al.* (1990). A monoclonal mouse anti-human IL-2 receptor antibody (anti-Tac) will recognise molecules on the surface of *Xenopus laevis* immunocytes which specifically bind rIL-2 and are only slightly larger than the human Tac protein. *Immunol. Letters* **24,** 117–126.
78. Haynes, L., Moynihan, J.A. and Cohen, N. (1991). A monoclonal antibody against the human IL-2 receptor binds to paraformaldehyde-fixed but not viable frog (*Xenopus*) splenocytes. *Immunol. Letters* **26,** 227–232.
79. Lallone, R.L., Chambers, M.R. and Horton, J.D. (1984). Changes in antibody and complement production in *Xenopus laevis* during postmetamorphic development revealed in a primary *in vivo* or *in vitro* antibody response. *J. Embryol. Exp. Morphol.* **84,** 191–202.
80. Sekizawa, A., Fujii, T. and Katagiri, C. (1984). Isolation and characterization of the third component of complement in the serum of the clawed frog, *Xenopus laevis*, *J. Immunol.* **133,** 1436–1443.
81. Grossberger, D. *et al.* (1989). The conservation of structural and functional domains in the third complement component C3 of *Xenopus* and mammals. *Proc. Nat. Acad. Sci. USA* **86,** 1323–1327.
82. Horton, J.D. (1988). *Xenopus* and developmental immunobiology: a review. *Dev. Comp. Immunol.* **12,** 219–229.
83. Canicatti, C. and Veuthey, F. (1987). The membrane attack complex of *Xenopus laevis* complement. *Experientia* **43,** 192–194.
84. Jensen, L.B. and Koch, C. (1991). An assay for complement factor B in species at different levels of evolution. *Dev. Comp. Immunol.* **15,** 173–179.
85. Cohen, N. (1971). Amphibian transplantation reactions: a review. *Amer. Zool.* **11,** 193–205.

86. Cohen, N. (1980). Salamanders and the evolution of the major histocompatibility complex. In *Contemporary Topics in Immunobiology, Vol. 9*, Marchalonis, J.J. and Cohen, N. (eds). Plenum Press, New York, pp. 109–139.
87. Obara, N., Kawahara, H. and Katagiri, C. (1983). Response to skin grafts exchanged among siblings of larval and adult gynogenetic diploids in *Xenopus laevis. Transplantation* **36,** 91–95.
88. Clark, J.C. and Newth, D.R. (1972). Immunological activity of transplanted spleens in *Xenopus laevis. Experientia* **28,** 951–953.
89. Nakamura, T. (1985). Lethal graft-versus-host reaction induced by parental cells in the clawed frog, *Xenopus laevis. Transplantation* **40,** 393–397.
90. Clothier, R.H. *et al.* (1989). Skin xenograft rejection in *Xenopus laevis*, the South African clawed toad. *Herpetopathologia* **1,** 19–28.
91. Jurd, R.D., Abdul-Salem, F.M.Y. and Maldonado, B.M. (1989). Delayed hypersensitivity phenomena in *Xenopus laevis* Daudin. *Herpetopathologia* **1,** 89–95.
92. Balls, M. *et al.* (1989). The incidence and significance of malignant neoplasia in amphibians. *Herpetopathologia* **1,** 97–104.
93. Clothier, R.H. *et al.* (1989). A transmissible lymphoblastic lymphoma in *Xenopus laevis*, the South African clawed toad. *Herpetopathologia* **1,** 7–11.
94. Picard, J.J. *et al.* (1982). A transplantable malignant adenocarcinoma induced in *Xenopus borealis* (Parker) by a B cell line transformed by *N*-methyl-*N*-nitro-*N*-nitroso-guanidine. In *First International Colloquium on Pathology of Reptiles and Amphibians*, Vago, C. and Matz, G. (eds). University of Angers Press, Angers, pp. 177–179.
95. Rollins, L.A. and Cohen, N. (1981). Effect of early thymectomy on development of renal tumors in Lucké tumor herpesvirus-infected leopard frogs. In *Aspects of Developmental and Comparative Immunology I*, Solomon, J.B. (ed.), Pergamon, Oxford, pp. 505–507.
96. Watkins, D. (1985). T cell function in *Xenopus*: studies on T cell ontogeny and cytotoxicity using an IL-2-like growth factor. PhD Thesis, University of Rochester, New York.
97. Ruben, L.N., Clothier, R.H. and Balls, M. (1986). Thymic involvement in memory responses after primary challenge with TNP-Ficoll in *Xenopus laevis*, the South African clawed toad. *Thymus* **8,** 341–348.
98. Du Pasquier, L. (1982). Antibody diversity in lower vertebrates—why is it so restricted? *Nature* **296,** 311–313.
99. Elkan, E. (1976). Pathology in the Amphibia. In *Physiology of the Amphibia III*, Lofts, B. (ed.). Academic Press, New York, pp. 273–312.
100. Cohen, N. *et al.* (1984). Identification and treatment of a lethal nematode (*Capillaria xenopodis*) infestation of the South African frog, *Xenopus laevis. Dev. Comp. Immunol.* **8,** 739–741.
101. Volpe, E.P. (1980). *The Amphibian Embryo In Transplantation Immunity*. Karger, Basel.
102. Volpe, E.P., Tompkins, R. and Reinschmidt, D. (1979). Clarification of studies on the origin of thymic lymphocytes. *J. Exp. Zool.* **208,** 57–66.
103. Smith, P.B., Flajnik, M.F. and Turpen, J.B. (1989). Experimental analysis of ventral blood island hematopoiesis in *Xenopus* embryonic chimeras. *Dev. Biol.* **131,** 302–312.
104. Maeno, M., Tochinai, S. and Katagiri, C. (1985). Differential participation of ventral and dorsolateral mesoderms in the hemopoiesis of *Xenopus*, as revealed in diploid-triploid or interspecific chimeras. *Dev. Biol.* **110,** 503–508.
105. Turpen, J.B. and Smith, P.B. (1986). Analysis of hemopoietic lineage of accessory cells in the developing thymus of *Xenopus laevis. J. Immunol.* **136,** 412–421.
106. Turpen, J.B. and Smith, P.B. (1989). Precursor immigration and thymocyte succession during larval development and metamorphosis in *Xenopus. J. Immunol.* **142,** 41–47.
107. Ohinata, H., Tochinai, S. and Katagiri, C. (1989). Ontogeny and tissue distribution of leukocyte-common antigen bearing cells during early development of *Xenopus laevis. Development* **107,** 445–452.
108. Nagata, S. and Cohen, N. (1984). Induction of T cell differentiation in early-thymectomized *Xenopus* by grafting adult thymuses from either MHC-matched or from partially or totally MHC-mismatched donors. *Thymus* **6,** 89–104.
109. Flajnik, M.F., Du Pasquier, L. and Cohen, N. (1985). Immune responses of thymus/lymphocyte embryonic chimeras: studies on tolerance and major histocompatibility complex restriction in *Xenopus. Eur. J. Immunol.* **15,** 540–547.
110. Arnall, J.C. and Horton, J.D. (1986). Impaired rejection of minor-histocompatibility-antigen-disparate skin grafts and acquisition of tolerance to thymus donor antigens in allothymus-implanted, thymectomized *Xenopus. Transplantation* **41,** 766–776.
111. Maeno, M. *et al.* (1987). Analysis of allotolerance in thymectomized *Xenopus* restored with semiallogeneic thymus grafts. *Transplantation* **44,** 308–314.

112. Horton, J.D. (1969). Ontogeny of the immune response to skin allografts in relation to lymphoid organ development in the amphibian *Xenopus laevis* Daudin. *J. Exp. Zool.* **170,** 449–466.
113. Cohen, N. (1966). Tissue transplantation immunity in the adult newt, *Diemictylus viridescens. J. Exp. Zool.* **163,** 173–189.
114. Du Pasquier, L. and Flajnik, M.F. (1987). *Xenopus* MHC class II antigens. *Basel Institute for Immunology, Annual Report* **47**.
115. Houssaint, E. and Flajnik, M.F. (1990). The role of thymic epithelium in the acquisition of tolerance. *Immunol. Today* **11,** 357–360.
116. Marrack, P. and Kappler, J. (1987). The T cell receptor. *Science* **238,** 1073–1079.
117. Plytycz, B. (1981). Ontogeny of transplantation immunity in the common frog, *Rana temporaria* L. *Differentiation* **20,** 71–76.
118. Du Pasquier, L. *et al.* (1986). Ontogeny of the immune system in anuran amphibians. *Progress in Immunology* **6,** 1079–1088.
119. Manning, M.J. and Al Johari, G.M. (1985). Immunological memory and metamorphosis. In *Metamorphosis*, Balls, M. and Bownes, M.E. (eds). Clarendon Press, Oxford, pp. 420–433.
120. Du Pasquier, L. (1970). Ontogeny of the immune response in animals having less than one million lymphocytes: the larvae of the toad *Alytes obstetricans. Immunology* **19,** 353–362.
121. Rollins-Smith, L.A. and Cohen, N. (1983). The Triplett phenomenon revisited: self-tolerance is not confined to the early developmental period. *Transplant. Proc.* **15,** 871–874.
122. Maeno, M. and Katagiri, C. (1984). Elicitation of weak immune response in larval and adult *Xenopus laevis* by allografted pituitary. *Transplantation* **38,** 251–255.
123. Triplett, E.L. (1962). On the mechanism of immunologic self recognition. *J. Immunol.* **89,** 505–510.
124. Cohen, N. *et al.* (1985). The ontogeny of allo-tolerance and self-tolerance in larval *Xenopus laevis.* In *Metamorphosis*, Balls, M. and Bownes, M.E. (eds). Clarendon Press, Oxford, pp. 388–419.
125. Du Pasquier, L. and Chardonnens, X. (1975). Genetic aspects of the tolerance to allografts induced at metamorphosis in the toad *Xenopus laevis. Immunogenetics* **2,** 431–440.
126. DiMarzo, J.J. and Cohen, N. (1982). Immunogenetic aspects of *in vivo* allotolerance induction during the ontogeny of *Xenopus laevis. Immunogenetics* **16,** 103–116.
127. Arnall, J.C. and Horton, J.D. (1987). *In vivo* studies on allotolerance perimetamorphically induced in control and thymectomized *Xenopus. Immunology* **62,** 315–319.
128. Horton, J.D. *et al.* (1989). Attempts to break perimetamorphically induced skin graft tolerance by treatment of *Xenopus* with cyclophosphamide and interleukin-2. *Transplantation* **47,** 883–887.
129. Nakamura, T. *et al.* (1987). Tolerance induced by grafting semi-allogeneic adult skin to larval *Xenopus laevis*: possible involvement of specific suppressor cell activity. *Differentiation* **35,** 108–114.
130. Manning, M.J. and Botham, P.A. (1980). The *in vitro* reactivity of lymphocytes in embryonically-induced transplantation tolerance. In *Development and Differentiation of Vertebrate Lymphocytes*, Horton, J.D. (ed.). Elsevier, Amsterdam, pp. 215–226.
131. Du Pasquier, L. and Bernard, C.C.A. (1980). Active suppression of the allogeneic histocompatibility reactions during the metamorphosis of the clawed toad *Xenopus. Differentiation* **16,** 1–7.
132. Barlow, E.H. and Cohen, N. (1983). The thymus dependency of transplantation allotolerance in the metamorphosing frog *Xenopus laevis. Transplantation* **35,** 612–619.
133. Ruben, L.N. *et al.* (1989). Amphibian metamorphosis: an immunologic opportunity. *Bioessays* **10,** 8–12.
134. Marx, M. *et al.* (1987). Compromised T-cell regulatory functions during anuran metamorphosis: the role of corticosteroids. In *Developmental and Comparative Immunology*, Cooper, E.L., Langlet, C. and Bierne, J. (eds). Alan R. Liss, New York, pp. 129–140.
135. Flajnik, M.F. *et al.* (1987). Changes in the immune system during metamorphosis of *Xenopus. Immunol. Today* **8,** 58–64.
136. Clothier, R.H. *et al.* (1992). Neuroendocrine regulation of immunity: the effects of noradrenaline in *Xenopus laevis*, the South African clawed toad. *Int. J. Neuroscience* **62,** 123–140.
137. Nagata, S. (1986). Development of T lymphocytes in *Xenopus laevis*: appearance of antigen recognized by an anti-thymocyte mouse monoclonal antibody. *Dev. Biol.* **114,** 389–394.
138. Koniski, A.D. and Cohen, N. (1991). Reproducible proliferative responses of axolotl lymphocytes cultured with phytohaemagglutinin-P (PHA). *Dev. Comp. Immunol.* **15** (Suppl. 1), S55.
139. Kerfourn, F. *et al.* (1992). T-cell-specific membrane antigens in the Mexican axolotl (urodele amphibian). *Dev. Immunol.* **2,** 237–248.
140. Sato, K. *et al.* (1993). Evolution of the MHC: iso-

lation of class II β-chain cDNA clones from the amphibian *Xenopus laevis*. *J. Immunol.* **150,** 2831–2843.

141. Flajnik, M.F. *et al.* (1991). Evolution of the MHC: molecular cloning of the major histocompatibility complex class I from the amphibian *Xenopus*. *Proc. Natl. Acad. Sci. USA* **88,** 537–541.
142. Flajnik, M.F. *et al.* (1991). Which came first, MHC class I or class II? *Immunogenetics* **33,** 295–300.
143. Flajnik, M.F. *et al.* (1991). Reagents specific for MHC class I antigens of *Xenopus*. *Amer. Zool.* **31,** 580–591.
144. Fellah, J.S. *et al.* (1992). Transient developmental expression of IgY and secretory component like protein in the gut of the axolotl (*Ambystoma mexicanum*). *Dev. Immunol.* **2,** 181–190.
145. Ibrahim, B. *et al.* (1991). Analysis of T cell development in *Xenopus*. *Fed. Proc.* **5,** 7651.
146. Fellah, J.S. *et al.* (1993). Conserved structure of amphibian T-cell antigen receptor β chain. *Proc. Natl. Acad. Sci. USA,* **90,** 6811–6814.
147. Du Pasquier, L. and Robert, J. (1992). *In vitro* growth of thymic tumour cell lines from *Xenopus*. *Dev. Immunol.* **2,** 295–307.
148. Rennie, J. (1993). No snake oil here. *Sci. Am.* **268,** 136–138.
149. Bechtold, T.E., Smith, P.B. and Turpen, J.B. (1992). Differential stem cell contributions to thymocyte succession during development of *Xenopus laevis*. *J. Immunol.* **148,** 2975–2982.
150. Rollins-Smith, L.A., Blair, P.J. and Davis, A.T. (1992). Thymus ontogeny in frogs: T-cell renewal at metamorphosis. *Dev. Immunol.* **2,** 207–213.
151. Horton, J.D., Horton, T.L. and Ritchie, P. (1993). Incomplete tolerance induced in *Xenopus* by larval tissue allografting: evidence from immunohistology and mixed leucocyte culture. *Dev. Comp. Immunol.* **17,** 249–262.
152. DuPasquier, L., Wilson, M. and Robert, J. (1994). The immune system of *Xenopus*: special focus on B cell development and immunoglobulin genes. In *The Biology of Xenopus*, Tinsley R.C. and Kobel, H.R. (eds.), Oxford University Press, in press.
153. Haynes, L. and Cohen, N. (1993). Further characterization of an interleukin-2-like cytokine produced by *Xenopus laevis* T lymphocytes. *Dev. Immunol.,* **3,** 231–238.
154. Hsu, E. and Du Pasquier, L. (1992). Changes in the amphibian antibody repertoire are correlated with metamorphosis and not with age or size. *Dev. Immunol.,* **2,** 1–6.

5

REPTILES AND BIRDS

R.D. Jurd

Department of Biology, University of Essex, Colchester, UK

5.1 INTRODUCTION

Reptiles and birds, together with mammals, are amniotes. Amniotes possess an amnion, a foetal membrane which normally fuses with the chorion to form a sac which contains a suspending amniotic fluid. Also present in amniotes is another foetal membrane, the allantois. This is formed from a diverticulum of the hindgut, and in reptiles and birds it provides a store for metabolic wastes. Later the outer mesodermal surface of the allantois fuses with the inner mesodermal surface of the chorion to form a richly vascularized chorioallantois which facilitates embryonic respiration. These features together allow the development of a closed, waterproof 'cleidoic' egg, and permit in amniotes a development and existence independent of an aquatic environment such as is not possible for amphibians (Figure 5.1).

The Reptilia is a rather heterogeneous class which now occupies a relatively minor place in the land fauna compared with that which it occupied in the Mesozoic era. Birds and mammals evolved from respectively different groups of reptiles. Modern reptiles are all poikilotherms but homoiothermy has evolved, presumably independently, in the birds and in the mammals. The birds have retained the egg-laying habit of the reptiles (although some reptiles, for example adders, are viviparous); with the exception of the monotremes, the mammals have evolved viviparity.

The reptiles evolved from labyrinthodont amphibians in the middle of the Carboniferous

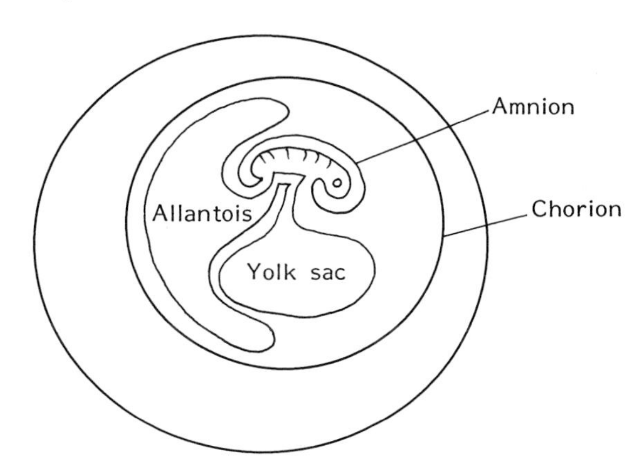

Figure 5.1. Avian foetal membranes.

Immunology: A Comparative Approach. Edited by R.J. Turner

period when there were profound changes in the terrestrial habitat. Swamps variously formed and dried out and land floras became established. The seasonal variations in climate characterized by periods of drought and flooding and the postulated prior evolution of the amniote egg must have favoured an increasingly terrestrial existence. There is a good fossil sequence from the Labyrinthodonta to the early reptiles, and forms such as *Seymouria* and *Solenodonsaurus* are continually having their status as amphibians or reptiles revised. Palaeontologists and zoologists classify reptiles on the basis of post-orbital vacuities in the dermal bone casings of the skull: none (anapsid), one (synapsid, parapsid and euryapsid) and two (diapsid).

The earliest reptiles were the Cotylosauria or stem reptiles: these anapsid reptiles are totally extinct. From the cotylosaurs it is believed that several reptile lines diverged (Figure 5.2).

One such line, the anapsid Chelonia, is represented today by about 250 species of tortoises, terrapins and turtles which demonstrate specialized body armour and jaw features but which are, in many other respects, relatively primitive.

Other lines, now extinct, gave rise to Mesozoic parapsid and euryapsid marine reptiles such as nothosaurs, mososaurs, plesiosaurs and the dolphin-contoured ichthyosaurs. Another extinct line, the Synapsida, is extremely important because from it evolved, through a sequence well-documented by fossils, the earliest mammals, which appeared in the early Triassic period.

The diapsid Lepidosauria is the reptile subclass

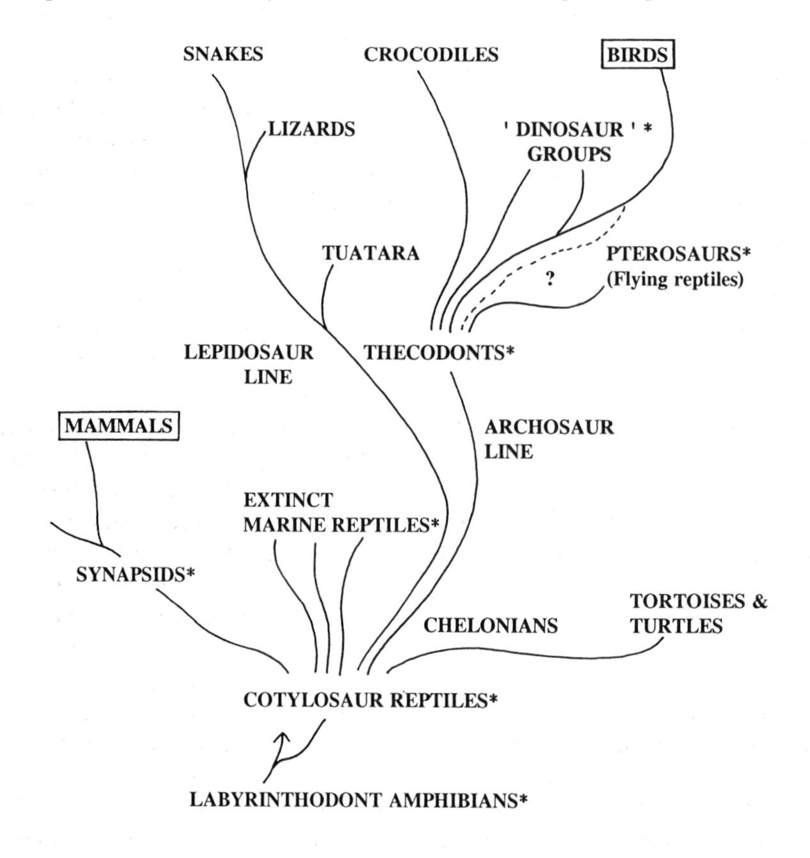

Figure 5.2. A simplified reptile family tree.

which contains most living forms. The primitive Rhynchocephalia is today represented solely by the tuatara, *Sphenodon punctatum*, a very rare 'living fossil' found on the Cook Straits islands off the South Island of New Zealand. Better known are the Lacertilia (the lizards) which have radiated widely, the Amphisbaenia (a small group of burrowing, lizard-like reptiles) and the Ophidia (the snakes). There are about 6000 living species of lepidosaurs and most studies of reptilian immunity have been carried out on members of this subclass.

The final reptile subclass is the diapsid Archosauria or 'ruling reptiles'. The ancestral archosaurs were the thecodonts of the Triassic. From the thecodonts may have arisen four orders. The Saurischia and the Ornithischia, grouped together (arguably artifically) as the 'dinosaurs', were the dominant land fauna of the Mesozoic era. They reached their apogee in the Cretaceous before dying out relatively suddenly at the end of that period in somewhat mysterious circumstances. Contemporary with the rise and demise of the 'dinosaurs' were the Pterosauria, the flying archosaurs which included the pterodactyls. The last order is the Crocodilia; some 22 species of crocodiles, caimans and alligators are the only archosaurs extant today.

The birds were once thought to have evolved from thecodonts too, and some still hold this view[1]. However, recent palaeontological finds show near identity in skeletal structure between groups of relatively small, bipedal, carnivorous saurischian 'dinosaurs' (such as the upper Jurassic coelurosaur, *Compsognathus*, and the early Cretaceous coelurosaur, *Deinonychus*) and the earliest known bird with feathers, *Archaeopteryx lithographica*, from the late Jurassic. Many workers therefore feel that these close similarities point to the origin of the birds lying with the bipedal saurischian carnivores (the theropods) (Figure 5.3). This debate is intensified by much discussion during recent years concerning whether the 'dinosaurs' (or at least some of them) exhibited a homoiothermic physiology. Bakker[2] provocatively suggests that the 'dinosaurs' and pterosaurs should be taken out of the class Reptilia and placed with the birds in a newly erected class, the Dinosauria. It might not be unreasonable, if this view is sustained, to speak of birds as 'feathered dinosaurs' and, indeed, Bakker asks his readers 'squarely [to] face the dinosaurness of the birds and the birdness of the Dinosauria. When the Canada geese honk their way northward, we can say: "The dinosaurs are migrating, it must be spring!"'

This view of reptile and bird classification shows that the Reptilia is a very diverse class and also that the birds are probably very closely related to one particular reptile subclass. Certainly, in terms of systematic zoology, the birds are a very homogeneous group. All modern birds are in one superorder, the Neognathae, which comprises nearly 30 orders. The flightless ratites (such as ostriches, emus and kiwis) were formerly considered more primitive than the flighted birds, the carinates. Young[3] suggests, however, that the ratites are descended from flighted birds and are not a natural group. Despite this, there is some evidence from studies on the embryology of ankle bones[4] for the traditional view that the ratites are a more primitive group.

The immunology of reptiles has not been studied in such depth as that of mammals and birds, or indeed of fishes and amphibians. There are several reasons for this. Reptiles are of relatively little economic importance and there is no impetus for research from interests comparable to fish-farming or poultry-keeping to promote studies of their immune systems. Generally reptiles are difficult to maintain in captivity and the conditions under which they are kept can affect profoundly their immune responses. Many reptiles are quite rare and workers are reluctant to work with them because of very proper conservation considerations; some reptiles are venomous (or dangerous in other ways) causing further problems best left to readers' imaginations! There are currently no real inbred strains of reptiles available for use by immunologists. Consequently the database of foundation information on which to build more advanced work is small. Reptile immunology currently consists of a disparate body of information from many species investigated in a hetero-

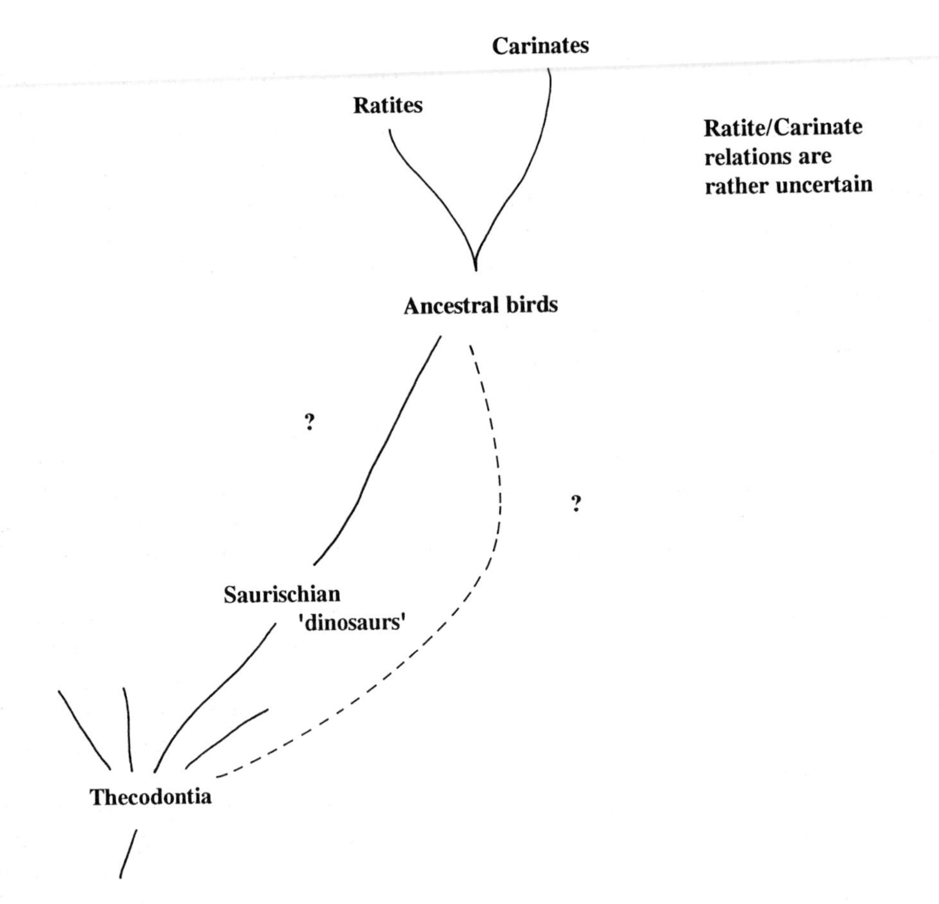

Figure 5.3. A view of bird phylogeny. The birds may be a polyphyletic group, several bird lines having evolved from reptile ancestors. There is debate about the relationship of ratite birds to carinates, and about which is ancestral or more 'primitive'.

geneous class of tetrapods. Nevertheless an overall picture of immunity in Reptilia is emerging.

Immunity in birds has been studied much more extensively, albeit mainly in a few species of economic importance such as the chicken, turkey and quail. Inbred strains of birds are becoming available, and the relative ease of breeding and maintenance of the birds has facilitated in-depth studies of developmental aspects of avian immunology and has permitted experimental intervention. The birds have proved particularly useful in helping our understanding of B-lymphocyte immunity.

From a zoologist's or an evolutionary biologist's point of view it is unfortunate that reptiles have been so little studied: a knowledge of reptilian immunity could provide valuable indicators to the evolution of features in the mammalian immune response (although it must always be remembered that modern reptiles may differ significantly in their immune systems from forms which were ancestral to mammals). Certainly reptiles are nearer to mammals than are birds, a point to be borne in mind when considering extrapolating avian findings to mammals. The zoological solecism of assuming that mammals evolved from birds is not unknown in immunological writing! Similarities between features of the avian and mammalian immune systems may indicate either a common ancestry for those features to be sought also in reptiles, or a remarkable degree of convergent evolution.

5.2 REPTILIAN IMMUNITY

5.2.1 The Cells of the Immune System

Although there is some confusion about how certain cell types, particularly granulocytes, are identified and named and to what extent they are homologous with similarly identified and named cells in mammals, it does seem that the Reptilia possess a full repertoire of leucocytes which mediate immunity. The topic is exhaustively reviewed elsewhere[5]. Suffice it here to say that eosinophilic, heterophilic and basophilic granulocytes are identifiable, as are cells in the monocyte-macrophage series. Lymphocytes and their descendants are seen. Numbers and relative proportions of different leucocytes reported in different reptile species vary greatly.

Granulocytes

'Heterophils' are phagocytic[6] and are associated with inflammatory responses to infective agents[7]: thus they parallel mammalian polymorphonuclear neutrophils. In the snapping turtle, *Chelydra serpentina*, 'basophils' with surface immunoglobulin (Ig) have been implicated in histamine release[8]. 'Eosinophils' are described in a variety of reptile species[9].

Monocytes and Macrophages

Mobile, phagocytic monocytes and macrophages may comprise over 20% of the circulating leucocytes in some snakes[10], although proportions are less in other reptiles[11]. The proportions of these cells, unlike some leucocytes, exhibit little variation with season[12].

Cells of the monocyte series are seen in inflammatory and granulomatous reactions to parasitic and bacterial infections[13,14]. Following challenge with antigen, monocytes and macrophages are prominently visible in the sinus walls and red pulp cords of the spleen of the turtle *C. serpentina*[15]. Monocytes show cytophilic surface Ig in antigen-specific reactions when *C. serpentina* is challenged with sheep erythrocytes[16]. The capacity of reptilian dendritic cells to trap antigen complexes has been demonstrated in the spleen of the snake, *Python reticulatus*[17], but whether reptile macrophages play a comparable role to those in mammals as antigen-presenting cells is not yet proven.

Lymphocytes

The lymphocyte is the most prevalent leucocyte in the peripheral blood of reptiles, comprising over 80% of the white cell count in some species[10,18]: both large and small lymphocytes can be discerned[10,18,19]. Overall numbers of lymphocytes vary according to sex[12], age[10], nutritional status[20], infection history[7,13] and particularly season of the year[12,21–24] where high lymphocyte counts, especially in the spleen and peripheral blood of tropical species and throughout the lymphoid system of temperate species, are associated with the summer months[21–25].

The stem cells which give rise to lymphocytes first appear in the embryonic yolk sac; the haemopoietic site then changes to the foetal spleen and then to the bone marrow which largely supplants the spleen in adult lizards[26]. Stem cells develop in the thymus to mature into T-lymphocytes. Embryonic T-cells bear a surface marker not found on adult T-cells; further maturation of the T-lymphocytes occurs in the spleen[27]. The most likely organ wherein B-lymphocyte differentiation occurs is the embryonic liver[28].

A complete range of effector lymphoid cells can be identified, including lymphoblasts and plasma cells[30], although these are somewhat rare in peripheral blood. Lymphocyte heterogeneity is discussed below.

5.2.2 Lymphoid Tissues

As vertebrates have become more 'advanced' and sophisticated in their physiologies their lymphoid tissues have tended to become more discrete and well-defined. Most prominent among reptile lymphoid tissues and organs are the thymus and the spleen; other lymphoid tissue is rather diffusely distributed.

Thymus

The thymus is normally markedly lobulated. Each lobule has a defined cortex with densely-packed lymphocytes and a medulla with fewer lymphocytes, together with many accessory cells of the monocyte-macrophage series and large myoid cells. The thymus differentiates early in development[31] and, as in other vertebrates, it involutes permanently as the reptile ages; it also involutes temporarily during disease or winter[20–24,32,33].

Spleen

The spleen in reptiles contains red and white pulp, although the distinction between the two is not always well defined. The white pulp has sheaths of lymphocytes surrounding the central arterioles and ellipsoids; discrete areas for functional B- or functional T-lymphocytes are not identifiable. Ellipsoids consist of a reticular tissue envelope enclosing cuboidal epithelium and a capillary: these are surrounded by a periellipsoidal lymphocyte sheath(s). Within the periellipsoidal sheaths immune complexes are trapped by dendritic cells. Follicles and germinal centres are not seen[34]. Antigen will stimulate lymphoproliferation in the periarteriolar sheaths: lymphoblasts migrate to the red pulp and here secrete antibody[15].

Other Lymphoid Tissue

Densely aggregated regions of packed lymphocytes are seen associated with all regions of the gut and in strategic sites such as the urinary system and the bladder[21–24,35–37]. In a number of Egyptian lizards and snakes it has been noted that the size and prominence of such lymphoid aggregations varies with the season of the year, declining during the winter[21–24,31]. Present at the hind end of the gut in most reptiles is a 'cloacal complex' which contains lymphocytes. However, although it is tempting to see this organ as an anatomical and functional homologue of the avian bursa of Fabricius, such a parallel has never been demonstrated[38]. Work on thymocyte ontogenesis in the lizard *Calotes versicolor*[39] suggests that the cloacal complex is a secondary lymphoid tissue within the gut-associated lymphoid tissue series: this would certainly be more in accordance with its embryological origins.

5.2.3 Lymphocyte Heterogeneity

A measure of how closely the reptile immune response parallels that in mammals might be provided by a consideration of the functional and morphological heterogeneity of the lymphocyte population. A rabbit antiserum to *C. versicolor* thymocytes was more effective against thymocytes than splenocytes and bone marrow and peripheral blood lymphocytes[40]. Similar results were obtained for other reptiles including the snapping turtle *Chelydra serpentina*[41] and the snake *Spalerosophis diadema*[42]. Thymectomy, not surprisingly, reduces the number of anti-thymocyte serum-sensitive cells[43]. From the *Calotes* and *Chelydra* work it seems that the reptiles investigated have T-lymphocytes (or functional T-lymphocytes) originating in the thymus and that these cells possess two antigens, one specific for thymocytes, and the other for thymocytes and splenocytes of the lymphocyte series. This is supported by studies with the lizard *Chalcides ocellatus* where such T-lymphocyte differentiation is first seen in the embryo[27].

Several workers have demonstrated surface immunoglobulin-positive populations of lymphocytes in reptiles, using xenogeneic anti-immunoglobulin sera: species investigated include the snake *S. diadema*[42], lizards such as *Agama stellio*[44] and the turtle *C. serpentina*[41]. Such studies showed surface immunoglobulin-positive lymphocytes in the spleen and peripheral blood but very few such cells in the thymus.

Investigation of lymphocyte heterogeneity is also possible using differential responses to mitogens, although such use of mitogens for fine discrimination of lymphocyte subsets has been likened to studying archaeology with a bulldozer! Peripheral blood lymphocytes from the archosaur *Alligator mississippiensis* were fractionated on glass-wool columns: non-adherent cells responded to the mitogens phytohaemagglutinin and concanavalin A (PHA and Con A: T-lymphocyte mitogens in mice); adherent cells responded to the murine B-lymphocyte mitogen *Escherichia coli* lipopolysaccharide (LPS)[45,46]. It is assumed that B-cells are adherent to glass-wool whereas T-cells are not. An anti-surface immunoglobulin antiserum negated the LPS response. Splenocytes from the lizards *Calotes versicolor* and *Lacerta viridis* responded to PHA, Con A, LPS and pokeweed mitogen (PWM), although stimulation indices were lower than for mammalian lymphocytes, even at high mitogen concentrations; *Calotes* responded more poorly than did *Lacerta*[47]. Lymphocytes from female *Psammophis sibilans* responded more vigorously to mitogens than did male cells[48]. In *Chalcides ocellatus* Con A responsiveness seems to develop early in embryogenesis[49].

Migration inhibition studies with nylon-wool fractionated *Calotes* lymphocyte subpopulations confirm the lymphocyte heterogeneity indicated by mitogen work[50,51]. Whether such populations of lymphocytes are functional analogues or homologues of mammalian lymphocyte populations is a key question, and the precise nature of such surface markers that they possess is not yet ascertained.

5.2.4 Antibody-mediated Immunity

All reptile groups which have been investigated exhibit antibody responses to antigens ranging from bacteria and xenogeneic erythrocytes to macromolecules such as proteins; foundation work is reviewed by Cohen[52]. All responses seem to be T-lymphocyte dependent[31]. Immunization with sheep erythrocytes leads to antigen-specific, rosette-forming cells in the *Calotes versicolor* spleen within 3 days; these peak by 7 days[53]. Antibody-secreting, plaque-forming cells are also discernible by 3 days and peak at day 14[54]. There is, however, great variation in the kinetics of reptile antibody production as reported in the literature.

Accelerated secondary antibody responses have been described in the lizard *Tiliqua rugosa*[55], although this phenomenon is apparently not demonstrable in *C. versicolor*[54].

5.2.5 Immunoglobulins

Two immunoglobulin (Ig) classes have been reliably described in Reptilia[56]. There is an 18S IgM and a 7S non-IgM Ig, variously designated 'IgG-like', IgRAA or IgY. IgY seems to be the term most accepted today. IgM and IgY have been found in all reptiles investigated, including the primitive lepidosaur *Sphenodon punctatum*[57]. In chelonians a 5.7S Ig has been described[58,59]; this 5.7S Ig probably contains a deleted IgY heavy chain. Such Igs with deleted heavy chains have been noted in other vertebrates including teleosts, dipnoans and mammals. Antigenic comparisons suggest that reptilian IgY may be the homologue of mammalian IgA[60].

In the tortoise *Agrionemys horsfieldii* 13–23% of peripheral blood lymphocytes bear surface IgM and 5–8% also bear IgY; splenic lymphocytes show a higher proportion of surface Ig-bearing cells[61].

Both IgM and IgY have antibody function, although some antigens seem capable of eliciting only an IgM response[55]. As in other poikilothermic vertebrates the low molecular weight antibody tends to supplement rather than supplant the 19S IgM antibody[62]. Antibody-secreting cells are detectable by plaque assays: such cells are found in the spleen, bone marrow and peripheral blood but not in other lymphoid tissues such as the thymus and the cloacal complex[53,54]. Splenectomy suppresses plaque-forming cells in the peripheral blood and the bone marrow in *Calotes versicolor* immunized with sheep erythrocytes, suggesting that the spleen is the main site for antibody-producing cell differentiation.

5.2.6 Immunoglobulin Genes

Studies on the crocodilian *Caiman crocodylus*[63,64], suggest that there is a 65–70% nucleotide sequence identity and a large degree of organizational homology when immunoglobulin V_H genes from the caiman and the mouse are compared. Sequences are 60–65% identical at the amino-acid level. Some parts of the V_H genes seem to be very highly conserved, suggesting minimal evolutionary change comparing the caiman and the mammal, which must have shared a cotylosaurian common ancestor as long ago as the Permian.

The turtle *Chelydra serpentina* has been found to possess a large V_H gene family[65], although many genes may be currently non-functional[66].

5.2.7 Cell-mediated Immunity

A variety of cell-mediated immune phenomena have been reported in a wide range of reptile species.

Graft Rejection

Until a decade ago it was generally considered that the processes of allogeneic or xenogeneic graft rejection were rather slow in reptiles. The

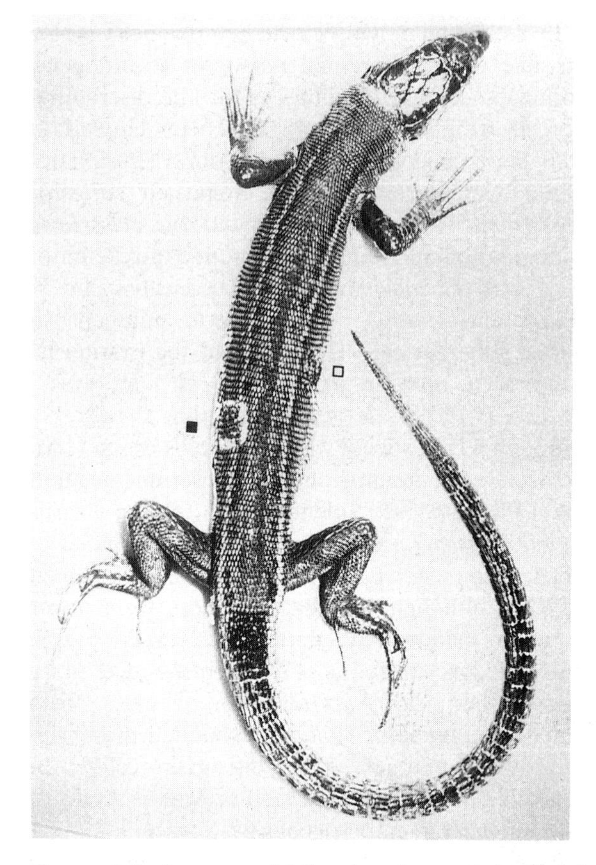

Figure 5.4. *Lacerta viridis*, showing autograft (□) and allograft (■) of skin/scales 42 days post-graft. Lizards kept indoors at 26 ± 4°C.

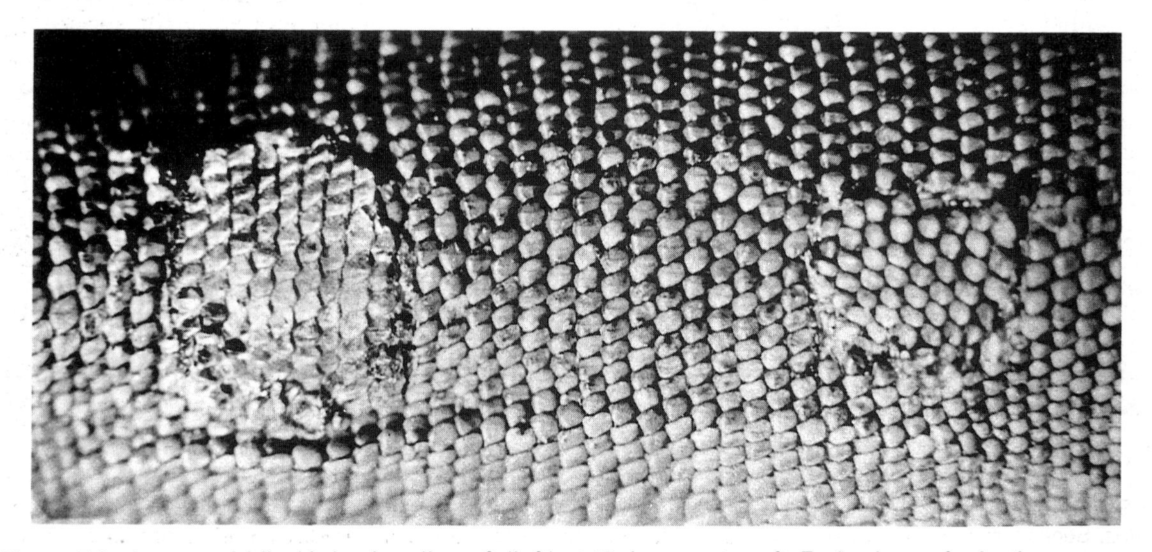

Figure 5.5. *Lacerta viridis* skin/scales allograft (left) at 48 days post-graft. Early signs of rejection are seen. Control autograft on right. Lizards kept indoors at 26 ± 4°C.

Figure 5.6. *Lacerta viridis* allograft rejection 201 days post-graft. Lizard kept indoors at 26 ± 4°C.

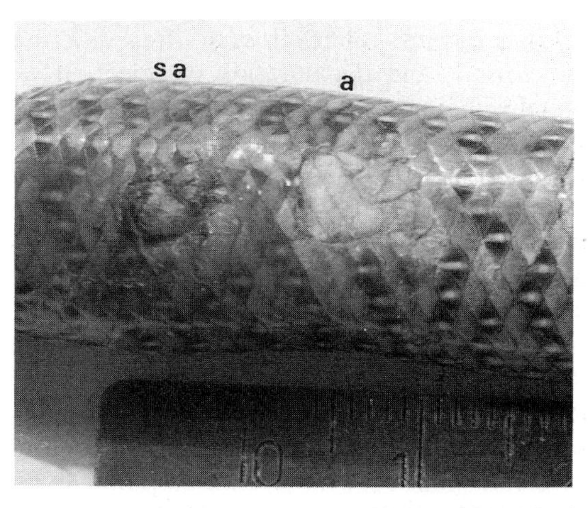

Figure 5.7. *Chalcides ocellatus*. Healthy skin autograft (a) and rejected second-set allograft (sa), the latter showing marked erythema. Photograph by courtesy of M. El Rushdy, A. Afifi and R. El Ridi, University of Cairo.

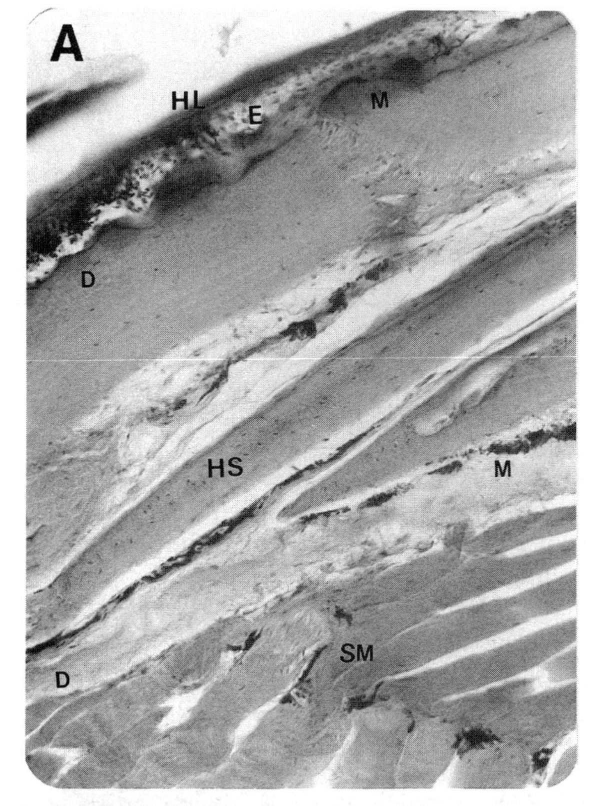

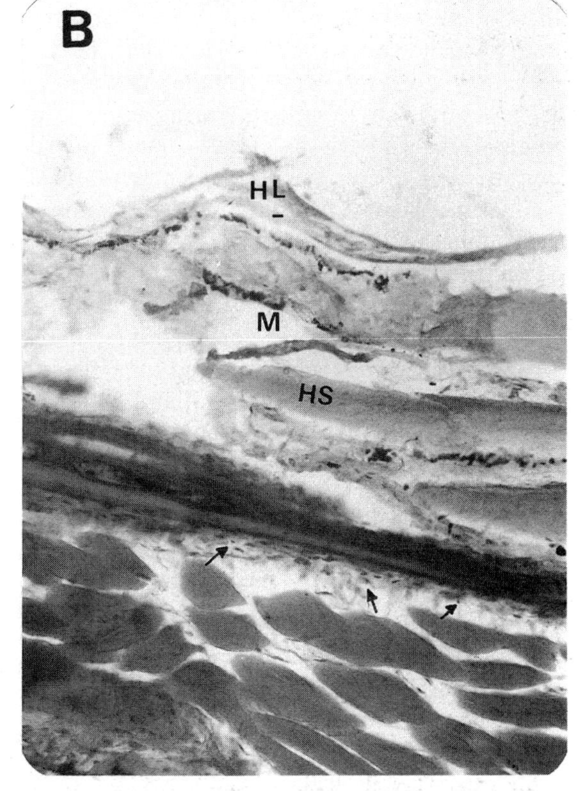

Figure 5.8. (A) *Chalcides ocellatus*. Section of a 5-day autograft showing structure of normal skin. Epidermis (E) with horny layer (HL) and dermis (D). Melanophores (M) lie in both the epidermis and dermis. Horny scales (HS) apparent. The dermis contains smooth muscle bundles (SM). Haematoxylin and eosin. (B) *Chalcides ocellatus*. Section of a 5-day first-set allograft of skin. Partial destruction of horny layer (HL), melanophores (M), and disturbance in the structure of the horny scale (HS). Mononuclear cell infiltration seen (arrows). Haematoxylin and eosin. Photographs by courtesy of M. El Rushdy, A. Afifi and R. El Ridi, University of Cairo.

kinetics of graft rejection were often described as 'chronic' and the duration of rejection was noted as being very variable[67,68]. Xenografts are rejected more speedily than are allografts, and second-set grafts are rejected more quickly than are first-set grafts. As with most immune phenomena in poikilotherms, the rejection process is very temperature-dependent, although there may be two separate processes: one, antigen recognition, being less temperature-sensitive than a second, rejection sequence[69]. Such chronic graft rejection (Figures 5.4–5.6) kinetics seemed puzzling in view of the acute or subacute rejections seen in anuran amphibians and teleost fishes[70]. It is possible that reptiles used in experiments were collected from local, inbred populations; alternatively it is possible that reptiles lack a major histocompatibility complex—this is discussed below. A third possibility relates to the fact that the conditions under which reptiles are maintained in the laboratory are often very dissimilar to those of their natural environments with respect to temperature and light cycles, local microhabitats and diets: such 'immunoecological' considerations may profoundly affect immune performance[47,68,175].

The picture outlined above has been complicated by findings during the 1980s in Cairo that acute allograft rejection is demonstrable in the snake, *Psammophis sibilans*, and the lizard, *Chalcides ocellatus*[71,72] (Figures 5.7 and 5.8). The implications of this are discussed below.

Other Indicators of Cell-mediated Immunity

Migration inhibition assays show that sensitized splenocytes from *Calotes versicolor* given skin allografts migrate from capillary tubes less effectively in the presence of donor spleen cell extracts, and that this migration inhibition is highly spleen cell specific[73].

The cells which mediate migration inhibition seem to be T cells[50]. Similar T-cell involvement is seen when migration inhibition studies are performed using spleen cells from *C. versicolor* injected with sheep erythrocytes. TNP-ovalbumin immunization results in inhibition of splenocyte migration in the presence of TNP-ovalbumin, but not in the presence of the TNP hapten conjugated to another carrier: here a classic helper T-cell effect is seen[74].

Two-way mixed lymphocyte reactions (MLRs) have been demonstrated with *Alligator mississippiensis* cells[46]. In *Calotes versicolor* and *Lacerta viridis* MLRs with splenocytes have also been seen, although the *Calotes* response is much less marked than that of *Lacerta*[47]. Vigorous MLRs are observed in the snake, *Psammophis sibilans*, and the lizard, *Chalcides ocellatus*[72,75]. Graft-versus-host (GVH) phenomena have been described in the turtle, *Chelhydra serpentina*, where, for example, adult allogeneic splenocytes can induce splenomegaly in explants of immature spleens[76]. Delayed hypersensitivity responses to the purified protein derivative of *Mycobacterium tuberculosis* have been seen in the lizard *Tarentola annularis*[77].

5.2.8 Major Histocompatibility Complex

The foregoing findings suggest a wide repertoire of T-lymphocyte activity in reptiles from all subclasses investigated. A key question remains: do reptiles possess a major histocompatibility complex (MHC)?

The earlier work showing chronic allograft rejection of skin grafts might point to the existence of only minor histocompatibility antigens in reptiles. However, the recent studies by the Cairo group of comparative immunologists, mentioned above, suggest that acute allograft rejection, strong MLRs and delayed hypersensitivity reactions in snakes and lizards favour the existence of antigens encoded by an MHC. Evidence supporting the latter view is obtained from normal lymphocyte transfer reaction studies where the kinetics of localized GVH reactions in *Tarentola annularis* resemble those in mammals[77]. Chronic skin allograft rejection may reflect the nature of reptile skin: allografting of other tissues to appropriate sites may be instructive[5].

The seasonal dimension may be very pertinent too. Corticosteroids are known to have a profound immunosuppressive effect, and they are known to occur at high levels in winter in *C. ocellatus*[78,79]. Such endocrine immunomodulation may have led to misleading results in early herpetoimmunological studies. Reptilian molecules similar to MHC molecules found in mammals, birds and amphibians have been described[80]. The case for a reptile MHC is thus strong but not proven because studies are hindered, partly because of a lack of inbred reptile strains and appropriate immunogenetic studies, and partly because of endocrinologically-mediated seasonal variations in the immune status of reptiles[81,82].

A monoclonal antibody to chicken class II MHC products has been shown to precipitate a corresponding caiman class II molecule. Chicken cDNA sequences for beta-microglobulin and class I MHC genes hybridize well with reptile DNA but show low homology with their mammalian counterparts, perhaps emphasizing the bird-archosaur relationship[83].

5.2.9 Cytokines

Soluble, physiologically-active mediators of immune function are assuming an increasingly important role in our understanding of immune phenomena. There is indirect evidence for the production of at least one pyrogenic factor by leucocytes from the lizard *Dipsosaurus dorsalis*[84,85]: whether such pyrogens are homologous or are identical to interleukin-1, TNF or interleukin-6 is not known[86]. Evidence for an interleukin-2 (IL-2) comes from work by El Ridi *et al.*[87]. Spring and autumn thymocytes and splenocytes from the snake *Spalerosophis diademia* respond well to Con A: supernatants from such cultures enhance proliferation of summer and winter thymocytes and will also favour growth of Con A-generated spleen cell lymphoblasts from autumn and spring spleens, even in the absence of Con A. Analysis of the active components in the culture supernatants shows two cytokine forms, which may be IL-2 homologues.

5.2.10 Hypersensitivity Reactions and Autoimmunity

A secondary challenge with an antigen can result in an immune reaction of such vigour that tissue damage results. This is known as an immune hypersensitivity reaction. Four (sometimes five) types of such reaction are described in mammals; this may be seen as the immune system over-reacting, the 'cost' of efficient immunity. In Reptilia little work has been done on hypersensitivity reactions but Type IV delayed hypersensitivity (a T-lymphocyte-mediated phenomenon) has been reported in the lizard *Tarentola annularis*[77]. Basophils with surface immunoglobulin have been described in the turtle *Chelhydra serpentina*[8]; these granulocytes are implicated in histamine release, implying parallels with Type I, IgE-mediated anaphylactic immediate hypersensitivity in mammals. Complement has been described in several species of snake[88], suggesting that complement-mediated Type II cytotoxic hypersensitivity may operate in these ophidians. Complement is also described in other reptile groups[89]. *Lacerta viridis* splenocytes will mediate antibody-dependent cellular cytotoxicity[90] in a chicken erythrocyte/anti-chicken erythrocyte system: this phenomenon has also been seen as an *in vitro* correlate of Type II hypersensitivity (Figure 5.9). Clearly, further investigations and experiments on hypersensitivity phenomena could be invaluable to ascertain the efficiency (or over-efficiency) of the reptile immune response.

A further indicator of the functioning of the reptilian immune system could be obtained from probing autoimmune phenomena; however, no such work seems to have been carried out.

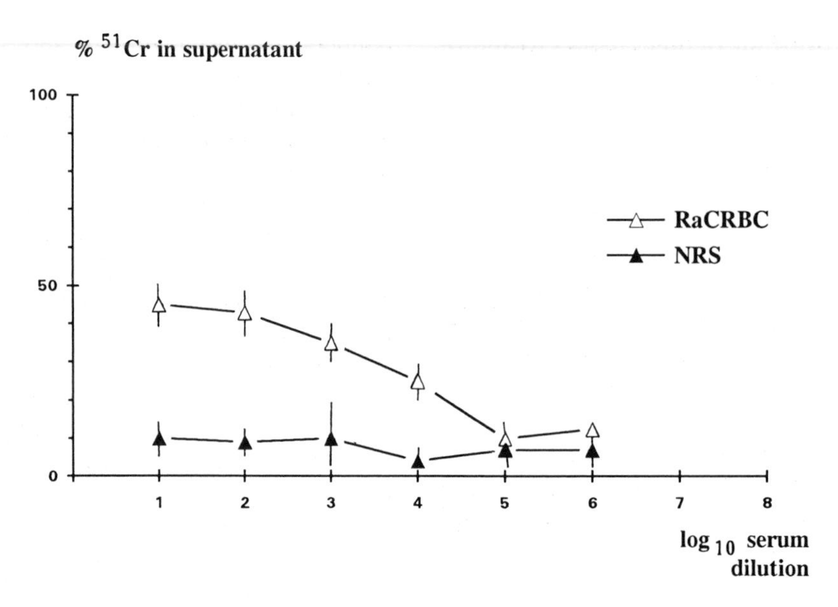

Figure 5.9. Chromium-release antibody-dependent cellular cytotoxicity assay using *Lacerta viridis* spleen cells in a system with chick erythrocytes as target cells and either a rabbit anti-chick erythrocyte serum (RaCRBC) or normal rabbit serum (NRS). See reference 90 for further details.

5.3 AVIAN IMMUNITY

Compared with the Reptilia, much more information is available about immunity in the Aves, the birds. Worthy of particular note are (i) the primary lymphoid organ, the bursa of Fabricius, which prepares B-lymphocytes and which is peculiar to the birds; (ii) the well-investigated major histocompatibility complex, known as the B system in chickens, whose antigens are conveniently expressed on the nucleated erythrocytes; and (iii) the development of interspecific chimaeras between the quail and the chicken[91], which have proved an exciting model for understanding aspects of avian immune system ontogenesis. Avian immunology is assisted by the availability of inbred strains and lines, including a number of specialities such as an obese chicken line which develops an autoimmune thyroiditis.

5.3.1 The Cells of the Immune System

A full range of leucocytes has been described in birds[92,93].

Granulocytes

'Heterophils' (or pseudoeosinophils) are the most common granulocytes (26% of peripheral blood leucocytes[94]) in the blood of chickens. They possess polymorphic nuclei, and three types of cytoplasmic granules have been identified. Interestingly the heterophils appear to lack peroxidase activity. Their role is bactericidal, being involved in phagocytosis during inflammatory reactions. Eosinophils account for 2.25% of peripheral blood leucocytes; these are most reliably identifiable by their enzyme activity. Eosinophils in birds are thought to have a role in modulating Type IV cell-mediated hypersensitivity reactions[95]. Basophils comprise 2.8% of the peripheral blood leucocyte count: several types of cytoplasmic granules are present, including some similar in structure to those found in human basophils associated with heparin and histamine. Such basophils release histamine and are involved in Type I anaphylactic hypersensitivity reactions[96]. There is a well-defined maturation process for avian granulocytes, as seen by light microscopy[92]; however, analysis of development using differentiation antigens on cell surfaces has yet to be fully explored.

Monocytes and macrophages

Chicken peripheral blood monocytes present an appearance not dissimilar to that characterizing mammalian monocytes. Their tissue equivalent, macrophages, can often be identified by the presence of digested inclusions, such as bacteria, in the cytoplasm. As in mammals, monocytes and macrophages carry major histocompatibility complex (MHC) class I and class II gene products on their surfaces. Monocytes/macrophages from marrow, thymus, spleen, bursa and peripheral blood of the chicken have been examined[97]. Macrophages appeared to be the predominant adherent cell population from these sources. B-L antigens were expressed by 60–80% of the macrophages (except in the thymus where the equivalent figure was 10%).

Monocytes and macrophages are phagocytic. It has been suggested that immunoglobulin receptors are important in the phagocytosis of antibody-coated allogeneic erythrocytes[98]; others deny a role for immunoglobulin or complement in macrophage phagocytosis of erythrocytes[99], although fibronectin or other serum factors may be implicated. T-lymphocyte activation of macrophages may explain their cytotoxic effects on allogeneic erythrocytes from donors previously used to sensitize the host whence the splenic macrophages were derived[100].

Lymphocytes

Both T-lymphocytes and B-lymphocytes are discernible in birds.

T-lymphocytes These are most easily seen packed in the thymic cortex. Like most lymphocytes they are small with a relatively large nucleus. Chicken thymocytes have numerous, small, Feulgen-positive dots when stained using the Feulgen-Rossenbeck technique for DNA; quail thymocytes have a single, large, Feulgen-positive mass and some smaller masses attached to the nuclear membrane[93]. This chicken/quail difference in nuclear histomorphology has provided a helpful marker system.

The absence of congenic bird strains has so far prevented extensive development of alloantisera directed against T-lymphocyte subpopulations and the elucidation of specific T-lymphocyte antigenic markers[101]. General anti-T-cell antisera show about half the splenic lymphocytes to be T-cells in 2–3-week-old chickens. Characteristic, unvarying proportions of T-lymphocytes have been demonstrated in various lymphoid organs from hatching up to 24 weeks[102]. T-lymphocytes were most common in the spleen and the peripheral blood, and were rather rare in Harder's gland. Several subsets of chicken T-lymphocytes have been distinguished using turkey and rabbit xenoantisera and fluorescence-activated flow cytometry[103]. These studies, in conjunction with mitogen response assays and the capacity to elicit graft-versus-host (GVH) reactions, suggest the existence of helper T-lymphocytes and suppressor/cytotoxic T-lymphocytes. This is explored further below.

Reciprocal immunization of inbred chicken lines produced antisera which discerned two T-lymphocyte antigens, styled Th-1 and Ly-4[104,105]. Anti-Th-1 antisera bind to 70–75% of thymocytes, and to a population of large and medium lymphocytes in the peripheral blood. Ly-4^{+} cells constitute only 4% of thymocytes but up to 70% of the peripheral blood lymphocytes, the latter almost exclusively small. A few Th-1^{+}, Ly-4^{+} cells were detected.

Monoclonal antibodies designated CT-1 and CT-1a have been shown to illumine two T-lymphocyte antigenic determinants on one molecule, with more than 90% of chicken thymocytes being positive. The CT-1 antigenic determinant was found on thymus cortex and medulla chicken lymphocytes, but on less than 5% of peripheral blood and spleen lymphocytes, and seemed to be absent from marrow and bursa cells. Interestingly CT-1 reacts with quail thymus cortex lymphocytes only, and the CT-1a antigenic determinants are present on about 5% of 11–12-day chicken embryos, increasing to adult proportions within 4 days[106]. Other studies tend to confirm the existence of subpopulations of T-lymphocytes with an ontogenetic succession of surface antigens and a differential distribution of the subpopulations in the lymphoid organs. Cooper *et al.*[176], reviewing recent studies on chicken lymphocyte development with monoclonal antibodies, remark on the similarities between mammalian and avian T-cell development and function.

B-lymphocytes These are distinguishable by their possession of surface IgM (sIgM): they develop in the bursa of Fabricius. Kincade *et al.*[107] injected anti-μ-chain antibody into 13-day embryos and then bursectomized the newly hatched chickens. The chickens became agammaglobulinaemic, suggesting that most if not all of the 13-day embryo's peripheral B-cells are $sIgM^+$. Seventeen-day *M-1a/M-1b* heterozygous chickens have B-lymphocytes expressing only one allotypic B-cell surface marker per cell, suggesting that allelic exclusion occurs: there is a 1:1 ratio of the two allotypic markers[108,109]. Gilmour *et al.*[104] showed the presence of another alloantigen on B-cells: using EL6 and EL7 inbred chicken lines these workers showed that there exists an autosomal locus *Bu-1* controlling alleles *Bu-1a* and *Bu-1b*.

Lymphocyte Heterogeneity

Evidence from surface marker studies which strongly favours the existence of T- and B-lymphocytes, and of subsets among them, is discussed above. Functional studies support these findings.

Mitogens As for mammals, the mitogens phytohaemagglutinin (PHA) and concanavalin A (Con A) are specific for T-lymphocytes. The responses appear to be under genetic control[110]. Differential PHA responses were obtained for lymphocytes from inbred chickens which were respectively susceptible or non-susceptible to Marek's disease, suggesting the existence of T-lymphocyte subpopulations. Ducks, which in the limited number of studies carried out on these birds, tend to be poor immune performers, have lymphocytes which seem to respond best to two or three mitogens given together[111].

Helper T-lymphocytes Helper T-cell function has been shown using hapten-carrier experiments. Normal birds were immunized with TNP-KLH and cyclophosphamide-bursectomized birds were immunized with BSA. Adoptive transfer experiments were performed: 'normal' splenic B-cells primed to make anti-TNP (the hapten), and 'bursectomized' splenic T-cells sensitized to the carrier KLH were used in these chicken experiments. If either cell type was transferred into bursectomized recipients with TNP-BSA, anti-TNP plaques were not induced. When a mixture of both cell types was transferred anti-TNP plaques were seen[112]. Generally, cell co-operation follows a pattern similar to that in mammals[113].

Suppressor function If 4-month old chickens are irradiated with X-rays, anti-KLH antibody formation is suppressed when bone marrow cells from agammaglobulinaemic chickens are given. Normal marrow cells have no such effect[112]. Suppressor T-cells (T_S) seem to be exerting this effect. Embryonic thymus cells (from days 16 to 20 of incubation) from isogeneic and allogeneic donors suppress graft rejection but do not seem to have a GVH effect[115]; adult thymus cells show the reverse effect. Thus two cell populations are probably implicated: an early T_s cell, probably present in the thymic cortex, and a late GVH-potentiating cell. Droege suggests that the former cell could explain embryonic tolerance and immunological non-reactivity[115]. Grebenau *et al.* investigated the role of T_S cells further[116]. Spleen cells from donors tolerized to human γ-globulin (HGG) do not prevent helper T-cell (T_H) activity nor will they prevent delayed hypersensitivity to HGG+complete Freund's adjuvant in normal or agammaglobulinaemic chickens, suggesting that tolerance is not (at least totally) a factor of T_S cells. T_S activity is affected by bursectomy[115], implying that bursa cells exert a feedback effect on thymic function: bursectomy reduced T_S activity and thymus size. Such T_S cells may be a $CT8^+$, $CT4^-$, $TCR1^-$ subset[117].

Cytotoxic T-cells Cytotoxic T-cells which respond to T-cell mitogens are found in the marrow, thymus and spleen (but not in the bursa)[118]. Bursectomy does not affect such cells' function. A population of NK cells is also described in birds[119]: at least some of these have the morphology of lymphocytes[120].

5.3.2 Lymphoid Organs

Bone Marrow

In the adult chicken the bone marrow is the site of granulopoiesis and lymphopoiesis: the spleen does not appear to fulfil these roles. This can be shown by sublethally irradiating a chicken and reconstituting it with bone marrow cells. No proliferating marrow cell colonies are seen in the spleen. Lymphopoiesis occurs in the extravascular spaces of the marrow: here the lymphoid tissue is discretely arranged and is characterized by the presence of germinal centres[121,122]. The lymphoid tissue has capillaries which contain lymphocytes; some of the latter are observed to be crossing the capillary walls.

Sorrel and Weiss found cells in the marrow of embryonic chickens at 12 days of incubation[123]. Already the cells were identifiable as belonging to lines such as erythrocytes or lymphocytes. Nevertheless, bone marrow cells are of extrinsic origin. In a quail/chick chimaeric system (see below) bone marrow in limb-bud grafts was populated by haemopoietic stem cells from the host species[124].

Thymus

The avian thymus is a lobulate organ ranged alongside the jugular veins on either side of the neck. A cortex densely packed with cells and a less densely packed medulla are discerned[122]. Cells seen include lymphoid cells, epithelial cells, macrophage/monocytes and myoid cells. Much of our understanding of avian thymus architecture and development comes from elegant experiments using chimaeras in which embryonic quail tissue is transplanted into chicken embryos or vice versa. Quail cell nuclei possess a distinctive heterochromatin mass which can be seen readily by light microscopy: hence the fates of transplanted and host cells and their progenies can be traced (Figure 5.10)[93].

If very young epidermal thymus rudiment tissue of one species is transplanted, it is found that the epithelial cells of the developing thymus are of donor origin (Figure 5.11), and that they express class II MHC antigens of the donor species[125]. Such cells are particularly prominent in the cortex where their long processes form a reticular framework. More compact epithelial cells are found in the corticomedullary junction regions. It is possible that their class II MHC antigens may play an educating role in connection with the phenomenon of MHC restriction. The ontogenesis of the thymus has been studied extensively. The thymus originates from the pharynx wall as an epitheliomesenchymal rudiment. The endoderm derives from the gut wall and the mesenchyme from neural crest cells of the mesectoderm. There is no ectodermal component. If quail rhombencephalon (hind brain) primordium is grafted into a 6–9 somite chick embryo a chimaeric thymus results. The connective tissue is quail whereas epithelial and endothelial tissue and lymphoid cells are chicken. It would seem that the mesectoderm contributes to myoid cells, cells lining the outer surfaces of the blood vessels and the interlobular connective tissue[126].

The bird chimaera system has enabled us to make sense of the long-standing problem concerning the origins of thymic lymphocytes, as to whether they are of epithelial or mesodermal origin. Early work[127,128] using parabiosed 4–5-day male/female chickens showed, by day 9, extensive chimaerism in the thymic lymphocyte populations. This suggested an extrinsic origin for such lymphocytes. If parabiosis is performed after

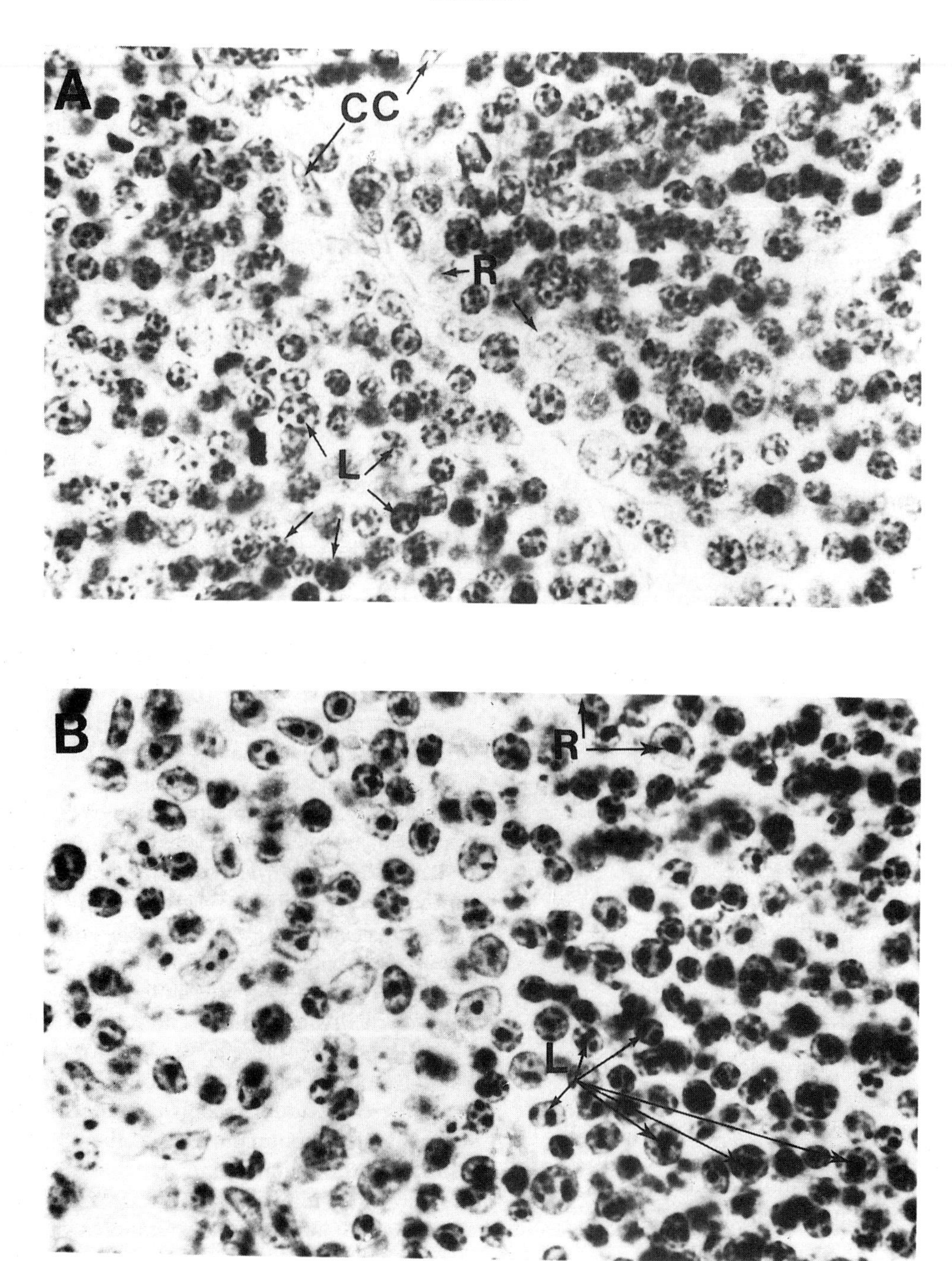

Figure 5.10. Normal thymus from an 18-day chick embryo (A) and a 16-day quail embryo (B). Feulgen & Rossenbeck staining; ×1200. R = reticular cells with clear nuclei containing an evenly dispersed network of chromatin for chick, or a large central DNA condensation for quail. L = lymphocyte nuclei with small, strongly Feulgen-positive chromocentres for chick, and a large heterochromatin mass, often attached to the nuclear membrane, plus small patches of chromatin for quail. CC = connective cells. Photographs by courtesy of Nicole Le Douarin, Institut d'Embryologie Cellulaire et Moléculaire, Nogent-sur-Marne.

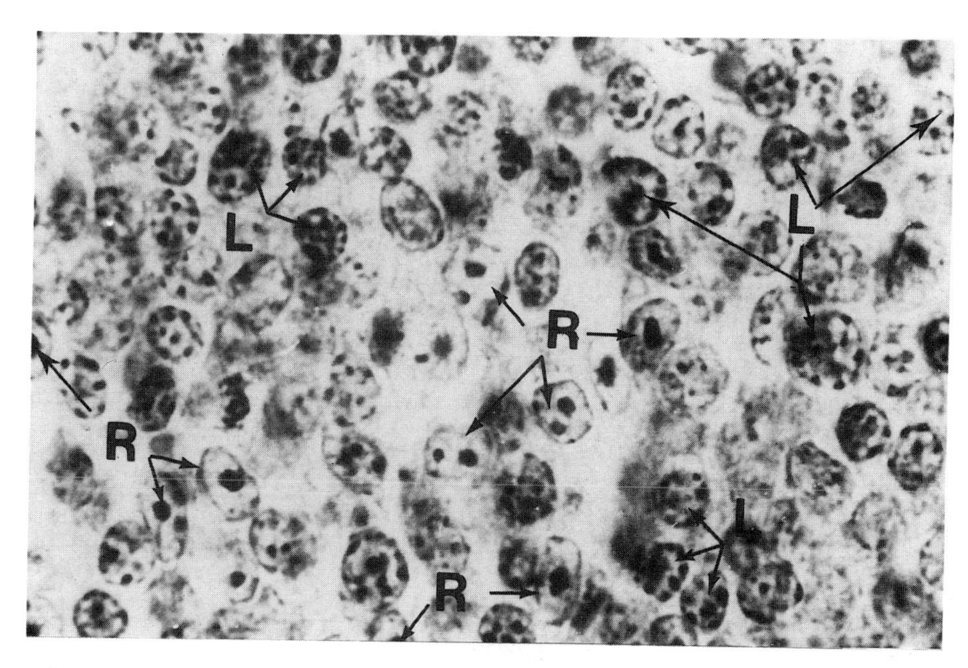

Figure 5.11. Cells of thymus, resulting from a graft of endomesodermal thymic rudiment of a 23-somite quail embryo into the somatopleurae of a 3-day-old chick. The reticular cells (R) are of quail origin, whereas the lymphocytes (L) are of host (chick) origin. Feulgen & Rossenbeck staining; ×1500. Photograph by courtesy of Nicole Le Douarin, Institute d'Embryologie Cellulaire et Moléculaire, Nogent-sur-Marne.

9 days little chimaerism is seen. This sets a timescale for thymus colonization. The quail/chicken model helps further here. Quail thymus anlagen were grafted to chickens (and vice versa) at various stages and were examined at 14 days of incubation. The thymus was invaded by stem cells after day 5 in the quail and day 6.5 (of incubation) in the chicken; the invasions lasted 24h and 36h respectively, after which the thymus entered a refractory period when it was not receptive to invading stem cells. Jotereau and Le Douarin indicated that the thymus later became receptive to the stem cells again, and a cycle of receptiveness and refractiveness was established[129]. They grafted a quail thymus, taken after the first stem cell colonization was complete, into a 3-day chicken embryo and then into a 3-day quail embryo. The thymus was left in the chicken for varying lengths of time. It was found that a second invasion of stem cells occurred at 11–12 days. Similar experiments using quail thymuses taken after the second invasion showed that a third influx of stem cells occurred between 17 and 18 days (2 days after the quail hatches). Further investigation beyond this time is impossible because the birds' maturing immune systems react to the xenogeneic cells in an *in vivo* mixed lymphocyte reaction (Figure 5.12).

In summary, colonization of the quail thymus is a cyclic process with at least three receptive 24h periods interspersed with 5-day non-receptive periods. Quail-quail parabiosis experiments confirm these findings and rule out criticisms that the observed process is artefactual due to the xenogeneic chimaeric regime[130]. A similar seeding pattern of the chicken thymus is seen when the reverse experiments are performed: here the receptive periods tend to be longer and later.

If a quail thymus is grafted onto a chicken embryo after the thymus has received its first (quail) lymphocyte immigration, chicken cells are observed in the grafted organ. Initially they are seen on the peripheries of the lobes and scattered in the medulla. For some 3 days after the 11–12

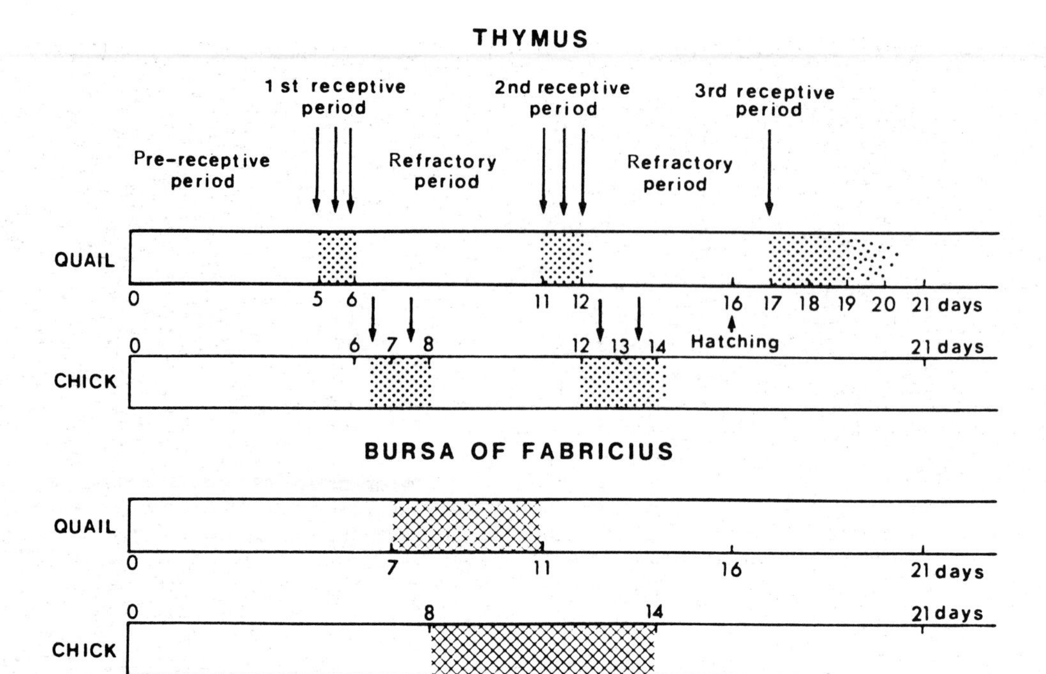

Figure 5.12. A comparison of the patterns of colonization of the avian primary lymphoid organs by stem cells (shaded areas). The thymus receives stem cells during short, recurrent periods which obey a precise rhythm. The rhythm is specific for each species. The bursa of Fabricius is colonized during a distinct period of embryonic life, before hatching. From reference 93 with permission.

day colonization period they are quiescent before a burst of mitotic activity results in numerous chicken cells throughout the lobe. It is noted that both the cortex and the medulla are simultaneously colonized. T-lymphocytes start to acquire their surface marker antigens by about 12 days of incubation, with almost all cells possessing them within 48h, before the start of the second wave of stem cell immigration: presumably this timing is of significance. Equivalent findings are reported in the quail thymus.

Bursa of Fabricius

The bursa is an organ which is unique to birds: no reptilian homologue of the organ has yet been satisfactorily described. Its immunological significance was realized in 1956 when Glick *et al.* removed the bursa and discovered that such bursectomized chickens failed to mount an immune response to bacteria[131]. The finding was instrumental to a realization that the immune system was dichotomous, with thymus (T) lymphocyte and bursa (B) lymphocyte arms.

The avian bursa is an oval, saccular diverticulum of the dorsal cloaca. Like the cloaca it is lined with pseudostratified columnar epithelium with secretory cells. The inner bursal surface is folded: these folds are full of follicles, about 10^4 overall (Figure 5.13). Between the follicles is sparse connective tissue. The follicles can be delineated into a cortex and a medulla, separated by a layer of reticular cells and a capillary bed (Figure 5.14). Epithelial cells are found in the follicular cortex and medulla. The cortex contains densely-packed lymphocytes, many of which are in mitosis. The medulla possesses more loosely packed medium and large lymphocytes, with large lymphoblasts near the edges. Bursa lymphocytes (Figure 5.15) seem to be generally larger than thymus lymphocytes.

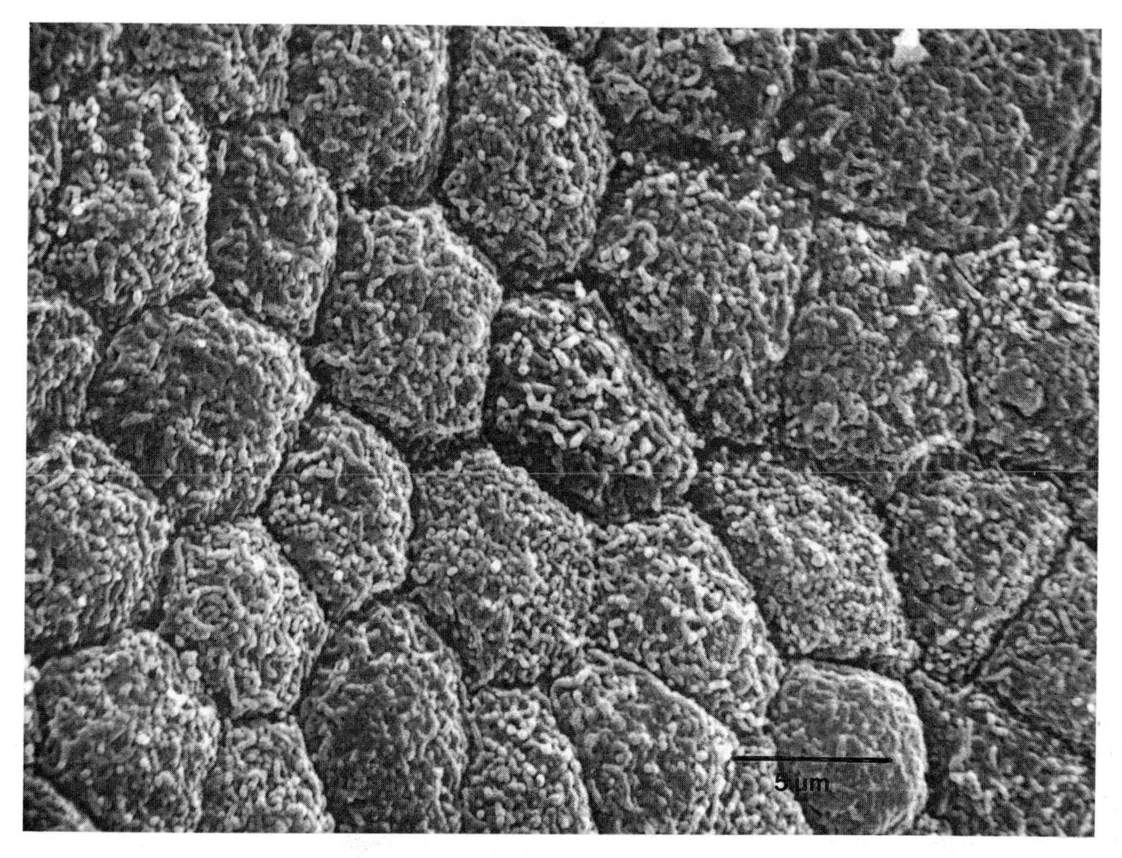

Figure 5.13. Scanning electron micrograph of the follicle-associated epithelium of the bursa of Fabricius of a 7-week chicken. Note microvilli. From reference 177 with permission.

Experimental manipulations have enabled studies of bursa function: the bursa has been ablated by surgery, often in conjunction with, or by the use instead of anti-immunoglobulin antisera, X-irradiation, or cyclophosphamide or testosterone treatment. Surgical bursectomy on hatching, coupled with the use of an anti-μ-chain antiserum leads to permanent agammaglobulinaemia[132]. However, bursectomy at 5 days of incubation seems to lead only to a hypogammaglobulinaemia: immunoglobulin is present, although the antibody repertoire seems to be absent or grossly depleted[133,134]. Quail/chicken chimaeras (wherein a quail bursa rudiment has been grafted to a chicken bursa) have been invaluable in the study of this organ too[135]. Quail cells express class II MHC antigens on their surfaces and also a unique quail antigen, styled MB1, which may be the quail IgM μ-chain[136]. The MB1 antigen is found on quail endothelial and haemopoietic cells, and later expresses itself on epithelial cells at the time that stem cells have invaded the organ. Class II antigens are found on the endothelial cells of the blood capillaries which surround the lymphocyte follicles where they probably have a recognition signalling role.

The drug cyclophosphamide leads to bursal atrophy. If the drug is given to chickens on hatching, the chickens can be reconstituted with immature, normal chicken bursa cells, and the morphology of the bursa is restored. A bursa environment is necessary in order that full immunocompetence may be restored, by providing a setting for bursa lymphocyte maturation and proliferation. For T/B lymphocyte co-operation to occur (and hence for the immune system

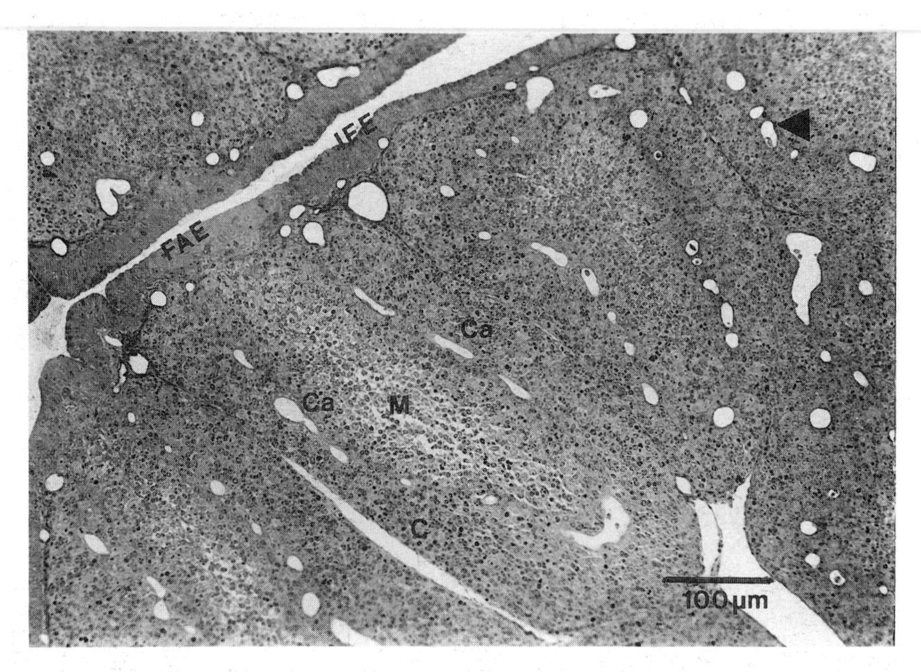

Figure 5.14. Chicken bursa of Fabricius, perfused. Capillaries (Ca) are visible within the cortex (C) of the bursal follicle (arrow). FAE = follicle-associated epithelium; IFE = interfollicular epithelium; M = medulla. From reference 177 with permission.

to be fully functional) the reconstituting bursa cells must share one MHC haplotype with their hosts' cells[137]. Attempts to reconstitute the structure of the bursa with older bursa cells are unsuccessful, but such cells are already mature and so considerable immune function is restored. Testosterone also has a bursectomizing effect: the bursal epithelium is apparently androgen-sensitive.

The bursa is first apparent in ontogeny at 5 days of incubation (in the chicken); epithelial folds are seen at 10 days, vessels at 11 days and lymphoid cell differentiation in follicles at 15 days. Cortical differentiation in the bursa occurs after hatching[93]. Granulopoiesis is seen to peak at 12–13 days in the mesenchyme within the bursal folds. The pattern in the quail is similar. The bursa continues to grow for about 1 month after hatching; soon afterwards involution occurs.

The lymphocyte precursors are basophilic cells seen in the peripheral mesenchyme at 7–9 days, and in the epithelium and capillary lumina from 10–12 days. Their appearance in epithelial follicles occurs at 14–15 days. Thereafter their incidence declines as lymphopoiesis progresses. Use of the quail/chicken chimaera model shows that the quail bursa is seeded by stem cells from 7–11 days, the chicken bursa from 8–14 days (Figure 5.12). Chimaeras prepared late will have mixed populations of lymphoid cells: individual follicles can have quail, chicken or both types of lymphocytes, suggesting that follicle seeding is not simultaneous or that multiseeding can take place[138,139].

Chicken/chicken bursa chimaeras with differing surface allotypic markers show that the chicken bursa is populated from outside at 8–15 days by less than 10^5 precursor cells lacking sIgM[140,141]. Later sIgM is seen on B-lymphocyte precursor cells which will proliferate in the bursa; the extrabursa precursors are undetectable in the hatched chicken. Such $sIgM^+$ B-lymphocyte precursors in the bursa later disappear and new B-lymphocytes are arguably generated outside of the bursa in older chickens.

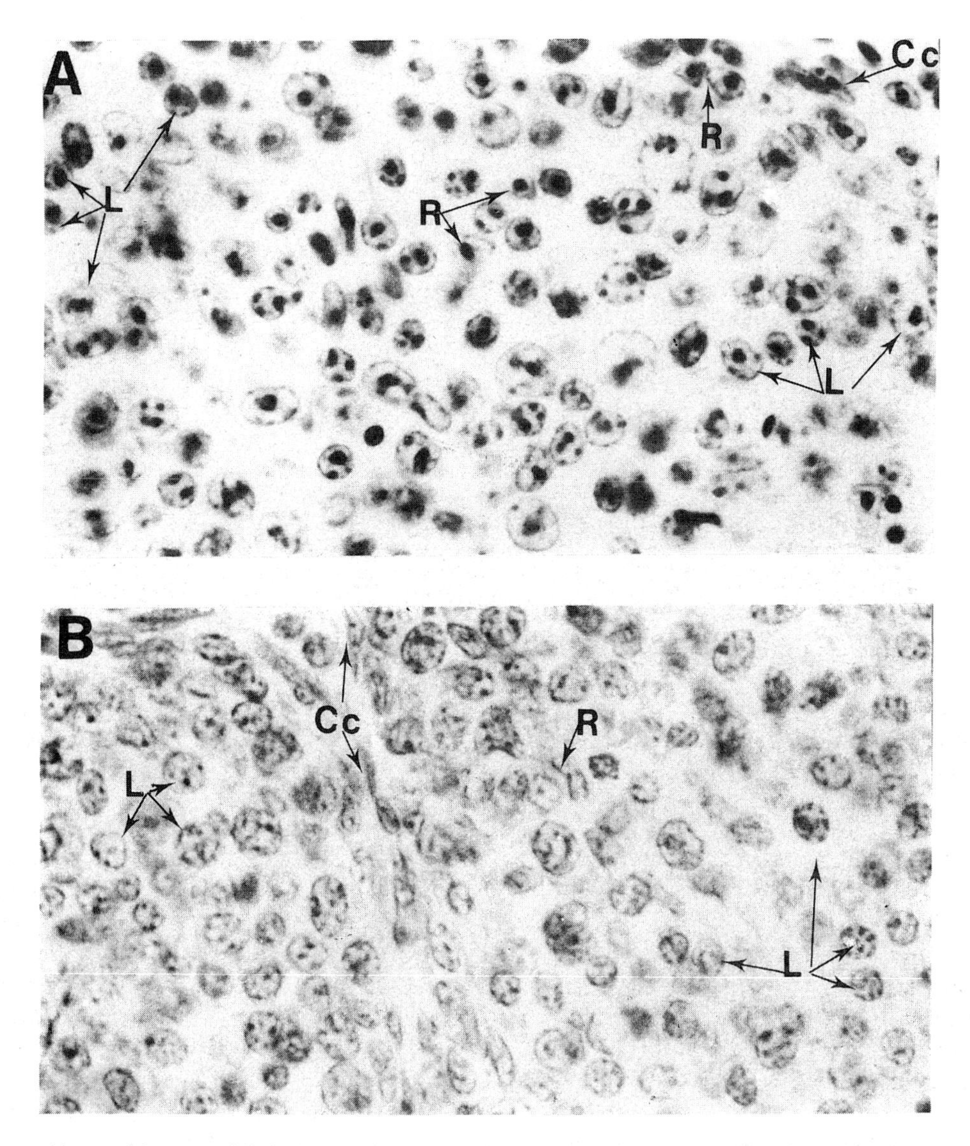

Figure 5.15. Normal bursa of Fabricius of a quail (A) and a chick (B) 3 days after hatching. Feulgen & Rossenbeck staining, ×1300. Note condensation of heterochromatin in the quail nuclei. R = reticular cells; L = lymphocytes; Cc = connective cells. Photographs kindly supplied by N. Le Douarin, by courtesy of E. Houssaint, Institut d'Embryologie Cellulaire et Moléculaire, Nogent-sur-Marne.

Spleen

The avian spleen is rather complex (Figure 5.16). Early in ontogenesis the spleen is primarily granulopoietic and erythropoietic. Its structure is complete by the end of embryonic life. A few lymphocytes are seen by day 5 of incubation in the quail and by day 12 of incubation in the chicken. Modest granulopoiesis continues after hatching but the predominant leucocytes in the post-hatch spleen are lymphocytes.

In the white pulp central arteries are surrounded by a periarterial lymphoid sheath (PALS) containing T-lymphocytes. At the end of the PALS the arteriole bends to become the pencilliform capillary which is surrounded by an ellipsoid (the

Figure 5.16. Section of chicken spleen, ×220. The central arteries (CA) have one muscle layer surrounded by a periarterial lymphatic sheath (PALS). At the end of the PALS there is a near right-angle curve in the vessel. The proximal portion of the pencilliform capillary (PC) does not have muscle cells and a PALS. The second, mid-portion of the PC is surrounded by the ellipsoid (E) which is embedded in the periellipsoidal white pulp (PWP) from reference 142 by permission of Wiley-Liss, a division of John Wiley and Sons, Inc.

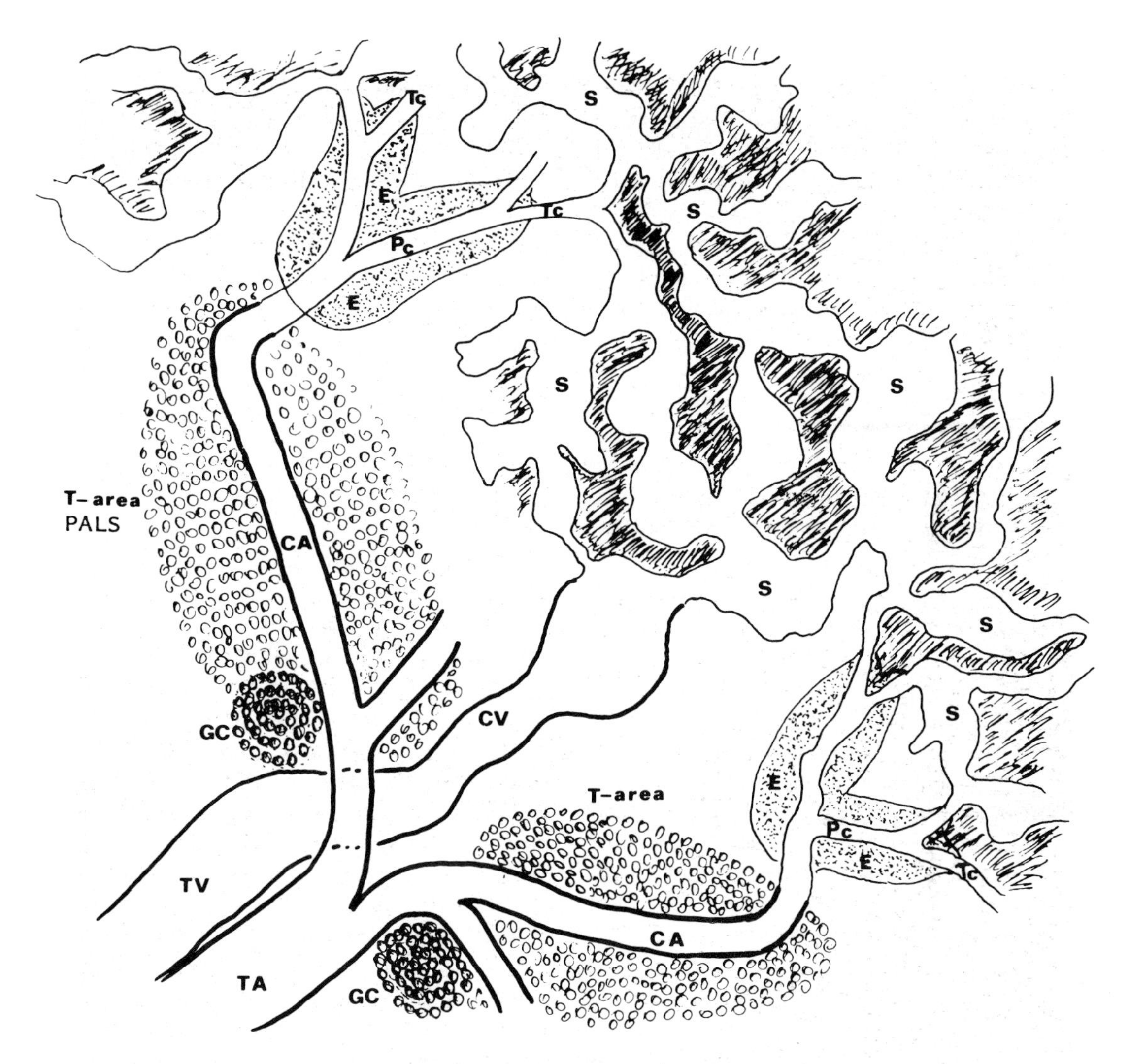

Figure 5.17. Diagram showing the organization of the chicken spleen and its circulation. The trabecular artery (TA) continues as a central artery (CA) surrounded by the T-dependent periarterial lymphatic sheath (PALS). The germinal centres (GC) are located at the beginning of the CA and PALS. The mid-portion of the pencilliform capillary (Pc) is encompassed by the ellipsoid (E) which is surrounded by the periellipsoidal white pulp. The distal portion of the Pc or terminal capillary (Tc) continues as the sinuses (S) of the red pulp. The sinuses are drained by the collecting veins (CV) which join to form the trabecular vein (TV). From reference 142 by permission of Wiley-Liss, a division of John Wiley and Sons, Inc.

Schweigger–Siedal sheath) which projects from the white pulp into the red pulp. Around the ellipsoid are regions of B-lymphocytes and dendritic cells. Beyond the ellipsoids the pencilliform capillary branches to form flattened terminal capillaries which drain into the red pulp sinuses. Germinal centres near the central arteriole have antigen-presenting cells and B-lymphocytes which have migrated from the ellipsoids. Ellipsoid-associated cells are also found, present on the surfaces of the ellipsoids. They can ingest colloidal carbon before migrating to the red pulp.

Other lymphoid tissues

Several other secondary lymphoid tissues have been described in birds[143]. The caecal tonsil in the proximal part of the caecal pouches contains lymphoid tissue with germinal centres. The proximal portion of the germinal centres possesses small lymphocytes and lymphoblasts and macrophages. The small lymphoblasts are actively dividing, particularly in the tonsils of younger birds. 'Thymic' and 'bursal' areas within the tonsils are reported, and the tonsils seem capable of producing antibodies. The Harderian gland near the nose and eye is rich in plasma cells in many species: these cells are derived from an early immigration of B-lymphocytes[144].

Lymph nodes are reported in a variety of aquatic and littoral species of birds, but not in chickens. All birds have lymphoid nodules formed from lymphocyte accretions in the walls of lymphatics. Antigenic stimulation increases their size, with greater numbers of germinal centres and plasma cells. Both B- and (fewer) T-lymphocytes are present in the nodules. Finally the pineal gland possesses numerous lymphocytes and germinal centres in 2–4 week chickens; later these disappear.

5.3.3 Antibody-mediated Immunity

Birds are very good responders to both particulate and soluble antigens, including T-independent antigens; antibody responses are under genetic control[145]. The involvement of B-lymphocytes processed in the bursa of Fabricius has already been noted. Generally antibody responses in birds tend to be more rapid and of shorter duration than those seen in Mammalia, a reflection, perhaps, of the rather high avian body temperature, which can exceed 40°C.

High and low molecular weight antibodies are produced to most antigens. Bacterial lipopolysaccharides, which are handled as T-independent antigens in Mammalia, elicit only IgM antibodies. T-dependent antigens elicit IgM antibody, later largely supplanted by low molecular weight IgY.

5.3.4 Immunoglobulin Classes

As in all craniates, IgM is found also in birds. In serum it exists mainly in a high molecular weight, pentameric 19S form. Monomeric IgM is found as surface immunoglobulin (Ig) on B-lymphocytes, and the use of an anti-μ-chain antibody indicates that IgM is found very early in ontogenesis, on developing B-lymphocytes within the bursa. Kincade *et al.* showed that all peripheral B-lymphocytes in a 13-day chicken embryo bear sIgM[107].

The low molecular weight non-IgM immunoglobulin found in serum, principally during secondary immune responses, is usually termed IgY[146]. It is arguably homologous with the IgY or 'IgRAA' found in Amphibia and Reptilia, and Ambrosius and Hädge suggest that avian (and reptilian) IgY may, on the basis of its antigenicity, be the homologue of mammalian IgA[60]. Certainly no γ-chain-containing IgG is found in birds. A 5.7S Ig has been described in ducks: it is suggested that this possesses a deleted IgY heavy chain.

Birds have a secretory Ig: it is the only Ig found in bile. This is termed IgA and it poses an intriguing evolutionary puzzle. If it is the homologue of mammalian IgA then either we could expect it to be found in reptiles (and none has been discovered, despite several searches), or the structural and functional similarities between IgA in birds and mammals are the result of a remarkable process of convergent evolution from a common ancestral IgA precursor. Hädge and Ambrosius suggest that in turkeys and in chickens the 'IgA' is a development unique to these galliform birds, and IgA should be renamed IgB[147]. In the anseriform ducks and geese the secretory 'IgA' may be IgM-related. There is a single report of an avian IgE with an ability to participate in passive cutaneous anaphylaxis reactions[148], and Leslie[149] makes a case for an IgD occurring on B-lymphocytes, sometimes on sIgM-positive cells. Avian

Igs have both κ and λ light chains; the former is the predominant isotype in chickens.

5.3.5 Transfer of Maternal Antibodies

In many placental mammals transfer of maternal IgG across the placenta confers passive immunity on the foetus. Embryonic birds are well-protected from antigenic insults by their cleidoic eggs, but maternal antibodies are nevertheless passed from the follicular epithelium of the ovary to the egg yolk. IgY seems to be passed preferentially, but IgM and IgA (?IgB) are also transmitted to the albumin as the egg descends the oviduct. Such antibody transfer seems to be inhibited in chickens with autosomal albinism, suggesting that it is under genetic control[150].

As the embryo develops it acquires IgY across the yolk sac. IgM and IgA (?IgB) enter the amniotic fluid and thence the alimentary canal.

5.3.6 Development of Immunocompetence

The passively acquired antibodies protect the avian embryo and fledgling, but then the bird needs to develop its own active immunity. Particulate matter is absorbed in the yolk sac by day 3 of chicken incubation, macrophage activity is noted by day 7, and the liver has a phagocytic role by day 10. By day 15 of incubation transplantation immunity is discernible, and injection of goat erythrocytes at this time elicits opsonizing antibodies within 1 week.

5.3.7 Immunoglobulin Genes

V_H, V_L and constant region genes have been investigated in the chicken[66]. This work has shown 61% homology between chicken and both mouse and human constant region light chain amino acid sequences[151]; on the other hand, the chicken V_λ sequence is nearer to human than to mouse $V_\lambda 1$. The overall gene seems to consist of single C_λ, J_λ and V_λ component genes, with somatic rearrangement of the V_λ and J_λ genes resulting in DNA deletion; a single base is added to the V_λ and J_λ coding segments. Non-functional flanking V_λ pseudogenes are involved in gene conversions and these help to generate antibody diversity[152].

Avian V_H genes belong to a single subgroup. Two D genes may join to give a D-D unit. A single chicken V_H gene with multiple pseudogenes on each side has been described[153]. Somatic changes similar to those described for the V_λ genes diversify the expression of the V_H gene: pseudogenes which include VD fusions are implicated in gene conversion. The composite picture seems therefore to portray a V_H gene and flanking pseudogenes, intimately linked to more than one D segment, a J_H and a constant region gene.

It is interesting that DNA probes discern major differences between Ig heavy chain gene rearrangements in bursa and erythrocyte DNA at a very early stage in B-cell ontogenesis[154]. This finding may help to explain why later constitution of bursectomized chickens with committed bursa cells does not require a bursa environment: perhaps relevant gene rearrangements have already taken place during the bursa-dependent phase of development.

It is tempting to speculate that the evolutionary homogeneity of the Aves is reflected by their paucity of Ig genes compared with the multiplicity of Ig genes in those mammals which have been investigated: the birds have generated antibody diversity by alternative genetic mechanisms such as somatic mutation.

5.3.8 Cell-mediated Immunity

The kinetics of allograft and xenograft rejection in birds is rapid and 'acute', as in mammals; good second-set memory is exhibited. Cytotoxic T-lymphocytes and their cytokines, and cytotoxic antibodies are involved. Graft rejection is under the control of an MHC (the B system in chickens) and also minor histocompatibility genes, including a sex-linked histocompatibility antigen.

In vitro correlates of graft rejection have been studied. Mixed lymphocyte reactions (MLRs) were reported by Alm[155]. Here MLRs were set up using spleen cells from outbred chickens; bursectomy did not affect the cells' ability to respond, suggesting T-lymphocyte involvement. Use of inbred, Prague strains of chickens shows the importance of the B complex in the control of MLRs[156], although other genetic determinants, not related to the B complex, together with the age of the stimulator and responder cells (according to strain) may also be critical. There is an endocrine link with MLR reactivity through the mediation of tri-iodothyronine[157]. Reactivity to class I (B-F) MHC antigen-disparity is not detectable before hatching, and embryonic thymocytes are not IL-2 sensitive[158].

Graft-versus-host (GVH) reactions are seen in birds. Implantation of allogeneic spleen cells into the chorioallantoic membrane of the egg results in a GVH reaction characterized by local lesions at the site of inoculation and by splenomegaly in the host. The phenomenon has proved to be a useful tool for testing immunocompetence in donor cells (allogeneic donor cells from 20-day of incubation embryos can potentiate a GVH reaction). GVH reactions are under the control of the MHC B-system. Some B alleles confer strong GVH reactivity (e.g. the B^{14} allele), others (e.g. the B^{2} allele) poor reactivity. Implantation of allogeneic cells and the induction of a GVH reaction is reported to accelerate the development of immunological maturity in the host[159].

Delayed hypersensitivity reactions are reported in birds. Human γ-globulin sensitization will result in a delayed hypersensitivity reaction when chickens are challenged with human γ-globulin plus complete Freund's adjuvant, even if the birds are agammaglobulinaemic[116].

5.3.9 Immune Tolerance

As in Mammalia, a state of specific immunological non-reactivity can be induced by parabiosing two chicken embryos (by fenestrating the egg-shells and allowing chorioallantoic blood vessels to anastomose). The two chickens will tolerate each other's cells. High-zone tolerance or immune paralysis can be induced by flooding the immune system with large doses of antigen, and low-zone tolerance is also demonstrable: here very small antigen doses will induce tolerance. Presumably this is how the bird learns self-tolerance *in ovo* as new antigens appear in very small amounts during ontogenesis. However, sensitization of an embryo with allogeneic erythrocytes will result in a second-set rejection of skin from the erythrocyte-donor when a graft is made after hatching, suggesting that the border between low-zone tolerance and immune reactivity is a very delicate one. Bursectomized birds can acquire tolerance, suggesting that blocking, enhancing antibodies are not implicated in the phenomenon. A T_S cell may be implicated in tolerance in the embryo, although this cannot be the whole story[115,116].

5.3.10 Major Histocompatibility Complex (MHC)

The chicken MHC and its gene products have been the subject of several review articles[82,160,161]. The chicken MHC is called the B complex and it encodes class I (B-F) and class II (B-L) gene products and also a multigene family (B-G or class IV) of cell surface, covalently linked dimer antigens. Class II gene products are identifiable on the non-lymphoid cells of the bursa, and then on thymus cells, B-lymphocytes and activated T-lymphocytes, antigen-presenting macrophages and dendritic cells. Class I gene products are found on most tissues including erythrocytes, while class IV (B-G) gene products are seen on erythrocytes, platelets, intestinal cells and arguably lymphocytes. The roles of the class I and class II molecules seem similar to those in mammals. Class I molecules are recognized by MHC-restricted cytotoxic T-lymphocytes specific for virally-infected cells; class II molecules potentiate MLRs and interactions between helper T-lymphocytes, B-lymphocytes and macrophages. No satisfactory roles have yet been found for the class IV (B-G) molecules; B-G genes and their products are reviewed by Kaufman and Salomonsen[162].

The B complex of genes is very highly compacted onto about one quarter of an 8 megabase microchromosome. The genes are very closely spaced with only very small introns. A B-F/B-L region has been seen: in the B12 haplotype four cosmid clusters in this region have been seen to possess, *inter alia*, five class II β genes, six class I genes and a class IV gene. Other class IV (B-G) genes map on cosmids in the B-G region. Generally the number of class I and class II β genes is low, although Miller questions this in the light of findings using more outbred chickens[161]. Class II genes may map outside of the B complex. No evidence for class III MHC genes (encoding for C4, C2, TNF, etc.) has yet been seen.

If chicken MHC gene products are compared with those of mammals, it is noted that homologous domains of the chicken class I heterodimer tend to show respectively differing rates of molecular divergence. By and large, intra- and interdomain contacts in binding sites are most conserved.

The detailed organizational minutiae of the B complex may suggest that the MHC isotypes evolved after the separation of the lines leading respectively to the birds and the mammals[160], even if the overall structure of the B complex, intron-exon arrangements and the division of class I and class II molecules into domains is conserved in homoiotherm evolution. One might expect to find such features in reptiles too.

5.3.11 Cytokines

The growing importance of cytokines to our understanding of immune function has already been noted for the reptiles. Cohen and Haynes[86] and Schat[163] review the research on avian cytokines.

Antiviral and tumoricidal interferons have been reported in a number of studies using chicken cells, although the type(s) of interferon is not fully defined. Similarly chickens and turkeys produce migration inhibiting factor (MIF) when sensitized with antigen; the MIF exhibits a degree of species-specificity[164,165]. Other cytokines reported include a lymphocyte inhibiting factor and a monocyte chemotactic factor. Pathogens such as avian influenza virus enhance lymphocyte activation and cytokine release[166].

Interleukin-1 (IL-1) has been implicated in a number of studies in which avian cells have been stimulated. Adherent spleen cells, elicited peritoneal macrophages and peripheral blood monocytes, for example (but not marrow macrophages), treated with stimulating agents such as *Escherichia coli* endotoxin produce a substance (or substances) in the supernatant which co-stimulates chicken thymus cells. This putative IL-1 seems arguably to be species-specific: mammalian IL-1 has no effect on chicken thymus lymphocytes, and in the reverse situation chicken IL-1 has only minimal stimulatory powers[167]. More work seems desirable to characterize further chicken IL-1, to rule out the possibility of inhibitors in the supernatants. T-lymphocytes show a greater dependence on macrophages than B-lymphocytes for the production of cytokines[168].

Mitogen stimulation of chicken spleen or peripheral blood lymphocytes results in two IL-2 type molecules being found in the supernatant: these will stimulate T-cells. Spleen T-lymphocytes have a receptor with a molecular weight of 45000 for such IL-2-type cytokines[169]. It has been shown that in ontogenesis the development of Con A responsiveness by T-cells is concomitant with their ability to produce IL-2; these events occur at 1 week post-hatching. The genetic control of Con A responsiveness (noted earlier) also relates to the ability of T-cells to produce IL-2[170]. As with IL-1, present evidence seems to suggest that chicken IL-2 is species- (or at least bird-) specific.

5.3.12 Hypersensitivity

The foregoing descriptions of avian immune phenomena suggest that birds possess a complex, sophisticated immune system. It is not unreasonable to expect them to 'pay the price' of this by immune over-reactions manifested as hypersensitivities or autoimmunity.

Anaphylactic reactions have been described in

chickens sensitized to and subsequently challenged with protein antigens: the sensitivity is transferable by serum, and the existence of basophils and a putative IgE[148] or an IgY subclass which is reaginic, suggests that Type I hypersensitivity reactions occur, with mechanisms at least partially similar to those seen in mammals. Histamine is involved in the reaction, since it can be blocked by antihistamines.

Figure 5.18. 10-week obese strain (OS) (left) and normal White Leghorn (right) chickens. The hypothyroid OS bird is clearly smaller and has ruffled feathers due to cold sensitivity, even at ambient temperature. Photograph kindly supplied by Georg Wick, University of Innsbruck.

Birds possess classical and alternative complement pathways, indicating that an avian, complement-mediated Type II cytotoxic hypersensitivity is feasible. The alternative complement pathway can be fixed by bacterial lipopolysaccharide or by cobra venom factor. Type IV delayed hypersensitivity (cell-mediated hypersensitivity) is also seen, as when sensitized, agammaglobulinaemic chickens are challenged with human γ-globulin in complete Freund's adjuvant[116]; the phenomenon is also described in the quail and the turkey[171].

5.3.13 Autoimmunity

The obese strain (OS) of chickens has been extensively studied[172] (Figures 5.18 and 5.19). It develops a spontaneous autoimmune thyroiditis, which has many close similarities with human

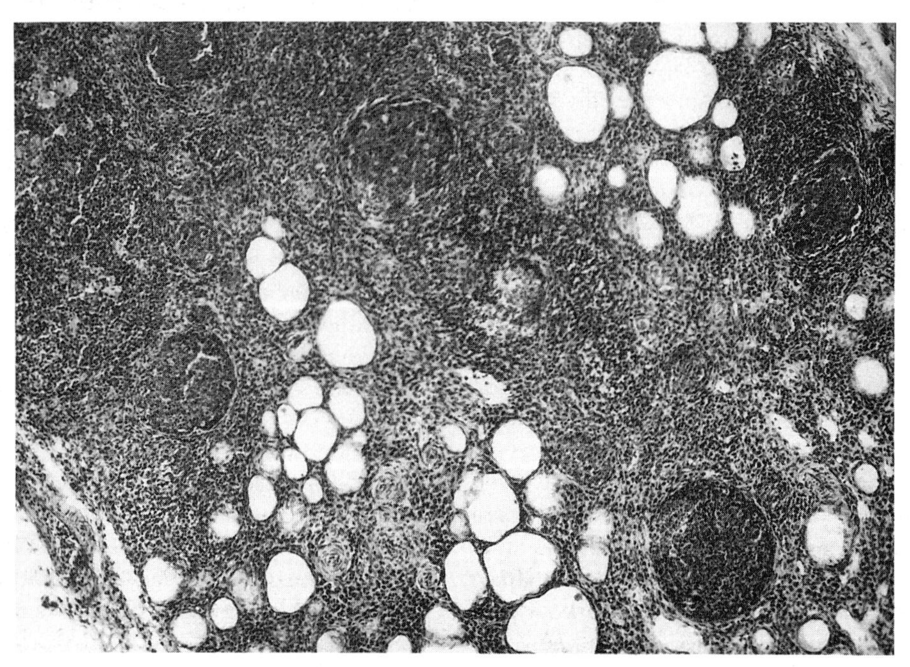

Figure 5.19. Thyroid gland of a 7-week OS chicken with nearly complete destruction by mononuclear cell infiltration and only a few intact follicles left. Note numerous well-developed germinal centres. ×250. Photograph kindly supplied by Georg Wick, University of Innsbruck.

Hashimoto's disease. OS chicken serum contains autoantibodies to thyroglobulin, to microsomes and to T_3 and T_4. Anti-thyroglobulin is produced by lymphocytes which actually infiltrate the thyroid and there form germinal centres. The autoantibody can be transferred from mother to egg. OS chickens also produce autoantibodies to a range of other organ and non-organ-specific self-antigens.

The disease seems to be initiated by T-lymphocytes which infiltrate the thyroid. Generally OS chickens have other hyperreactive immune features, perhaps associated with a defect in suppressor activity and imbalances in the neuroimmunoendocrine axis. It is noted that they have enhanced IL-2 and IL-2 receptor levels[173], the possible result of a deficiency in an IL-2 inhibitory factor in the serum.

Other autoimmune conditions in chicken include the Smyth–Dam line chicken and the UCD line 200 chicken which exhibit diseases resembling human vitiligo and scleroderma respectively[174].

5.4 CONCLUSIONS

The foregoing survey of reptilian and avian immunity is interesting on several counts. First, reptilian immunity is still not nearly so well researched as is mammalian, avian or even amphibian and piscine immunity. Much of the work is anecdotal or isolated; concentrated, in-depth studies of particular, defined aspects of reptilian immunity are few. The lack of genetically-defined inbred strains and lines of reptiles hampers research. Nevertheless, the picture which increasingly develops is of a system which shows signs of considerable sophistication and finesse, with possession of an MHC, a separate low molecular weight immunoglobulin, hypersensitivity reactions and so on. To date studies have not been sufficiently exhaustive or rigorous to say whether particular orders of reptiles (e.g. the Chelonia or the Crocodilia) show distinctive immunological features: all reptiles would seem to have broadly similar immune responses on the evidence so far collected. The importance of the external and the internal environment on reptilian immune performance is, however, a matter for note[5,175].

The avian immune response is much more fully investigated in terms of depth, but almost all the work has been done on the chicken, with minor studies on the quail, turkey and duck, all species of economic importance. It is frustrating not to have data about immunity in other bird species in what is admittedly an evolutionarily very homogeneous class: for instance, it would be instructive to have some knowledge of immunity in ratites or in passeriform birds. Galliform birds such as the chicken and the quail may be atypical, or show atypical responses due to long years of inbreeding and domestication.

On the present evidence, birds do seem to have a very full repertoire of immune responses and to be excellent immune performers although ducks, interestingly, are noted to have poor immune responses[111].

There are some remarkable parallels between avian and mammalian immune phenomena. If we bear in mind that the bird and the mammal lines diverged at the level of the Cotylosauria, it is not unreasonable to suppose that common features of the two classes' immune systems might be found in today's reptiles, albeit somewhat subdued by the reptiles' poikilothermic physiology. It must also be acknowledged that modern reptiles may have lost ancestral immunological features which have been conserved in birds and in mammals. If only we could look at immunity in fossils such as captorhinomorphs, pelycosaurs, thecodonts and 'dinosaurs'! A further possibility is that a striking degree of parallel or convergent evolution in birds and mammals has occurred in their immune systems.

Notwithstanding the similarities between avian and mammalian immunity, it must also be admitted that birds have some features in their immune systems which are unique: the bursa, IgY rather than IgG, the arrangement of the immunoglobulin and MHC genes. Here we see, perhaps, the birds evolving their own refinements to the basic immune system model.

5.5 REFERENCES

1. Tarsitano, S. and Hecht, M. (1980). A reconsideration of the reptilian relationships of *Archaeopteryx*. *Zool. J. Linn. Soc.* **69**, 149–182.
2. Bakker, R. (1986). *The Dinosaur Heresies—a Revolutionary View of Dinosaurs*. Longman, Harlow, UK.
3. Young, J.Z. (1981). *The Life of Vertebrates*, 3rd edn. Clarendon Press, Oxford.
4. McGowan, C. (1984). Evolutionary relationships of ratites and carinates: evidence from ontogeny of the tarsus. *Nature* **307**, 733–735.
5. Sypek, J. and Borysenko, M. (1988). Reptiles. In *Vertebrate Blood Cells*, Rowley, A.F. and Ratcliffe, N.A. (eds). Cambridge University Press, pp.211–256.
6. Efrati, P., Nir, E. and Yaari, A. (1970). Morphological and cytochemical observations on cells of the hemopoietic system of *Agama setllio* (Linnaeus). *Israel J. Med. Sci.* **6**, 23–31.
7. Jacobson, E.R., Gaskin, J.M., Page, D., Iverson, W.O. and Johnson, J.W. (1981). Illness asssociated with paramyxo-like virus infection in zoologic collection of snakes. *J. Amer. Vet. Med. Assoc.* **179**, 1227–1230.
8. Sypek, J.P., Borysenko, M. and Findley, S.R. (1984). Anti-immunoglobulin induced histamine release from naturally abundant basophils in the snapping turtle, *Chelydra serpentina. Dev. Comp. Immunol.* **8**, 359–366.
9. Caxton-Martins, A.E. (1977). Cytochemistry of blood cells in peripheral smears of some West African reptiles. *J. Anat.* **125**, 393–400.
10. Pienaar, U. de V. (1962). *Haematology of some South African reptiles*. Witwatersrand University Press, Johannesburg.
11. DeKumar, T. and Maiti, B.R. (1981). Differential leukocyte count in both sexes of an Indian soft-shelled turtle (*Lissemys punctata punctata*). *Z. Mikrosk-Anat. Forsch.* **95**, 1965–1969.
12. Duguy, R. (1970). Numbers of blood cells and their variation. In *Biology of the Reptilia—Vol. 3. Morphology*, Gans, C. and Parsons, T.S. (eds). Academic Press, New York, pp.93–109.
13. Wolke, R.E., Brooks, D.R. and George, A. (1982). Spirorchidiasis in loggerhead sea turtles (*Caretta caretta*): pathology. *Wildlife Dis.* **18**, 175–185.
14. Evans, R.H. (1983). Chronic bacterial pneumonia in free-ranging eastern box turtles (*Terrapene carolina carolina*). *J. Wildlife Dis.* **19**, 349–352.
15. Borysenko, M. (1976). Changes in spleen histology in response to antigenic stimulation in the snapping turtle, *Chelydra serpentina. J. Morphol.* **149**, 223–242.
16. Mead, K.F. and Borysenko, M. (1984). Surface immunoglobulin on granular and agranular leukocytes in the thymus and spleen of the snapping turtle, *Chelydra serpentina. Dev. Comp. Immunol.* **8**, 109–120.
17. Kroese, F., Leceta, J., Dopp, E., Herraez, M., Nieuwenhuis, P. and Zapata, A. (1985). Dendritic immune-complex trapping cells in the spleen of the snake, *Python reticulatus. Dev. Comp. Immunol.* **9**, 641–652.
18. Ryerson, D.L. (1949). A preliminary survey of reptilian blood. *J. Ent. Zool.* **41**, 49–55.
19. St. Girons, M.C. (1970). Morphology of circulating blood cells. In *Biology of the Reptilia–Vol. 3. Morphology*, Gans, C. and Parsons, T.S. (eds). Academic Press, New York, pp.73–91.
20. Borysenko, M. and Lewis, S. (1979). The effect of malnutrition on immunocompetence and whole body resistance to infection in *Chelydra serpentina. Dev. Comp. Immunol.* **3**, 89–100.
21. Hussein, M.F., Badir, N., El Ridi, R. and Akef, M. (1978). Differential effect of seasonal variation on lymphoid tissue of the lizard *Chalcides ocellatus. Dev. Comp. Immunol.* **2**, 297–310.
22. Hussein, M.F., Badir, N., El Ridi, R. and Akef, M. (1979). Lymphoid tissues of the snake, *Spalerosophis diadema* in the different seasons. *Dev. Comp. Immunol.* **3**, 77–88.
23. Hussein, M.F., Badir, N., El Ridi, R. and Charmy, R. (1979). Natural heterohemagglutinins in the serum of the lizard *Agama stellio. Dev. Comp. Immunol.* **3**, 643–652.
24. Hussein, M.F., Badir, N., El Ridi, R. and El Deeb, S. (1979). Effect of seasonal variation on the immune system of the lizard *Scincus scincus. J. Exp. Zool.* **209**, 91–96.
25. Wright, R.K. and Cooper, E.L. (1981). Temperature effects on ectotherm immune responses. *Dev. Comp. Immunol.* **5** (Suppl. 1), 117–122.
26. El Deeb, S., Zada, S. and El Ridi, R. (1985). Ontogeny of hemopoietic and lymphopoietic tissues in the lizard *Chalcides ocellatus. J. Morphol.* **185**, 241–253.
27. El Deeb, S., El Ridi, R. and Zada, S. (1986). The development of lymphocytes with T- or B-membrane determinants in the lizard embryo. *Dev. Comp. Immunol.* **10**, 353–364.
28. El Deeb, S.O. and Saad, A.-H.M. (1990). Ontogenic maturation of the immune system in reptiles. *Dev. Comp. Immunol.* **14**, 151–159.
29. Desser, S.S. (1979). Haematological observations on a hibernating tuatara, *Sphenodon punctatum. N.Z. J. Zool.* **6**, 77–78.

30. Borysenko, M. (1976). Ultrastructural analysis of normal and immunized spleen of the snapping turtle, *Chelydra serpentina*. *J. Morphol.* **149**, 243–264.
31. El Ridi, R., Zada, S., Afifi, A., El Deeb, S., El Rouby, S., Farag, M. and Saad, A.-H. (1988). Cyclic changes in the differentiation of lymphoid cells in reptiles. *Cell Different.* **24**, 1–8.
32. Bockam, D.E. (1970). The Thymus. In *Biology of the Reptilia—Vol. 3. Morphology*, Gans, C. and Parsons, T.S. (eds). Academic Press, New York, pp.111–133.
33. Leceta, J. and Zapata, A. (1985). Seasonal changes in the thymus and spleen of the turtle *Mauremys caspica*. A morphometrical, light microscopical study. *Dev. Comp. Immunol.* **9**, 653–668.
34. Kröese, F.G.M. and Van Rooijen, N. (1983). Antigen trapping in the spleen of the turtle, *Chrysemys scripta elegans*. *Immunology* **49**, 61–68.
35. Zapata, A. and Solas, M.T. (1979). Gut-associated lymphoid tissue (GALT) in Reptilia: structure of mucosal accumulations. *Dev. Comp. Immunol.* **3**, 477–487.
36. Jacobson, E.R. and Collins, B.R. (1980). Tonsil-like esophageal lymphoid structures of boid snakes. *Dev. Comp. Immunol.* **4**, 703–711.
37. El Ridi, R., El Deeb, S. and Zada, S. (1980). The gut-associated lymphoepithelial tissue (GALT) of lizards and snakes. In *Aspects of Developmental and Comparative Immunology—Vol. 1*. Solomon, J.B. (ed.). Pergamon Press, New York, pp.233–239.
38. Solas, M.T., Leceta, J. and Zapata, A. (1981). Structure of the cloacal lymphoid complex of *Mauremys caspica*. *Dev. Comp. Immunol.* **5** (Suppl. 1), 151–156.
39. Pitchappan, R.M. and Muthukkaruppan, V.R. (1977). *In vitro* properties of heterologous anti-lizard thymocyte serum. *Proc. Ind. Nat. Acad. Sci. B* **85**, 1–12.
40. Pitchappan, R.M. and Muthukkaruppan, V.R. (1977). Analysis of the development of the lizard, *Calotes versicolor*. II. Histogenesis of the thymus. *Dev. Comp. Immunol.* **1**, 217–230.
41. Mead, K.F. and Borysenko, M. (1984). Turtle lymphocyte surface antigens in *Chelydra serpentina* as characterized by rabbit anti-turtle thymocyte sera. *Dev. Comp. Immunol.* **8**, 351–358.
42. Mansour, M.H., El Ridi, R. and Badir, N. (1980). Surface markers of lymphocytes in the snake, *Spalerosophis diadema*. I. Investigation of lymphocyte surface markers. *Immunology* **40**, 605–613.
43. Manickasundari, M. and Pitchappan, R.M. (1987). Characteristics of heterologous anti-lizard (*Calotes versicolor*) thymocyte serum which identify the functions of thymus-derived cell lineage. *Dev. Comp. Immunol.* **11**, 605–612.
44. Negm, H. and Mansour, M.H. (1982). Phylogenesis of lymphocyte diversity. I. Immunoglobulin determinants on the lymphocyte surface of the lizard *Agama stellio*. *Dev. Comp. Immunol.* **6**, 519–532.
45. Cuchens, M.A. and Clem, L.W. (1979). Phylogeny of lymphocyte heterogeneity. III. Mitogenic responses of reptilian lymphocytes. *Dev. Comp. Immunol.* **3**, 287–297.
46. Cuchens, M.A. and Clem, L.W. (1979). Phylogeny of lymphocyte heterogeneity. IV. Evidence for T-like and B-like cells in reptiles. *Dev. Comp. Immunol.* **3**, 465–475.
47. Worley, R.T.S., Jurd, R.D. and Muthukkaruppan, V.R. (1989). Lymphocyte behaviour in two species of Sauria. *Herpetopathologia* **1**, 41–47.
48. Saad, A.H. (1989). Sex-associated differences in the mitogenic responsiveness of snake blood lymphocytes. *Dev. Comp. Immunol.* **13**, 225–229.
49. El Deeb, S. and Saad, A.H. (1987). Ontogeny of Con A responsiveness and mixed leucocyte reactivity in the lizard *Chalcides ocellatus*. *Dev. Comp. Immunol.* **11**, 595–604.
50. Manickasundari, M., Selvaraj, P. and Pitchappan, R.M. (1984). Studies on T-cells of the lizard, *Calotes versicolor*: adherent and non-adherent populations of the spleen. *Dev. Comp. Immunol.* **8**, 367–374.
51. Manickasundari, M. and Pitchappan, R.M. (1988). Structural, functional relationship of thymus derived cells of the lizard *Calotes versicolor*. *Dev. Comp. Immunol.* **12**, 603–610.
52. Cohen, N. (1971). Reptiles as models for the study of immunity and its phylogenesis. *J. Amer. Vet. Med. Assoc.* **159**, 1662–1671.
53. Pillai, P.S. and Muthukkaruppan, V.R. (1977). The kinetics of rosette-forming cell response against sheep erythrocytes in the lizard. *J. Exp. Zool.* **199**, 97–104.
54. Kanamkambika, P. and Muthukkaruppan, V.,R. (1972). Immune response to sheep erythrocytes in the lizard, *Calotes versicolor*. *J. Immunol.* **109**, 415–419.
55. Wetherall, J.D. and Turner, K.J. (1972). Immune response of the lizard *Tiliqua rugosa*. *Austral. J. Exp. Biol. Med. Sci.* **50**, 79–95.
56. Jurd, R.D. (1984). Immunoglobulin diversity: an evolutionary perspective. In *Aspects of Developmental and Comparative Immunology—Vol. 2*, Cooper, E.L. and Wright, R.K. (eds). Pergamon Press, New York, pp.145–152.

57. Marchalonis, J.J., Ealey, E.H.M. and Diener, E. (1969). Immune response of the tuatara, *Sphenodon punctatum. Austral. J. Exp. Biol. Med. Sci.* **47**, 367–380.
58. Benedict, A.A. and Pollard, L.W. (1972). Three classes of immunoglobulin in the sea turtle, *Chelonia mydas. Folia Microbiol.* **17**, 75–78.
59. Leslie, G.A. and Clem, L.W. (1972). Phylogeny of immunoglobulin structure and function. VI. 17S, 7.5S and 5.7S anti-DNP of the turtle *Pseudemys scripta. J. Immunol.* **198**, 1656–1664.
60. Ambrosius, H. and Hädge, D. (1983). Phylogeny of low molecular weight immunoglobulins. *Dev. Comp. Immunol.* **7**, 721–724.
61. Andreas, E.-M. and Ambrosius, H. (1989). Surface immunoglobulin on lymphocytes of the tortoise *Agrionemys horsfieldii. Dev. Comp. Immunol.* **13**, 167–175.
62. Lykakis, J.J. (1968). Immunoglobulin production in the European pond tortoise, *Emys orbicularis*, immunized with serum protein antigens. *Immunology* **14**, 799–808.
63. Litman, G.W., Berger, L., Murphy, K., Litman, R.T., Hinds, K.R., Jahn, C.L. and Erickson, B.W. (1983). Complete nucleotide sequence of an immunoglobulin V_H gene homologue from *Caimen*, a phylogenetically ancient reptile. *Nature* **303**, 349–352.
64. Litman, G.W., Murphy, K., Berger, L., Litman, R.T., Hinds, K.R. and Erickson, B.W. (1985). Complete nucleotide sequences of three V_H genes in *Caimen*, a phylogenetically ancient reptile: evolutionary diversification in coding segments and variation in the structure and organization or recombination elements. *Proc. Nat. Acad. Sci. US* **82**, 844–845.
65. Litman, G.W., Chisholm, R. and Erickson, B.W. (1984). Structure and organization of immunoglobulin genes in phylogenetically diverse species: studies at the DNA level. In *Aspects of Developmental and Comparative Immunology—Vol. 2.* Cooper, E.L. and Wright, R.K. (eds). Pergamon Press, New York. pp.131–138.
66. Litman, G.W., Varner, J. and Harding, F. (1991). Evolutionary origins of immunoglobulin genes. In *Phylogenesis of Immune Functions*, Warr, G.W. and Cohen, N. (eds). CRC Press, Boca Raton, pp.172–189.
67. Borysenko, M. (1970). Transplantation immunity in Reptilia. *Transplant. Proc.* **2**, 299–306.
68. Worley, R.T.S. and Jurd, R.D. (1979). Graft rejection in *Lacerta viridis* (the green lizard). *Dev. Comp. Immunol.* **3**, 653–665.
69. Borysenko, M. (1979). Evolution of lymphocytes and vertebrate alloimmune reactivity. *Transplant Proc.* **11**, 1123–1130.
70. Jurd, R.D. (1985). Specialisation in the teleost and anuran immune response: a comparative critique. In *Fish Immunology*, Manning, M.J. and Tatner, M.F. (eds). Academic Press, London, pp.9–28.
71. Farag, M.A. (1987). A contribution to the study of the MHC in snakes. PhD Thesis, University of Cairo.
72. Saad, A.-H. and El Ridi, R. (1984). Mixed leukocyte reaction, graft-versus-host-reaction, and skin allograft rejection in the lizard *Chalcides ocellatus. Immunobiology* **166**, 484–493.
73. Jayaraman, S. and Muthukkaruppan, V.R. (1977). *In vitro* correlate of transplantation immunity: spleen cell migration inhibition in the lizard *Calotes versicolor. Dev. Comp. Immunol.* **1**, 133–144.
74. Ramila, G. (1978). Studies on cellular interactions in the immune response and antigenic competition in the lizard *Calotes versicolor*. PhD Thesis, University of Madurai, India.
75. Farag, M.A. and El Ridi, R. (1985). Mixed leucocyte reaction (MLR) in the snake *Psammophis sibilans. Immunology* **55**, 173–181.
76. Sidky, Y.A. and Auerbach, R. (1968). Tissue culture analysis of immunological capacity of snapping turtles. *J. Exp. Zool.* **16**, 187–196.
77. Badir, N. Afifi, A. and El Ridi, R. (1981). Cell-mediated immunity in the gecko, *Tarentola annularis. Folia Biol.* **27**, 28–36.
78. Saad, A.-H., El Ridi, R., El Deeb, S. and Soliman, M.A.W. (1986). Effect of hydrocortisone on immune system of the lizard, *Chalcides ocellatus*. III Effect on cellular and humoral immune responses. *Dev. Comp. Immunol.* **10**, 235–246.
79. Saad, A.-H., El Ridi, R., El Deeb, S. and Soliman, M.A.W. (1987). Corticosteroids and immune system in the lizard *Chalcides ocellatus*. In *Developmental and Comparative Immunology*, Cooper, E.L., Langlet, C. and Bierne, J. (eds). Alan Liss, New York, pp.141–151.
80. Kaufman, J., Skjoedt, K., Salomonsen, J., Simonsen, M., Du Pasquier, L., Parison, R. and Riegart, P. (1990). MHC molecules in some non-mammalian vertebrates can be detected by some cross-reactive xenoantisera. *J. Immunol.* **144**, 2258–2272.
81. Kaufman, J., Skjoedt, K. and Salomonsen, J. (1990). The MHC molecules of non-mammalian vertebrates. *Immunol. Rev.* **90**, 83–117.
82. Kaufman, J., Flajnik, M. and Du Pasquier, L. (1991). The MHC molecules of ectothermic vertebrates. In *Phylogenesis of Immune Functions*,

Warr, G.W. and Cohen, N. (eds). CRC Press, Boca Raton, pp.125–169.
83. Kaufman, J., Thorpe, D., Skjoedt, K. and Salomonsen, J. (1989). MHC proteins and genes in birds and reptiles. *Dev. Comp. Immunol.* **13**, 374–375.
84. Bernheim, H.A. and Kluger, M.J. (1976). Fever and antipyresis in the lizard *Dipsosaurus dorsalis*. *Amer. J. Physiol.* **231**, 198–203.
85. Bernheim, H.A. and Kluger, M.J. (1977). Endogenous pyrogen-like substance produced by reptiles. *J. Physiol.* **267**, 659–666.
86. Cohen, N. and Haynes, L. (1991). The phylogenic conservation of cytokines. In *Phylogenesis of Immune Function*, Warr, G.W. and Cohen, N. (eds). CRC Press, Boca Raton, pp.241–268.
87. El Ridi, R., Wahby, A.F., Saad, A.-H. and Soliman, M.A.-W. (1987). Concanavalin responsiveness and interleukin-2 production in the snake *Spalerosophis diadema*. *Immunobiology* **174**, 177–189.
88. Kawaguchi, S., Maramatsu, S. and Mitsuhashi, S. (1978). Natural hemolytic activity in snake serum. I. Natural antibody and complement. *Dev. Comp. Immunol.* **2**, 287–296.
89. Koppenheffer, T.L. (1987). Serum complement systems of ectothermic vertebrates. *Dev. Comp. Immunol.* **11**, 279–286.
90. Jurd, R.D. and Doritis, A. (1977). Antibody-dependent cellular cytotoxicity in poikilotherms. *Dev. Comp. Immunol.* **1**, 341–352.
91. Le Douarin, L.M. (1973). A Feulgen-positive nucleolus. *Exp. Cell Res.* **77**, 459–468.
92. Lucas, A.M. and Jamroz, C. (1961). *Atlas of Avian Hematology*. US Dept. of Agriculture, Washington, D.C.
93. Dieterlen-Lièvre, F. (1988). Birds. In *Vertebrate Blood Cells*, Rowley, A.F. and Ratcliffe, N.A. (eds). Cambridge University Press, pp.257–336.
94. Hodges, R.D. (1974). *The Histology of the Fowl.* Academic Press, New York.
95. Maxwell, M.H. (1984). Histochemical identification of tissue eosinophils in the inflammatory response of the fowl (*Gallus domesticus*). *Res. Vet. Sci.* **37**, 7–11.
96. Fox, A.J. and Solomon, J.B. (1981). Chicken non-lymphoid leukocytes. In *Avian Immunology*, Rose, M.E., Payne, L.E. and Freeman, M.B. (eds). Poultry Science, Edinburgh, pp.135–166.
97. Peck, R., Murthy, K.K. and Vaino, O. (1982). Expression of B-L (Ia-like) antigens on macrophages from chicken lymphoid organs. *J. Immunol.* **129**, 4–5.
98. Grecchi, R., Saliba, A.M. and Mariano, M. (1980). Morphological changes, surface receptors and phagocytic potential of fowl mononuclear phagocytes and thrombocytes *in vivo* and *in vitro*. *J. Pathol.* **130**, 23–31.
99. Van Alten, P.J. (1982). Ontogeny of RES function in birds. In *The Reticuloendothelial System: A Comprehensive Treatise. Vol. 3: Ontogeny and Phylogeny*, Cohen, N. and Sigel, M.M. (eds). Plenum Press, New York, pp.659–685.
100. Palladino, M.A., Chi, D.S., Blyznak, N., Paolino, A.M. and Thorbecke, G.J. (1980). Cytotoxicity to allogeneic cells in the chicken. I. Role of macrophages in the cytotoxic effect on ^{51}Cr-labelled red blood cells by immune spleen cells. *Dev. Comp. Immunol.* **4**, 309–322.
101. Chen, C.H., Pickel, J.M., Lahti, J.M. and Cooper, M.D. (1990). Surface markers on avian cells. In *Avian Cellular Immunology*, Sharma, J.M. (ed.). CRC Press, Boca Raton, pp.1–22.
102. Albini, B. and Wick, G. (1975). Ontogeny of lymphoid cell surface determinants. *Int. Arch. Allergy Appl. Immunol.* **48**, 513–529.
103. Traill, K.N., Bock, B., Ratheiser, K. and Wick, G. (1984). Ontogeny of surface markers on functionally distinct T cell subsets in the chicken. *Eur. J. Immunol.* **14**, 61–67.
104. Gilmour, D.G., Brand, A., Donnelly, N. and Stone, H.A. (1976). Bu-1 and Th-1, two loci determining surface antigens of B or T lymphocytes in the chicken. *Immunogenetics* **3**, 549–563.
105. Fredericksen, T.L., Longenecker, B.M., Padzerka, F., Gilmour, D.G. and Ruth, R.F. (1977). A T-cell antigen system of chicken: Ly-4 and Marek's disease. *Immunogenetics* **5**, 535–553.
106. Chen, C.-L.H., Chanh, T.C. and Cooper, M.D. (1984). Chicken thymocyte-specific antigen identified by monoclonal antibodies: ontogeny, tissue distribution and biochemical characterization. *Eur. J. Immunol.* **14**, 385–391.
107. Kincade, P.W., Lawton, A.R., Bockman, D.E. and Cooper, M.C. (1970). Suppression of immunoglobulin G synthesis as a result of antibody-mediated suppression of immunoglobulin M synthesis in chicken. *Proc. Nat. Acad. Sci. US* **67**, 1918–1925.
108. Ivanyi, J. and Hudson, L. (1979). Allelic exclusion of M1 (IgM) allotype of the surface of chicken B cells. *Immunology* **35**, 941–945.
109. Ratcliffe, M.J.H. and Ivanyi, J. (1979). Allelic exclusion of surface IgM allotypes on spleen and bursal B cells in the chicken. *Immunogenetics* **9**, 149–156.
110. Fredericksen, T.L. and Gilmour, D.G. (1983). Ontogeny of Con A and PHA responses of chicken blood cells in MHC-compatible lines 6_3 and 7_2. *J. Immunol.* **130**, 2528–2533.

111. Higgins, D. (1991). Duck lymphocytes. IV. Collective effect of mitogens. *Dev. Comp. Immunol.* **15**, 357–368.
112. Weinbaum, F.I., Gilmour, D.G. and Thorbecke, G.J. (1973). Immunocompetent cells of the chicken. III. Co-operation of carrier-sensitized T cells from agammaglobulinemic donors with hapten immune B cells. *J. Immunol.* **110**, 1434–1438.
113. Veromaa, T. and Toivanen, P. (1990). Cellular co-operation in immune response and role of MHC. In *Avian Cellular Immunology*, Sharma, J.M. (ed.). CRC Press, Boca Raton, pp.23–33.
114. Blaese, R.M., Weiden, P.C., Koshi, I. and Dooley, N. (1974). Infectious agammaglobulinaemia: transmission of immunodeficiency with grafts of agammaglobulinaemic cells. *J. Exp. Med.* **140**, 1097–1101.
115. Droege, W. (1976). The antigen-inexperienced thymic suppressor cells: class of lymphocytes in the young chicken thymus that inhibits antibody production and cell-mediated responses. *Eur. J. Immunol.* **6**, 279–287.
116. Grebenau, M.D., Chi, D.S. and Thorbecke, G.J. (1979). T-cell tolerance in the chicken. II. Lack of evidence for suppressor cells in tolerant agammaglobulinemic and normal chickens. *Eur. J. Immunol.* **9**, 477–485.
117. Quere, P. and Thorbecke, G.J. (1991). Suppressor cells for antibody production *in vivo*, induced by bursa cell injection into agammaglobulinemic chickens, belong to a $CT4^-$, $CT8^+$, $TCR1^-$ subset of T cells. *Dev. Comp. Immunol.* **15**, 73–82.
118. Kirchner, H. and Blaese, R.M. (1973). Pokeweed mitogen, concanavalin A- and phytohaemagglutinin-induced development of cytotoxic effector lymphocytes. *J. Exp. Med.* **138**, 812–824.
119. Sharma, J.M. and Schat, K.A. (1990). Natural immune functions. In *Avian Cellular Immunology*, Sharma, J.M. (ed.). CRC Press, Boca Raton, pp.51–70.
120. Sieminski-Brodzina, L.M. and Marshaly, M.M. (1991). Characterization by scanning and transmission electron-microscopy of avian peripheral blood mononuclear cells exhibiting natural killer-like (NK) activity. *Dev. Comp. Immunol.* **15**, 181–188.
121. Campbell, F. (1967). Fine structure of the bone marrow of the chicken and pigeon. *J. Morphol.* **123**, 405–440.
122. Payne, L.N. and Powell, P.C. (1984). The lymphoid system. In *Physiology and Biochemistry of the Domestic Fowl*, Freeman, B.M. (ed.). Academic Press, London, pp.277–321.
123. Sorrell, J.M. and Weiss, L. (1980). Cell interactions between hematopoietic and stromal cells in the embryonic chick bone marrow. *Anat. Rec.* **197**, 1–19.
124. Jotereau, F.V. and Le Douarin, N.M. (1978). The developmental relationship between osteocytes and osteoclasts: a study using the quail-chick nuclear marker in endochondral ossification. *Dev. Biol.* **63**, 253–265.
125. Guillemot, F.P., Oliver, P.D., Péault, B.M. and Le Douarin, N.M. (1984). Cells expressing Ia antigens in the avian thymus. *J. Exp. Med.* **160**, 1803–1819.
126. Nahamura, H. and Ayer-Le Lievre, C. (1986). Neural crest and thymic myoid cells. *Curr. Topics Dev. Biol.* **20**, 111–115.
127. Moore, M.A.S. and Owen, J.J.T. (1965). Chromosome marker studies on the development of the haemopoietic system in the chick embryo. *Nature* **208**, 989–990.
128. Moore, M.A.S. and Owen, J.J.T. (1967). Experimental studies on the development of the thymus. *J. Exp. Med.* **126**, 715–723.
129. Jotereau, F.V. and Le Douarin, N.M. (1982). Demonstration of a cyclic renewal of the lymphocyte precursor cells in the quail thymus during embryonic and perinatal life. *J. Immunol.* **129**, 1869–1877.
130. Le Douarin, N.M., Dieterlen-Lièvre, F. and Oliver, P.D. (1984). Ontogeny of primary lymphoid organs and lymphoid stem cells. *Amer. J. Anat.* **170**, 261–299.
131. Glick, B., Chang, T.S. and Jaap, R.C. (1956). The bursa of Fabricius and antibody production. *Poultry Sci.* **35**, 224–225.
132. Toivanen, P., Toivanan, A. and Good, R.A. (1972). Ontogeny of bursal function in the chicken. III. Immunocompetent cells for humoral immunity. *J. Exp. Med.* **136**, 816–831.
133. Jalkanen, S., Granfors, K., Jalkanen, M. and Toivanen, P. (1983). Immune capacity of the chicken bursectomized at 60 hours of incubation: failure to produce immune, natural and auto-antibodies in spite of immunoglobulin production. *Cell Immunol.* **80**, 363–373.
134. Jalkanen, S., Jalkanen, M., Granfors, K. and Toivanen, P. (1984). Defect in the generation of light chain diversity in bursectomized chickens. *Nature* **311**, 69–71.
135. Belo, M., Martin, C., Corbel, C. and Le Douarin, N.M. (1985). A novel method to bursectomize avian embryos and obtain quail-chick bursal chimeras. I. Immunocytochemical analysis of such chimeras by using species-specific monoclonal antibodies. *J. Immunol.* **135**, 3785–3794.
136. Péault, B.M., Thiéry, J.P. and Le Douarin, N.M. (1983). Tissue distribution and ontogenetic

emergence of differentiation antigens on avian T cells. *Proc. Nat. Acad. Sci. US* **80**, 2976–2980.
137. Toivanen, A. and Toivanen, P. (1977). Histocompatibility requirements for cellular co-operation in the chicken: generation of germinal centers. *J. Immunol.* **118**, 431–436.
138. Le Douarin, N.M., Houssaint, E., Jotereau, F.V. and Belo, M. (1975). Origin of hemopoietic stem cells in embryonic bursa of Fabricius and bone marrow studied through interspecific chimeras. *Proc. Nat. Acad. Sci. US* **72**, 2701–2705.
139. Houssaint, E., Belo, M. and Le Douarin, N.M. (1976). Investigations on cell lineage and tissue interactions in the developing bursa of Fabricius through interspecific chimeras. *Dev. Biol.* **53**, 250–264.
140. Pink, J.R.L., Ratcliffe, M.J.M. and Vainio, O. (1985). Immunoglobulin-bearing stem cells for clones of B (bursa-derived) lymphocytes. *Eur. J. Immunol.* **15**, 617–620.
141. Pink, J.R.L., Vainio, O. and Rijnbeck, A.-M. (1985). Clones of B lymphocytes in individual follicles of the bursa of Fabricius. *Eur. J. Immunol.* **15**, 83–87.
142. Olah, I. and Glick, B. (1982). Splenic white pulp and associated vascular channels in chicken spleen. *Amer. J. Anat.* **165**, 445–480.
143. Glick, B. (1982). RES structure and function of the Aves. In *The Reticuloendothelial System: A Comprehensive Treatise. Vol. 3: Phylogeny and Ontogeny*, Cohen, N. and Sigel, M.M. (eds). Plenum Press, New York, pp.509–540.
144. Gallego, M. and Glick, B. (1988). The proliferative capacity of the cells of the avian Harderian gland. *Dev. Comp. Immunol.* **12**, 157–166.
145. Martin, A., McNabb, F.M.A. and Siegel, P.B. (1988). Thiouracil and antibody titers of chickens from lines divergently selected for antibody response to sheep erythrocytes. *Dev. Comp. Immunol.* **12**, 611–619.
146. Leslie, G.A. and Clem, L.W. (1969). Phylogeny of immunoglobulin structure and function. III. Immunoglobulins of the chicken. *J. Exp. Med.* **130**, 1337–1352.
147. Hädge, D. and Ambrosius, H. (1988). Comparative studies of the structure of biliary immunoglobulins of some avian species. II. Antigenic properties of the biliary immunoglobulins of chicken, turkey, duck and goose. *Dev. Comp. Immunol.* **12**, 319–329.
148. Burns, R.B. and Maxwell, M.H. (1981). Probable occurrence of IgE in the adult domestic fowl (*Gallus domesticus*) after horse serum stimulation. *Vet. Res. Commun.* **5**, 67–72.
149. Leslie, G.A. (1980). Idiotypy and IgD isotypy of chicken immunoglobulins. In *Phylogeny of Immunological Memory*, Manning, M.J. (ed.). Elsevier/North Holland, Amsterdam. pp.253–263.
150. Pardue, S.L. (1990). Autosomal albinism affects immunocompetence in the chicken. *Dev. Comp. Immunol.* **14**, 105–112.
151. Reynaud, C.-A. Dahan, A. and Weill, J.-C. (1983). Complete sequence of a chicken lambda-light chain immunoglobulin derived from the nucleotide sequence of its mRNA. *Proc. Nat. Acad. Sci. US* **80**, 4099–4103.
152. Reynaud, C.-A. Anquez, V., Grimal, H. and Weill, J.-C. (1987). A hyperconversion mechanism generates the chicken light chain preimmune repertoire. *Cell* **48**, 379–388.
153. Reynaud, C.-A., Dahan, A., Anquez, V. and Weill, J.-C. (1989). Development of the chicken antibody repertoire. In *Immunoglobulin Genes*, Honjo, T., Alt, F. and Rabbitts, T.H. (eds). Academic Press, San Diego, pp.151–162.
154. Weill, J.-C., Reynaud, C.-A., Lassila, O. and Pink, J.R.L. (1986). Rearrangement of chicken immunoglobulin genes is not an ongoing process in the embryonic bursa of Fabricius. *Proc. Nat. Acad. Sci. US* **83**, 3336–3340.
155. Alm, G.V. (1979). *In vitro* studies of chicken lymphoid cells III. The mixed spleen leucocyte reaction with special reference to the effect of bursectomy. *Acta Path. Microbiol. Scand.* **79**, 359–365.
156. Miggiano, V., Birgen, I. and Pink, J.R.L. (1974). The mixed leukocyte reaction in chickens. Evidence for control by the major histocompatibility complex. *Eur. J. Immunol.* **4**, 397–401.
157. Erf, G.F. and Marsh, J.A. (1989). Effect of dietary triiodothyronine on mixed-lymphocyte responsiveness in young male chickens. *Dev. Comp. Immunol.* **13**, 177–186.
158. Lehtonen, L., Vainio, O. and Toivanen, P. (1989). Ontogeny of alloreactivity in the chicken as measured by mixed lymphocyte reaction. *Dev. Comp. Immunol.* **13**, 187–195.
159. Desveaux-Chabrol, J., Gendreau, M. and Dieterlen-Lièvre, F. (1991). Serial transfer of GVH-R splenomegaly in chicken embryos. *Dev. Comp. Immunol.* **15**, 341–347.
160. Kroemer, G., Bernot, A., Béhar, G., Chaussé, A.-M., Gastinel, L.-N., Guillemot, F., Park, I., Thoraval, P., Zoorob, R. and Auffray, C. (1990). Molecular genetics of the chicken MHC: current status and evolutionary aspects. *Immunol. Rev.* **113**, 119–145.
161. Miller, M.M. (1991). The major histocompatibility complex of the chicken. In *Phylogenesis of Immune Function*, Warr, G.W. and Cohen, N. (eds). CRC Press, Boca Raton, pp.151–170.

162. Kaufman, J. and Şalomonsen, J. (1992). B-G: we know what it is, but what does it do? *Immunol. Today* **13**, 1–3.
163. Schat, K.A. (1990). T-cell immunity: mechanisms and soluble mediators. In *Avian Cellular Immunology*, Sharma, J.M. (ed.). CRC Press, Boca Raton, pp.35–50.
164. Vlaovic, M.S., Buening, G.M. and Loan, R.W. (1975). Capillary-tube leukocyte migration inhibition as a correlate of cell-mediated immunity in the chicken. *Cell. Immunol.* **17**, 335–341.
165. Trifonow, S., Solirow, N. and Filchev, A. (1977). *In vitro* migration of peritoneal and spleen cells and its inhibition in some avian species. *Cell. Immunol.* **21**, 361–369.
166. Holt, P.S. (1990). Enhancement of chicken lymphocyte activation and lymphokine release by avian influenza virus. *Dev. Comp. Immunol.* **14**, 447–455.
167. Klasing, K.C. and Peng, R.K. (1987). Influence of cell sources, stimulating agents and incubation conditions on release of interleukin-1 from chicken macrophages. *Dev. Comp. Immunol.* **11**, 385–394.
168. Joshi, P. and Glick, B. (1990). The role of avian macrophages in the production of avian lymphokines. *Dev. Comp. Immunol.* **14**, 319–325.
169. Lee, T.-H. and Tempelis, C.H. (1991). Characterization of a receptor on activated chicken T cells. *Dev. Comp. Immunol.* **15**, 329–339.
170. Schauenstein, K., Kromer, G., Fassler, R. and Wick, G. (1987). Implications of IL-2 in normal and disturbed immune functions in the chicken. In *Developmental and Comparative Immunology*, Cooper, E.L., Langlet, C. and Bierne, J. (eds). Alan Liss, New York, pp.69–77.
171. Matsuda, H. and Mikami, T. (1990). Features of cellular immune systems in avian species other than chickens. In *Avian Cellular Immunology*, Sharma, J.M. (ed.). CRC Press, Boca Raton, pp.133–138.
172. Wick, G., Brezinschek, H.P., Hála, K., Dietrich, H., Wolf, H. and Kroemer, G. (1989). The obese strain of chickens: an animal model with spontaneous autoimmune thyroiditis. *Adv. Immunol.* **47**, 433–500.
173. Kroemer, G., Schauenstein, K., New, N., Stricker, K. and Wick, G. (1985). *In vitro* T cell hyperreactivity in obese strain (OS) chickens is due to a defect in non-specific suppressor mechanism(s). *J. Immunol.* **135**, 2458–2463.
174. Kaplan, M.H., Sundrich, R.S. and Rose, N.R. (1990). Autoimmune diseases. In *Avian Cellular Immunology*, Sharma, J.M. (ed.). CRC Press, Boca Raton, pp.183–197.
175. Muthukkaruppan, V.R., Borysenko, M. and El Ridi, R. (1982). RES structure and function of the Reptilia. In *The Reticuloendothelial System: A Comprehensive Treatise. Vol. 3: Phylogeny and Ontogeny*, Cohen, N. and Sigel, M.M. (eds). Plenum Press, New York, pp.461–508.
176. Cooper, M.D., Chen, C.-L., Bucy, R.P. and Thompson, C. (1991). Avian T Cell ontogeny. *Adv. Immunol.* **50**, 87–117.
177. Glick, B. (1983). Bursa of Fabricius. In *Avian Biology, Vol. VII*, Farner, D.S., King, R. and Parkes, K.C. (eds). Academic Press, New York, pp.443–500.

6

MAMMALS

R.J. Turner

Institute of Biological Sciences, University of Wales, Aberystwyth, UK

6.1 INTRODUCTION

Mammals have solved the problems of living on land in particularly sophisticated and integrated ways, involving close control of body temperature and chemistry, elevated metabolic rate, and elaborate parental care. The most striking differences among the three main evolutionary groups are seen in their reproductive systems, although all of them, by definition, provide their young with milk. Thus the monotremes lay eggs instead of giving birth to live young; marsupials have a very brief gestation, after which the youngsters migrate to a pouch; and eutherian mammals show a characteristically long and intimate connection between the developing foetus and its mother.

The evolution of viviparity and the provision of milk raise important immunological questions. Medawar[1] suggested, some 40 years ago, that in outbred populations the relationship between a foetus and its mother is akin to a foreign graft. However, the factors which determine foetal survival versus 'rejection' have still not been fully determined: apart from the sheer complexity of the system in eutherians, one of the difficulties is that different 'model' species show differing degrees of invasion, destruction and interaction at the interface between the uterine wall and the trophoblast formed around the embryo. Marsupials could provide simpler models here, but investigations are at an early and controversial stage: authors are unable to agree whether marsupials even possess a trophoblast[2,3]. Nevertheless, continued efforts to identify the key immunological mechanisms which assist the survival of sperm and the foetus in allogeneic females could be rewarded by improved treatments of infertility, new methods of contraception, and an increased battery of natural immunosuppressive agents available for clinical use.

The eutherians, of course, are the most successful and diverse group of mammals, accounting for nearly a thousand living genera, or 93% of the total in the Class[2]. However, immunological studies have focused on a small number of favoured models, notably the laboratory mouse. The latter's cheapness, small size, short life span,

Immunology: A Comparative Approach. Edited by R.J. Turner

ready availability, large number of inbred strains, plethora of anti-mouse reagents and (one suspects) a relative lack of public sympathy for it, have all contributed to its success in this regard. This has meant that the mouse is used as the main immunological yardstick for comparative work. Unfortunately, in evolutionary terms the rodents are isolated, and attempts to relate their ancestry to other groups of mammals have been in vain[4]. A representative of the Insectivora would have provided a more suitable yardstick in view of their pivotal position in eutherian evolution (Figure 6.1), but this Order has received little immunological attention. Rodent models, on the other hand, exist for a wide variety of immunological functions and malfunctions, and an attempt will be made in this chapter to illustrate the strengths and weaknesses of such models: where possible, detailed comparisons will be made with other types of mammal, including man and domestic animals. Immunological studies on the latter are usually pigeonholed in the 'veterinary' section[5]. However, these animals deserve a wider audience since their large size makes them suitable for *in vivo* manipulations that would be extremely difficult in laboratory rodents. Moreover, detailed investigation (e.g. on the Peyer's patches of sheep and the endometrial cups of horses) has revealed some sharp contrasts with the 'typical' eutherian immune system as portrayed by rodents.

A consequence of the early birth of marsupials is that many developmental events which occur *in utero* in the eutherians, take place instead in the pouch. Although the newborn marsupial is tiny (less than a gram body weight in all species), its accessibility makes it very useful for studying the exposure of immunologically immature mammals to antigenic stimulation, and the effects of removing the thymus gland at a 'foetal' stage. The immunological roles of antibodies and leucocytes from the marsupial mammary gland are also of particular interest, since the lactation period lasts over 6 months in many species and accounts for

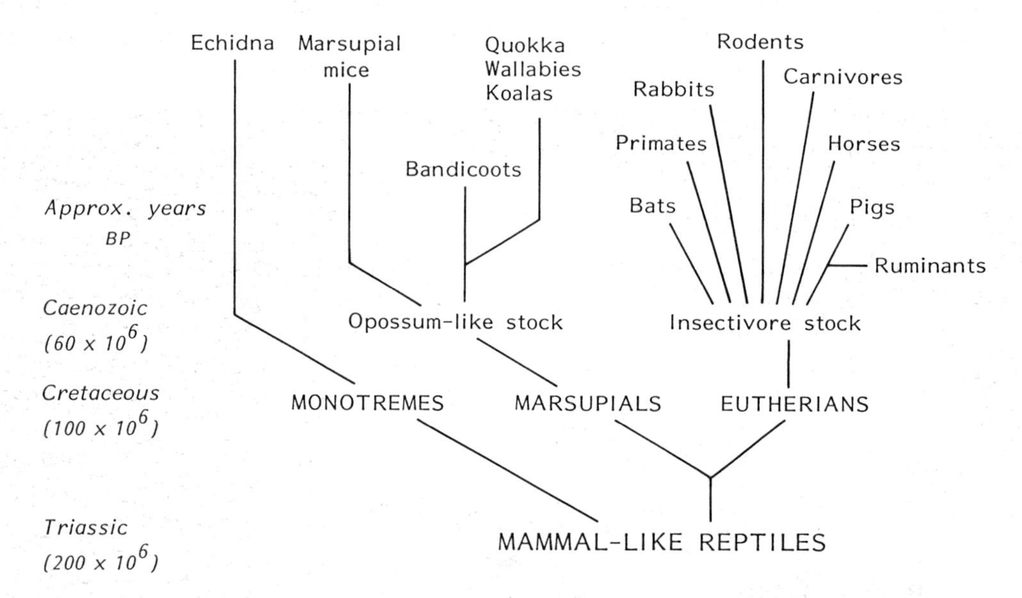

Figure 6.1. A simplified view of mammalian evolution, showing relationships of the different species used for immunological research. The eutherian evolutionary 'tree' is more comparable to a great bush: the different groups are the product of a very rapid evolutionary change beginning in the Caenozoic, and their affinities are difficult to delineate with any authority[8]. Similarly, the classification of marsupials has posed major difficulties for taxonomists, because the fossil record for early marsupials is poor, and comparative anatomy has produced inconsistencies.

most of their morphological development[2].

In view of the merits of marsupials as immunological models for reproductive and developmental studies, it is disappointing that, after a minor flurry of papers in the 1960s and early 70s, research on them has failed to keep pace with eutherian work[183]. One reason for this was the reliance on wild animal populations[6]. However, four marsupial species are now established as self-sustaining colonies, thus providing a good supply of laboratory animals of known source and genetic constitution. These are the grey short-tailed opossum, *Monodelphis domestica* (Figure 6.2), the tammar or dama wallaby, *Macropus eugenii*, another wallaby (Bennett's or red-necked), *Macropus rufogriseus*, and the fat-tailed dunnart (a marsupial mouse), *Sminthopsis crassicaudata*[6]. To the author's knowledge, two other marsupials which figure prominently in earlier immunological work, namely the North American (Virginia) opossum, *Didelphis virginiana*, and the Australian quokka, *Setonix brachyurus*, have not yet been established in self-sustaining colonies.

The availability of several marsupial species with known genetic constitution should broaden mammalian comparative studies, for example on the major histocompatibility complex (MHC). The improving prospects for immunological research on opossums are especially welcome, since this group most closely resembles the 'stock' marsupial (Figure 6.1) and may be the least modified of all viviparous mammals[2].

Finally, the study of monotremes offers a different kind of evolutionary insight. These Australasian creatures comprise a fascinating and bizarre mix of mammalian, reptilian and specialist features[7], and most workers believe that they represent an ancient and separate line (or lines) of descent from the mammal-like reptiles (Figure 6.1). Thus they could suggest to us what the immune systems of mammals were like before viviparity had evolved in the Class, but after internal control of body temperature had been at least partially attained. Not surprisingly, however, the limited availability of the monotremes at the present time has ensured that only a few privileged immunologists have been able to work on them: experimental studies to date consist of a few papers on just one species, the echidna, *Tachyglossus aculeatus*[183].

Figure 6.2. Female opossum with her young. The species shown (*Monodelphis domestica*) is pouchless. It is indigenous to Brazil, but has adapted well to laboratory conditions. Photograph by courtesy of W.H. Stone.

6.2 IMMUNOLOGICAL ASPECTS OF MAMMALIAN REPRODUCTION

6.2.1 Survival of Sperm

The survival of allogeneic sperm in the female tract is not a problem faced only by mammals—all terrestrial vertebrates and many aquatic ones fertilize internally— but it is in the higher mammals (especially rodents, rabbits and primates) that explanations have been sought[9,10].

There are conflicting reports over the degree of expression of MHC antigens by sperm, and this may differ from species to species[9]. Sperm-specific differentiation antigens are more widely accepted as a potential target and these, unlike the MHC antigens, can provoke immune responses in the male producing them as well as in sexual partners: since these autoantigens appear only at puberty, they will not be available to tolerize the host at the same time as his other 'self' components. In the testis of the normal male, however, the risk of autoimmunity is reduced by tight junctions of Sertoli cells which line the seminiferous tubules; these keep the developing sperms and immunocompetent cells apart. Further along, the male reproductive tubes are more permeable to antigen and cells, and here anti-sperm reactions are limited by suppressor T-cells in the linings. Deliberate breakdown of the blood–testis barrier as a means of male contraception has been achieved in domestic ruminants, carnivores and rabbits by local injection of a vaccine consisting of a mycobacterial extract; this allows leucocytes to invade and inhibit sperm production without causing damage to other components of the reproductive system[11].

Studies on different eutherian species have consistently shown that the female reproductive tubes contain antibodies, complement and leucocytes, and can produce immune responses. Fertilization of eggs can be blocked *in vitro* by coating sperm with antibody, and *in vivo* by systemic immunization with sperm antigens. However, such responses are obviously not the norm when sperm are introduced sexually (and account for only some cases of female infertility). It has been speculated that survival of sperms is achieved partly by minimizing their access to the uterus except at the most propitious times—humoral defences in the human cervical mucus are lowest at ovulation—and partly by phagocytosis of many sperms when they do reach the uterus. Thus, it is argued, the middle and upper reaches of the reproductive tract will remain poorly sensitized (though it seems reasonable to suppose that at least some of these sperm-consuming phagocytes in the uterus will act as antigen-presenting cells). A better substantiated explanation is that seminal fluid has potent immunosuppressive properties, which could act in the distal male ducts as well as in the female. The molecules responsible have a wide range of molecular weights and stabilities, and they include zinc-containing compounds, proteins, polyamines and prostaglandins of the E series. Some appear to act by masking or modifying antigens on the sperm surface, some by interfering with leucocyte function, others by inhibiting antibodies or complement[10,12]. Small vesicles (prostasomes) with immunosuppressive effects have also been identified in the fluid[174]. Although species cross-reactivity has been described, the components of this suppressive cocktail differ in detail for different animals: it has been reported, for example, that there is an immunosuppressive uteroglobin in the prostatic fluid of rabbits but not in other species[13]. It seems likely, therefore, that reported species variations in the potency of semen for artificial insemination[9,14] depend not only on the concentration of sperm and their viability after storage, but also on differences in their immunogenicity and the immunological make-up of the fluid surrounding them.

6.2.2 Survival of the Foetus

This topic has produced a voluminous and often conflicting immunological literature, and many questions are still unresolved[9,11,15–19,167]. Studies on humans have been complemented by animal models, chiefly mice: the latter allow detailed examination of the uterus, trophoblast and embryo at any chosen stage of gestation; manipulation of the genetic relationship between mother and foetus; and the use of females with selected immunodeficiencies. In common with humans, rodents develop an invasive, destructive trophoblast leading to the formation of a 'thin' discoid placenta. On the other hand, the uterine decidual reaction is more extensive in rodents, their embryonic yolk sac is relatively large and used for the transmission of maternal antibodies to the foetus, and their young are born at a less advanced stage of immunological development (see Sections 6.2.3 and 6.3).

The trophoblast is undoubtedly a key structure in determining the success or failure of pregnancy. In both rodents and humans it presents a low antigenic profile, and is resistant to immunological attack. In rats and mice, polymorphic MHC antigens of class 1 but not 2 have been detected on the trophoblast at low levels, although reports do not agree on their distribution within the structure[20]. In mice, although there is equal transcription of paternally and maternally derived genes, paternally inherited allotypes are sometimes more difficult to detect serologically, perhaps because of masking by maternal alloantibody. There is also a differential expression of murine MHC class 1 genes, which suggests a complex pattern of gene regulation during gestation[21].

In humans, no classical MHC antigens are detectable on that part of the trophoblast which is in closest contact with maternal tissues. This absence would account for the failure to sensitize maternal T- and B-cells against paternal antigens in most women but would, by the same token, present a target for natural killer (NK) cells. However, a non-classical, non-polymorphic MHC antigen (HLA-G) is expressed on the human trophoblast, and it has been proposed that a population of NK-like CD56^{+} large granular lymphocytes which appear in large numbers in the uterine lining only during the early stages of pregnancy interact with this unusual MHC molecule in controlling the degree of invasion of maternal tissues by the trophoblast[22]. More HLA-G expression by the foetal side would aid invasion; more activation of the NK-like cells by maternal cytokines would limit it.

It is not clear whether rodents and other mammals operate a similar controlling mechanism. Comparisons of the human MHC with that of cotton-top tamarin monkeys suggest that evolutionary pressures adapted a classical class 1 molecule for a role in placentation at the level of the higher primates by inducing a loss of polymorphism[19]. However, it has been reported that rats have an HLA-G analogue (Pa/K) and uterine NK-like cells. Similar cells, described as granulated endometrial gland (GMG) cells or endometrial granulocytes, occur in various other rodents (including mice), bats, horses and the elephant shrew (an insectivore)[23], but in some species the GMG cells are commonly seen in virgin uteri and do not appear to influence trophoblast invasion[175].

Although the rodent and human trophoblasts seem designed to minimize the likelihood of maternal T- and B-cell sensitization by allogeneic foetal material, the occasional presence of cytotoxic T-cells or antibodies with anti-paternal reactivity indicates that either the barrier does not maintain a uniformly low profile throughout gestation, or that foetal cells or antigens can escape across it into the maternal circulation. In horses, unlike rodents or humans, there is a consistent and strong cytotoxic antibody response by females during normal first pregnancy, probably induced by high levels of classical MHC-1 antigens expressed on trophoblast cells before and during uterine invasion. After invasion, these cells become isolated and transform into ulcer-like endometrial cups, which are peculiar to the horse family (Figure 6.3A). The cups seem to act as a target for maternal leucocytes (Figure 6.3B), although by this stage their MHC antigens are difficult to detect serologically[20]. Anti-MHC antibodies also appear frequently in parous sheep and cattle, though not with the same regularity as horses[61,62].

The horse and ruminant findings illustrate that immune responses do not normally jeopardize pregnancy. Similarly, cytotoxic anti-HLA antibodies in pregnant women are not associated with spontaneous abortion, stillbirths or developmental abnormalities[24]. These cytotoxic antibodies should perhaps be regarded as a controllable nuisance (except by the tissue typist), but other immune responses generated may actually

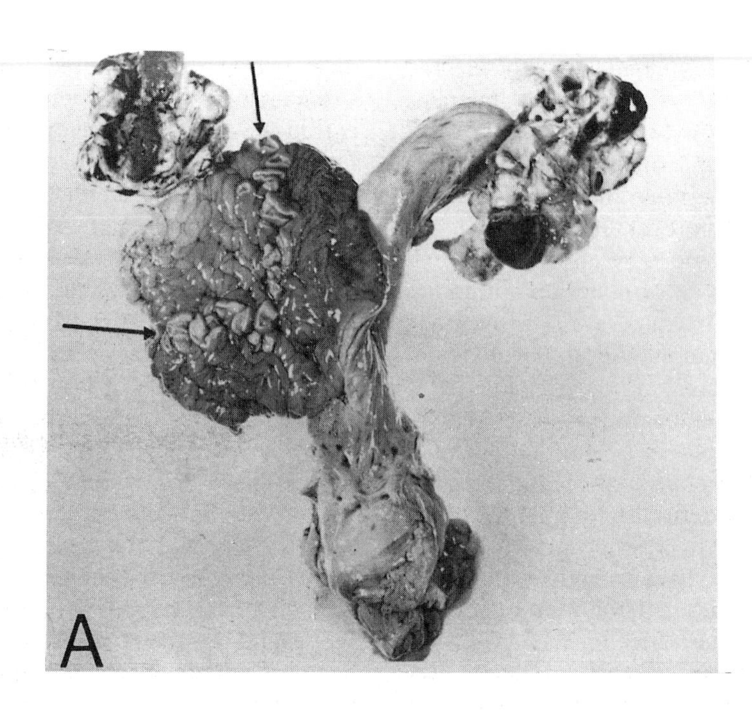
A

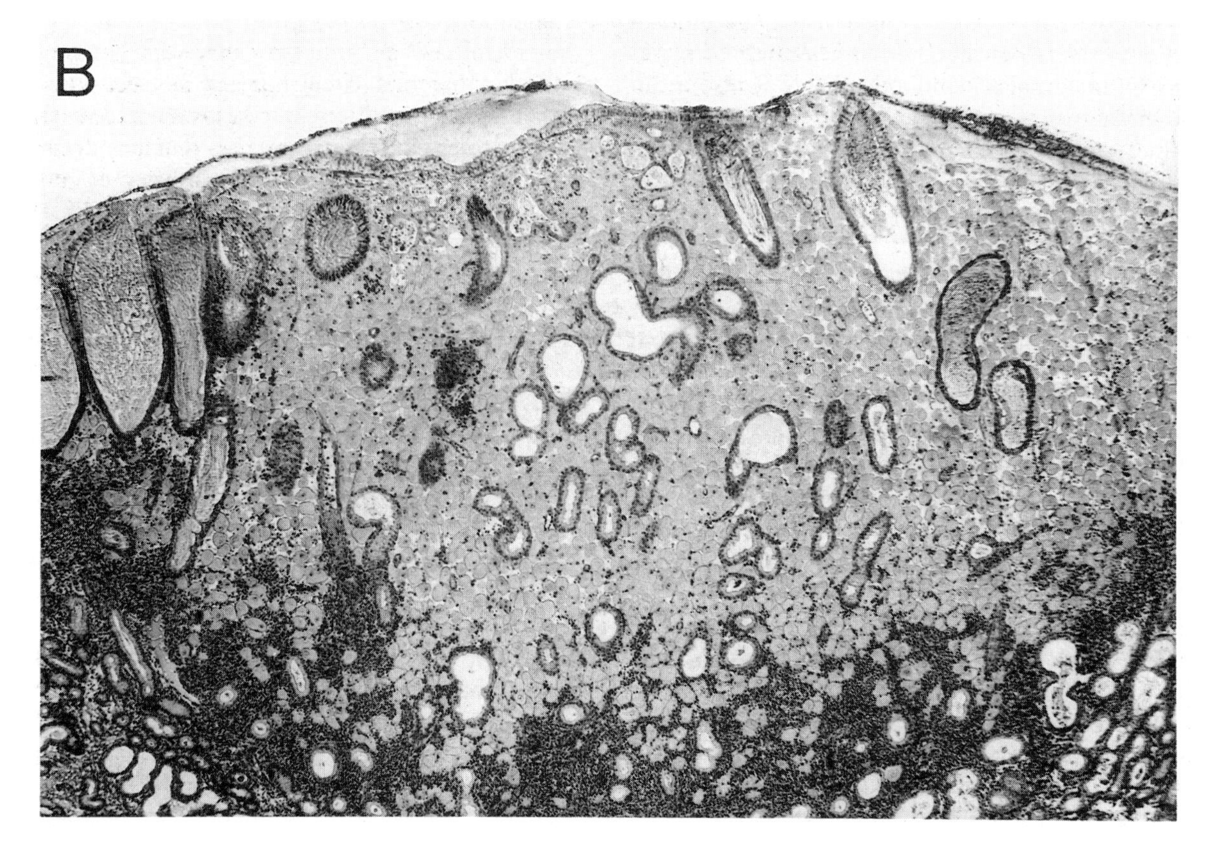
B

improve the chances of reproductive success. Antibodies which do not fix complement, for example, seem to mask antigens or block the function of maternal leucocytes in certain strain combinations of mice; and the risk of spontaneous abortion is greater in women who lack blocking antibodies, though these antibodies are not essential for normal pregnancy[25].

Studies on mice begun over 20 years ago showed that allogeneic pairings produced more invasive trophoblasts and larger placentas as well as larger foetuses compared with syngeneic pairings[15]. Moreover, the outcome could be altered by deliberate preimmunization or tolerization of the mother with paternal antigens. Similarly, it has been reported that women who suffer repeated abortions are more likely to share HLA antigens with their husbands than women who give birth successfully; administration of pooled allogeneic leucocytes was found to improve their prospects of completing a pregnancy[25], though this has to be balanced against the risk of infections and graft-versus-host reactions. It has been suggested that activated immune cells may assist trophoblastic growth through the secretion of cytokines; however, *in vitro* studies have identified inhibitory cytokines also, and any net effect is likely to depend on an array of interactions[17,18,21,167,176,177].

Where antigenic differences between mother and foetus are large, gestation sometimes proceeds but does not go to completion. This has been demonstrated in experiments on xenogeneic combinations[17,182]. After transfer of pre-implantation embryos from a species of wild mouse, *Mus caroli*, to the closely related laboratory mouse, *Mus musculus*, the embryos develop normally to mid-gestation, and are then infiltrated and destroyed by lymphocytes. Similarly, most donkey embryos transferred to horse mares grow normally then die, with intense accumulations of lymphocytes appearing along the entire foetal–maternal interface. These deaths are due to trophoblast failure: *M. caroli* embryos will only complete development in *M. musculus* if they are first given a 'host' type trophoblast, and the same is true for embryo transfers between sheep and goats[26]. Female equids are more accommodating: apart from the donkey-to-horse example, they show a remarkable ability to carry embryos of different species as well as their own (Figure 6.4), but this ability is still linked to the successful development of endometrial cups from the trophoblast (Table 6.1). Table 6.1 also shows that these embryo transfer techniques in equids can be used to assist the recovery of endangered species[27]. Immunization of surrogate mares with donkey lymphocytes improves the survival rate of donkey embryos, thus substantiating the results of immunization in mice and humans described above[17,182].

There is evidence in some mammals that populations of suppressor leucocytes are generated during normal pregnancy, presumably to counter any destructive immune responses that may occur. Small non-T granulated lymphocytes with suppressor activity appear in the uterine lining of mice which successfully carry allogeneic foetuses, but they are missing in the DBA/2J × CBA/J combination and the *M. caroli* × *M. musculus* xenogeneic combination which both spontaneously abort[17,28]. Depending on the paternal/maternal strain combination, lymph nodes draining the uterus undergo hypertrophy during allogeneic pregnancies, and their removal can result in impaired fertility[9]. However, claims that these nodes are responsible for generating the suppressor cells

Figure 6.3 (opposite). Endometrial cups of the horse. (A) Pregnant uterus at day 63 of gestation. The body and gravid horn have been opened and the conceptus removed to reveal the line of raised, ulcer-like cups (arrowed) forming a circle around the conceptus. The cups reach their maximum size at this stage. (B) Low-power (×36) section of a cup recovered from a mare at day 73 of gestation. The cup consists of hormone-secreting foetal trophoblast cells interspersed with distended endometrial glands and occasional blood vessels. Dense accumulations of maternal leucocytes are seen at the edge of the cup (bottom of picture). $CD4^+$ and $CD8^+$ lymphocytes occur in similar numbers in these accumulations, but more detailed characterization of the cells present awaits the development of additional equine leucocyte markers. Photographs by courtesy of W.R. Allen[164].

Figure 6.4. (A) A 6-day-old horse foal (*Equus caballus*) with its surrogate donkey mother (*E. asinus*), following extraspecific transfer of the early blastocyst on the 7th day after ovulation. (B) A 2-day-old Grant's zebra foal (*E. burchelli*) with its surrogate domestic horse mother. The embryo was again transferred 7 days after ovulation. Photos by courtesy of W.R. Allen, reproduced from references **165, 166** with permission.

have been disputed[29]. Large and small suppressor cells have also been detected in the human uterus[17]. In horses, an immunosuppressive function was suggested for the dense accumulations of leucocytes which form around the endometrial cups during successful pregnancies, but more recent work has been unable to confirm this (W.R. Allen, personal communication).

The normal foetus does not rely only on maternal mechanisms for its defence. Diverse humoral factors are produced in the placenta, including pregnancy-associated proteins, steroids, prostaglandins and cytokines[9,15–18], which could exert immunosuppressive effects near their site of production. Moreover, immunosuppressive cells of foetal origin have been identified, for example human Ts cells with an unusual phenotype ($CD4^{+}8^{-}$)[15]. The trophoblast has also been likened to a sponge or decoy, describing its ability to mop up or deflect attack by anti-foetal cells or antibodies. The latter pose the greater potential danger in the sense that they can pass with relative ease into the foetus (depending on species and isotype—see Table 6.2 and Section 6.2.3). Complement-mediated attack against the trophoblast itself may be inhibited by membrane-bound complement modulatory proteins: two of these, membrane cofactor protein (MCP) and

Table 6.1 Successful pregnancies in the horse family.

Foetal genotype	Mare genotype	Endometrial cup development	Usual outcome of pregnancy
Intraspecific mating			
Horse	Horse	+	+
Donkey	Donkey	+	+
Zebra	Zebra	+	+
Interspecific mating			
Mule	Horse	+	+
Hinny	Donkey	+	+
Extraspecific embryo transfer			
Przewalski's horse	Horse	+	+
Horse	Donkey	+	+
Horse	Mule	+	+
Donkey	Mule	+	+
Zebra	Donkey	+	+
Zebra	Horse	+	+
Donkey	Horse	−	−

Data from W.R. Allen *et al.*, 1987[182]
+=successful; −=unsuccessful

decay accelerating factor (DAF), have been identified on human trophoblast; they may also function in protecting sperm[18].

According to some authors, maternal T-cells which manage to cross the placenta and encounter an immunoincompetent foetus (e.g. in mouse or human) cause graft-versus-host disease. However, foetal rhesus monkeys halfway through gestation are not runted by injection of allogeneic bone marrow cells[48]. Studies on the effects of cytotoxic murine T-cells *in vitro* and *in vivo* suggest that neither placental nor foetal cells are susceptible to attack because they do not express paternal histocompatibility antigens in an 'appropriate' format[30].

In view of the work described above, it is interesting that agammaglobulinaemic women and mice, thymusless nude mice and mice with severe combined immunodeficiency (given a germ-free environment) are all capable of producing young successfully[9,16,21]. These observations put immunological aspects of eutherian pregnancy into perspective: conventional immune responses attributable to T- and/or B-cells are not central to its success; instead they appear to be engaged in an important and evenly balanced side show.

From a very early stage of development, the 'typical eutherian' (i.e. mouse) trophoblast is distinguishable as an external sphere which completely encloses the new embryo. It implants early into the uterine wall and continues to nurture and protect the embryo proper over a prolonged period. At no stage is there any shell membrane separating the maternal and foetal systems. Marsupials, in contrast, show no clear delineation between cells that are destined to form the embryo and those with extra-embryonic fates, and a maternally derived shell membrane (homologous to those seen in other amniotes) persists until late gestation[2].

Although there is no disagreement about the shell membranes, some authorities argue that the above description overstates eutherian and marsupial disparity since not all eutherian trophoblasts develop as in the mouse, and there are sufficient functional similarities in the contribution of this external layer to implantation, placental membranes, maternal–foetal exchange and steroid metabolism in the two groups to expect immunological similarities as well[3,31].

After intrauterine 'hatching', many marsupials, including opossums and wallabies, attach only tenuously to the uterine wall via a yolk sac placenta. Bandicoots, on the other hand, resemble eutherians by producing an invasive chorioallantoic placenta[32]. For this reason they could provide a good model for comparative immunological studies, although unlike eutherians their 'trophoblast' component fuses with maternal cells and therefore disappears as a functional layer.

Although the duration of marsupial gestation shows considerable variation among species, the 'naked' period without the shell membrane is short. Moreover, the placental attachments shown by the bandicoots are among the most transient (Figure 6.5). It has therefore been suggested that the marsupial foetus does not linger in the uterus because, unlike eutherians, it has no device to counteract immune responses by the mother against paternal type antigens. Attempts to test this hypothesis are few, and have focused on tammar wallabies and the grey short-tailed opossum. Their foetuses do present a potential target since the MHC is polymorphic, but large numbers of maternal serum samples all failed to show cytotoxic antibody[33,34]. This suggests a failure to trigger the B-cell compartment, although other types of antipaternal antibody need to be sought. Nevertheless, the results contrast with eutherians, where cytotoxic antibodies are occasionally or regularly induced by pregnancy.

Female tammar wallabies mated repeatedly with the same male or sensitized to his tissues by prior skin grafting do not show a reduction in their fertility[3]. It follows (from the latter procedure at least) that the foetus is somehow invulnerable to a cell-mediated attack, provided that residual effector cells were still active during late gestation or that foetal exposure to maternal immunocompetent cells was of sufficiently long duration to allow time to generate and deliver an anamnestic response. There is some doubt that the latter does occur in this animal, since its yolk sac placenta is in place for only 8 days of gestation (Figure 6.5), whilst second-set skin grafts take 10 days to be rejected[33]. This question might

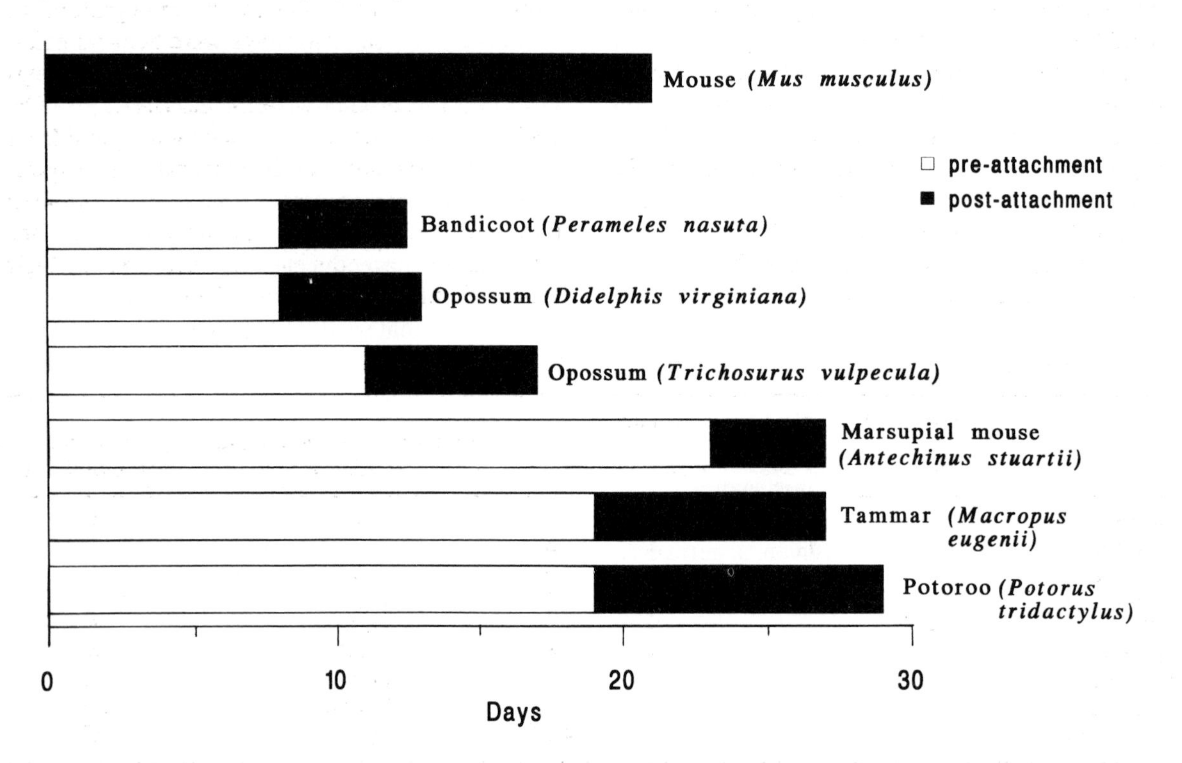

Figure 6.5. Some mammalian gestation times. Marsupial data from reference 3.

be resolved by examining the effect of injecting *in vitro*-expanded hyperimmune cytotoxic T-cells into pregnant wallabies as carried out in mice[30]. An extension of such studies to bandicoots would also be of interest.

6.2.3 Lactation

Immunoglobulin Transfer

A young mammal emerging from its clean and stable uterine environment has to cope with a sudden influx of potential pathogens from the outside world. Marsupials are especially vulnerable: at birth they crawl unaided across the mother's fur in order to latch onto a teat, and in so doing become colonized with bacteria[35]; in the absence of lymphocytes and organized lymphoid tissues at this stage (see Section 6.3), the baby is unable to initiate its own immune responses[6]. Eutherian neonates are more advanced than this but are nevertheless immunologically naive and would need time before primary responses could be mustered. In both groups, therefore, maternal protection is afforded by antibodies transferred to the offspring *in utero* and/or post-natally via the colostrum and milk, depending on species (Table 6.2). Milk also contains various other protective humoral components, including complement, lactoperoxidase and lactoferrin[36,37].

In humans, monkeys, rabbits and rodents, antibody is transmitted before birth. In the primates, circulating IgG is selectively taken up by the placenta, whereas rabbits and rodents use the yolk sac route. Selective recognition occurs via Fc receptors on the foetal membranes[38]. In all of these groups, IgG with anti-paternal reactivity has to be somehow prevented from reaching or damaging the foetus, while specificities directed against microorganisms and parasites previously encountered by the mother are allowed through.

After birth, the offspring of these mammals continue to receive antibody, but now mainly as IgA: it accounts for 97% of human colostral protein (120mg IgA per ml colostrum) and 10–25% of milk protein[37,39], whereas IgG is only a minor constituent. Similar changes are seen in rodents and rabbits (Table 6.2). Although some immunoglobulin is acquired from the circulation, the presence of large numbers of plasma cells in the mammary gland of rodents and primates suggests an important role in immunoglobulin secretion for the gland itself, either as a result of local antigenic stimulation of lymphocytes or immigration of immunocompetent cells which have been stimulated in lymphoid tissues associated with the linings of the gut and bronchi (GALT and BALT)[9,10]. The traffic of lymphocytes to the mammary gland from other mucosal sites (as demonstrated in experimental rodents) would explain why women exposed to respiratory or gut infections produce antibodies of the appropriate specificities in their milk without showing them in the blood.

Table 6.2 Maternal provision of immunoglobulins (main isotypes).

Species	Pre-natal transfer:		Presence in:		Postnatal transfer across neonatal gut (days)	Protection of neonatal gut
	Placenta	Yolk sac	Colostrum	Milk		
Man	G		A	A	–	A
Rabbit		G	A	A	–	A
Mouse/rat		G	A	A	G (16/21)	A
Sheep/cow	–	–	G	G	G (1)	M
Pig	–	–	G	A	G (1)	A
Horse	–	–	G	A	G (1)	A
Tammar	G			A/G?	G (150)	A?
Quokka	–	–		A/G?	G (180)	A?
Opossum	–	–		A/G?	G (100)	A?

Data from references 6, 9, 38, 39, 43, 46, 47.

Little or no antibody in human milk reaches the baby's circulation. However, it does protect the linings of the gut and respiratory tubes from infection, thus lending support to arguments for breast- versus bottle-feeding in infants[37]. It is not clear whether the risk of neonatal respiratory infections is reduced by milk spreading onto the respiratory mucosa during sucking, or whether migrating milk leucocytes are responsible (see below).

In rodents, unlike rabbits or humans, IgG reaching the neonatal gut in the colostrum and milk is transferred across it into the circulation. This absorption of antibody occurs for 2–3 weeks post-partum until changes in gut acidity and proteolysis occur[40]. The transfer is selective and highly efficient: in the rat, IgG from the milk binds at acidic pH to an Fc receptor on the surface of epithelial cells in the anterior small intestine; the IgG is then transported and released at higher pH (7.4) into the blood and tissue fluids[38,41]. The receptor protein responsible (FcRn) has aroused special interest because it is structurally very similar to MHC class 1 molecules yet has a role unrelated to T-cell recognition[41].

Significant quantities of IgD in rat milk (100 times the levels found in maternal serum) have also attracted attention[42]. IgD synthesis (like IgA) occurs in the mammary gland and there is no transfer across the neonatal gut. However, serum levels of IgD in the neonate are higher than expected, which implies synthesis *de novo* or transfer *in utero*. The role of this humoral IgD is not established, though a regulatory function in the development of the immune system and/or protective IgA-like functions in the neonatal gut have been suggested.

In contrast with the mammals described above, domestic ungulates receive no antibodies before birth (Table 6.2). The newborn therefore risk infection unless protective antibodies are transfered rapidly and effectively. The predominant immunoglobulin in the colostrum of these species is IgG, occurring at high concentrations (over 100 mg/ml in sheep) as a result of uptake by the mammary gland from the maternal circulation. In ruminants, this process begins about a month before parturition and is known to occur via high-affinity Fc receptors on mammary epithelial cells[36,39]. IgA and IgM are also taken up across the epithelial membranes, but appear to need the attachment of a secretory component. The influx of immunoglobulins into the gland is complemented by secretion from plasma cells close to epithelial surfaces. These cells are more evident in pigs than in ruminants, and some at least appear to be the progeny of gut-associated lymphocytes: exposure of sows to intestinal pathogens induces specific IgA and IgM activity in milk. Sows also show a rapid decline in IgG output from the high concentrations seen in colostrum, so that by about 3 days post-partum, levels have fallen below IgA[9]. The mucosal immune system therefore appears to be more important in the lactation of pigs, rodents and primates than in ruminants, where serum-derived IgG continues to dominate in terms of relative concentration (Table 6.2). It has been suggested that the low amounts of IgA delivered to lambs and calves allows the establishment of ruminal microbes; the IgM they receive appears to have the dominant role in gut protection[9].

Unlike rodents, new-born ungulates absorb colostral immunoglobulins unselectively across the small intestine[9,38]. Moreover, absorption ceases a day or so after birth when immature cells lining the small intestine are replaced. In the absence of prenatal transfer, this means that the occasional but sometimes grave risk posed by destructive antibodies from a mother sensitized to paternal antigens is very brief: newborn foals or piglets susceptible to haemolytic disease can be kept alive by withholding maternal colostrum[40].

Studies on the maternal provision of immunoglobulins in marsupials are limited. Reports on the quokka and three species of opossum all suggest that transfer occurs only after birth, but there is some evidence in the tammar for pre-natal transfer of IgG via the placenta as well as post-natal transfer of this isotype[34,43]. Absorption of maternal immunoglobulin (mainly G) across the gut appears to take place for a prolonged period after birth as in rodents but unlike ungulates; leisurely post-natal absorption may therefore be associated with short gestation times. The IgG content in the milk of the hill kangaroo (*Macropus robustus*) actually increases later on in lactation when the joey begins to take milk only intermittently, perhaps reflecting a second surge in its likely exposure to pathogens[6]. Marsupial IgA does not appear to cross either the placenta or the neonatal gut, and presumably remains in the gut to protect its lining as in eutherians.

If the eutherian condition is a guide, future research on marsupials is likely to reveal further species differences in the balance, timing and mode of delivery of different maternal immunoglobulin isotypes, although the disparity reported above for quokkas and tammars—both members of the kangaroo family—is surprising.

Cell Transfer

It has been known for years that leucocytes are present in normal colostrum and milk (Figure 6.6), but their functional significance is not established. On average 2 million leucocytes per ml occur in human colostrum and 1.4 million per ml in human milk. Similar concentrations have been reported in rat, guinea pig and cow's milk, with lower amounts in mares (8×10^3 cells per ml)[40]. Within a species, however, concentrations vary considerably according to the individual, stage of lactation and even the part of the gland from which samples were taken.

The macrophage is a prominent cell type in human, rat, cow, sheep and quokka milk, but polymorphs are sometimes present in greater concentrations, particularly in early lactation and in infections of the mammary gland[9,36,44]. The remaining leucocytes are mainly T- and B-cells, at least in cows, sheep, rats and humans. *In vitro* tests on these cells have sometimes indicated a reduced defensive capability compared with their counterparts in the blood. Milk lymphocytes, for example, show poor responses to mitogens; phagocytes show a reduced antibacterial activity, possibly because they become laden with debris

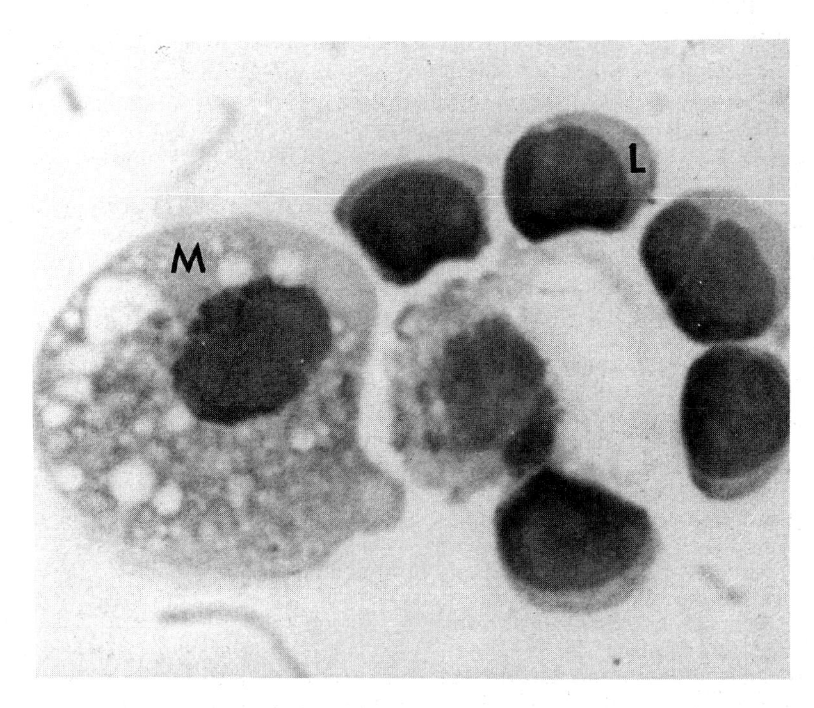

Figure 6.6. Leucocytes from cow's milk (×2000). The cytoplasm of the macrophage (M) is laden with lipid. Note also lymphocytes (L). Photo by courtesy of J.M. Finch.

and lipid from the milk[9,36] (Figure 6.6). On the other hand, studies in rats indicate that milk leucocytes are immunologically active in the recipient: protection against bacterial infection in the neonatal gut can be conferred by milk macrophages, whilst milk lymphocytes can leave the gut and migrate to the draining lymph nodes[9,45]. The latter might suggest another mechanism by which the lactating mother confers immunity to its young, but instead destructive graft-versus-host reactions have been reported in infant rats feeding on allogeneic milk, leading to runting and death[40]. Although allogeneic lymphocytes are also capable of movement from the gut lumen into the draining lymph nodes of lambs[45], there is no evidence that fostering causes graft-versus-host problems here, nor in humans. The most obvious explanation for these species differences is that neonatal rats are much less developed immunologically and therefore less able to counteract hostile maternal cells than newborn lambs or humans, but fostering of baby marsupials is also successful[6]. Alternatively, the cells responsible may be unable to migrate in large enough numbers, may be in too poor a condition or somehow too suppressed to cause damage in species other than rats.

Finally, attempts have been made to manipulate mammary immunoglobulin output and leucocyte function in order to protect the gland itself

Table 6.3 Maturation of the ovine immune system *in utero*.

Gestation (day)	Event/characteristic
16	Haemopoiesis begins in yolk sac
21	Haemopoiesis begins in liver (= main site in foetus)
24	First leucocytes (macrophages) in circulation
39	Complement (C1) in blood
40	Lymphocytes in thymus
48	Spleen is identifiable
50	Lymphocytes (mostly T-derived) appear in blood
55	First lymph node appears (prescapular)
60	Thymus a major lymphopoietic organ
60	Lymphocyte circulation between blood and lymph is established
60	Lymphoid sheaths in splenic white pulp
66	Antibodies to ferritin (in FCA*) and bacteriophage ϕX174 detected (IgM)
70	Haemopoiesis in bone marrow (minor role)
70	B-cells in lining of rectum and colon
72	Antibodies to chicken red blood cells
74	Antibodies to polymeric flagellin
75	Lymphopoiesis in jejunum
76	Antibodies to ovalbumin (in FCA)
77	Skin allografts are rejected as in adults
78	Antibodies to monomeric flagellin
82	Antibodies to chicken gamma-globulin (in FCA)
85	Lymphopoiesis in ileal Peyer's patches
87	Synthesis of IgG to antigens
91	Antibodies to *Brucella ovis*
95	Antibodies to bluetongue virus
100	Ileal Peyer's patches a major lymphopoietic organ (B-cells)
112	Mature eosinophils
116	Antibodies to *Brucella abortus*
123	Mature neutrophils
123	All complement factors present
150	BIRTH

*FCA=Freund's complete adjuvant
Compiled from findings in references 47–53.

more effectively[36]. This work is of considerable interest to the dairy industry which has suffered enormous losses from bacterial mastitis. However, a vaccine or dietary treatment which reliably and effectively boosts antibacterial immunity within the gland is still awaited.

6.3 DEVELOPMENT OF IMMUNOCOMPETENCE

Mammals are difficult subjects in which to investigate immune capabilities of the young: events *in utero* are obviously more difficult to monitor and manipulate than those in free-living animals; passively acquired maternal antibodies interfere with the development of active immunity in the offspring; and maternal cells and proteins which cross the placenta and/or form part of the milk can themselves provide an antigenic stimulus. The present section will focus on two types of mammal which overcome some of these problems: marsupials, where the post-natal phase accounts for most of development; and sheep, where large size has allowed early grafting, cannulation and organ removal, and the placenta is impermeable to immunoglobulins (Table 6.2). Experiments on sheep have done much to change earlier convictions that eutherian foetuses and neonates are immunologically incompetent (Table 6.3).

The first leucocytes appear in the sheep foetus 24 days after fertilization. These cells are

Table 6.5 Lymphoid maturation in young quokkas.

Age (days)	Event/characteristic
0	BIRTH
3	Lymphocytes appear in cervical thymus
4	Lymphocytes appear in thoracic thymus
4	Lymphocytes appear in blood
5	Rudimentary lymph nodes
5	Lymphocytes in spleen
10	Antibody responses to sheep erythrocytes can be initiated
10	Hassall's corpuscles in cervical thymus
60	Hassall's corpuscles in thoracic thymus
90	Plasma cells and germinal centres first seen in lymph nodes
120	Cervical and thoracic thymuses achieve adult structure
150	Lymph nodes achieve adult structure

Compiled from findings in references 6 and 46.

Table 6.4 Maturation of the opossum (*D. virginiana*) immune system.

Age (days)	Event/characteristic	
0	BIRTH	
2	Mature neutrophils in blood	
2	Lymphocytes appear in thymus (= thoracic)	
3	First lymph node rudiments appear	
4	Eosinophils in blood	
6	Lymphocytes appear in blood and lymph nodes	
5	Injection of bacteriophage f2	elicits specific antibodies
8	Injection of *Salmonella* flagella	elicits specific antibodies
10	Injection of sheep erythrocytes	elicits specific antibodies
11	Injection of bacteriophage ϕX174	elicits specific antibodies
12	Injection of DNP-bovine albumin	elicits specific antibodies
17	Skin allografts applied are eventually rejected	
17	Lymphocytes in spleen (white pulp)	
32	Thymus achieves adult structure	
65	Mature germinal centres and plasma cells in lymph nodes	
90	Skin allografts rejected at 'adult' speed	
100	Tonsils and Peyer's patches appear	

Compiled from findings in references 6 and 46.

macrophages, and can ingest carbon particles and bacteria[50]. However, their efficiency at this stage is likely to be impaired by the absence of opsonins and key intracellular enzymes[52]. Lymphocytes are found in the thymus gland at 40 days, soon after the organ can be identified, though presumably their stem cells are present in yolk sac and liver before this. Lymphopoiesis increases in intensity in the thymus throughout gestation, and most circulating foetal lymphocytes are T-derived: thymectomy at 55–60 days gestation causes an 80% reduction in numbers[49].

The acquisition of foetal immunocompetence follows the appearance of lymphocytes in the circulation and the establishment of routes through the secondary lymphoid tissues (Table 6.3). Foetal lambs can reject skin allografts about halfway through gestation and, depending on the antigen, produce specific antibodies at this time. Reported differences in the timing of humoral responses could be due to the staggered maturation of different antigen-recognizing clones or accessory cells, but technical factors (e.g. sensitivity of different assays) may be partly responsible. Foetal antibodies are principally IgM; different classes and larger amounts of antibody (especially IgG) are inducible in older foetuses, but in the absence of congenital infections or deliberate antigenic stimulation, overall levels remain low[47,51,52]. The late maturation of a full complement system and effector granulocytes (Table 6.3) will limit the defensive value of antibodies induced by pathogens *in utero*.

As in sheep and cattle, thymic differentiation and the production of lymphocytes occur at an early stage in human foetuses, and there is evidence of cellular and humoral immunocompetence shortly thereafter (from *c.* 10 weeks of gestation). These conclusions are based on functional tests of human foetal lymphoid cells *in vitro*, and on observations of foetal responses (especially IgM) to natural transplacental infections[47,48]. The foetuses of several other species, including monkeys, pigs and dogs, are also capable of producing specific antibodies following an appropriate antigenic stimulus. This capability appears at different times, but once present it is similar to that elicited in adults[47]. Assessments of T-cell-mediated responses *in utero* have been harder to obtain, although dog and rabbit foetuses grafted with foreign skin shortly before birth can reject it[48]. Rats and mice are unrepresentative of eutherians in that functional maturation of the immune system occurs late in relation to their time of birth. Moreover, immunological tolerance can easily be induced by experimental means in neonatal rats or mice, but not in foetal sheep or calves[49,50].

Whilst the bone marrow has been identified as a major source of B-cells in humans and rodents, ileal Peyer's patches appear to function as the bursa equivalent in sheep and perhaps also in pigs, foxes and mink[50,54,55,77]. In late foetal and neonatal life the patches contain large populations of B-cells; most of these cells die soon after they are generated, while the rest eventually seed the secondary lymphoid tissues[56]. Surgical removal of most of the patches from foetal or neonatal sheep or induced premature involution causes widespread B-cell deficiencies. Normal involution begins 2–4 months after birth and is virtually complete by 15 months[54].

Although these developmental events in ileal Peyer's patches show clear parallels with other primary lymphoid organs, the fact that antibody production precedes the appearance of the patches in the foetal lamb (Table 6.3) means there must be an alternative source of B-cells (possibly the spleen, jejunum or large intestine)[50,53]. The same could be true for ovine T-cells: early thymectomy does not prevent mixed lymphocyte reactions or graft rejection; the operated animals survive and grow and after a time regenerate Ig^{-ve} lymphocytes[49,50]. While it is possible that removal of the thymus was incomplete or too late, the results suggest that other sites are capable of supporting T-cell maturation. The recovery by T-cells after foetal thymectomy is restricted to the $\alpha\beta$ T-subset; $\gamma\delta$ T-cells remain barely detectable[57].

As in other eutherians, the ruminant thymus reaches its greatest size in early postnatal life, and shows the usual distribution of lymphocytes and MHC-rich antigen-presenting cells[49]. As expected, the $\gamma\delta$ T-subset is a minority population in the thymus (up to 4% of thymocytes)[57]. What is unusual is that $\gamma\delta$ T-cells are the dominant mononuclear cells in the blood of neonatal lambs and calves, and they remain numerous (but no longer dominant) in older animals[57]. This therefore suggests that ruminants rely to a greater extent on $\gamma\delta$ T-cells for immunity against infection than humans or mice, possibly because they are more prone to attacks by pathogens on their mucosal surfaces and skin from an early age. The cells are evident in linings of the gut, trachea, bladder and bare parts of the skin (Figure 6.7), recirculate through the blood, lymph, lymphoid tissues and epithelia, and are selectively recruited to sites of active inflammation[57]. Bovine $\gamma\delta$ T cells bind to an endothelial adhesion protein (E-selectin) which is found in other mammals, but via a novel glycoprotein receptor[178].

The proportion of ovine T- and B-cells recirculating between the blood and the lymph remains steady during gestation at approximately 24T:1B, but after birth the numbers and ratios change (3T:1B), peripheral lymphoid tissues complete their maturation, and some sites assume a new significance[50]. The prescapular lymph node is the largest and first to develop in the foetus but its importance declines after birth. In contrast, a dramatic increase is seen in the numbers and movement of lymphocytes associated with the gut. The latter has been attributed to a rapid influx and colonization of the gut by microbes after birth, the onset of digestion and absorption, and the release by ileal Peyer's patches of B-cells which they had previously held back after production[50].

Unlike the ileal Peyer's patches, lymphoid follicles in the jejunum and large intestine do not completely involute later in young lambs, implying a greater influence of luminal antigen on their lymphocyte production as in the Peyer's patches

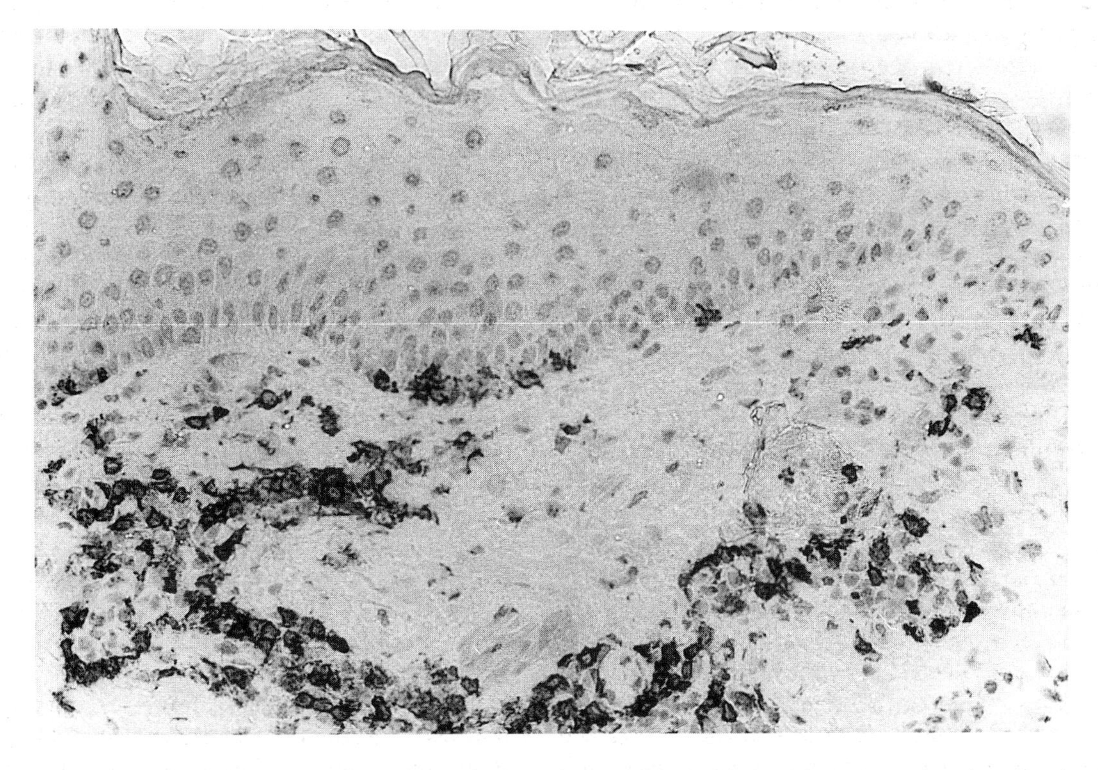

Figure 6.7. Section of bovine skin (×200) stained by immunoperoxidase using monoclonal antibody specific for ruminant $\gamma\delta$ T-cells. Large, dense-staining accumulations of these cells are seen predominantly in the subepidermal region, with scattered cells adjacent to and sometimes invading the basal layer of the epidermis. Photo by courtesy of W.R. Hein.

of rodents[53]. Specialized epithelial cells (M-cells) which overlie the domed follicular areas are generally thought to sample antigen from the intestinal lumen for presentation to the immune system: in rats, mice and sheep, the domes are rich in dendritic cells, macrophages and lymphocytes, although the balance of T- and B-cells differs from species to species[58]. Lymphocytes, sometimes in clusters, can be seen enclosed by the M-cells (Figures 6.12 and 6.13), but the extent of their local migration and proliferation is not yet clear. The envelopment of lymphocytes by jejunal M-cells in foetal sheep as well as in older animals suggests that these events are not always antigen driven.

Most ontogenetic studies of marsupial immunity have been made on quokkas and the Virginia opossum *Didelphis virginiana*. Unlike eutherians, they show little or no immunological development before birth (compare Tables 6.3 and 6.4).

The lymphatic system of *D. virginiana* is first seen on the 11th day of gestation, only 2 days before term, and the first lymph node primordia to develop (i.e. the deep cervical nodes) do so on the third post-natal day[46]. At birth the thymus gland of this opossum is identifiable only as a cluster of epithelial cells; small lymphocytes can be seen there 2 days later. This contrasts with the eutherian thymus, which shows lymphocytes and cortico-medullary differentiation at birth, even in those species such as the mouse and Syrian (golden) hamster where gestation is brief[48]. *D. virginiana* lymphocytes are seeded into the blood and lymph nodes 6 days after birth, whilst circulating neutrophils and eosinophils appear on days 2 and 4 respectively (Table 6.4).

On the evidence available, the newborn *D. virginiana* is incapable of mounting immune or inflammatory responses, and its survival seems to depend on maternal milk and structural barriers: a minor wound in the skin of a newborn opossum results in a rapid and fatal invasion of bacteria[46]. However, judging by findings in quokkas, defences based on macrophages may already be in place to counter infections of the respiratory tract[44]. Maturation and distribution of lymphocytes occur rapidly in both species after birth, no doubt to meet the new challenge from pathogens. Humoral reactivity is manifest shortly after lymphocytes appear in the thymus, and long before the lymphoid tissues reach their mature condition (Tables 6.4 and 6.5). As in foetal sheep, antibodies produced by young opossums are initially of low titre, and are detectable at different times depending on the antigen (Table 6.4). Pouch young produce an 'embryonic' 13S immunoglobulin which is gradually replaced by the adult forms[6]. First-set skin allografts placed on *D. virginiana* less than 12 days old are permanently accepted, although some of these animals can afterwards reject second-set grafts; first-set grafts given to 17-day-old animals are rejected, albeit slowly[46].

Studies on a second (more tractable) species of opossum, *Monodelphis domestica*, tend to support the above observations. At birth its thymus gland is very small, lacks a clear-cut cortex and medulla and is devoid of Hassall's corpuscles (Figure 6.8A). These features of the fully functional thymus become evident over the next 1–2 weeks (Figure 6.8B)[59]. Immunoglobulins are again absent in the neonate[34]. Neoplastic mouse cells (melanoma) grow in *M. domestica* injected from 12 days of age, but tumour growth could rarely be initiated in animals over 24 days old. This change coincides with the onset of homoiothermy[60]. It was also reported that established transplanted tumours disappear during the weaning period (days 45–62). Thus there may be two boosts to active immunity in opossums as development proceeds.

Whilst the thymus of the two opossum species described above is located in the thoracic region (Figure 6.8A), koalas and wombats have a cervical thymus, and kangaroos and quokkas have both[6,46,47]. The cervical thymus of the developing quokka is much larger than its thoracic counterpart at all stages, and its differentiation occurs more quickly (Table 6.5). It has been suggested that a cervical thymus is more important to species such as the quokka which possess a deep moist pouch, since the microflora to which the young are exposed are likely to be more diverse and concentrated than in other marsupials like opossums where the pouch is shallow or absent;

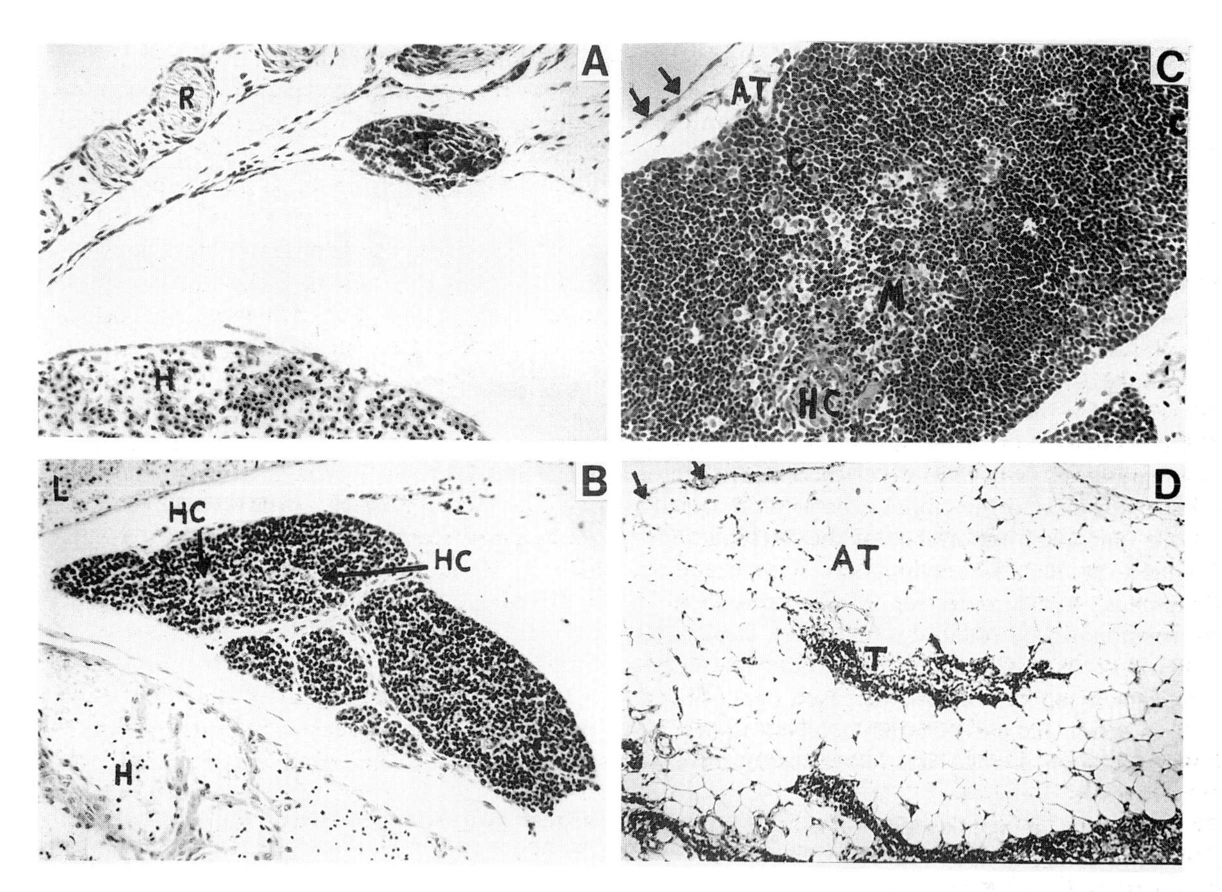

Figure 6.8. Development and ageing of the thymus gland in an opossum, *Monodelphis domestica* (H & E stain). (A) In a day-old animal the gland (T) is very small and is not differentiated into cortex and medulla. Note its thoracic location (heart H, ribs R) (×70). (B) At day 7 the cortex (C) is developing, and Hassall's corpuscles (HC, arrowed) can be seen (×70). (C) This thymus from a 15-month-old opossum is showing maximal development. M = medulla; AT = adipose tissue; thymic capsule is arrowed (×70). (D) Atrophic thymus of a 28-month-old animal. Small, poorly defined islands of thymic lymphoid tissue (T) are surrounded by abundant adipose tissue (×28). Photos by courtesy of W.H. Stone, reproduced from reference 59 with permission.

to meet this immunological threat a gland in the neck can enlarge more easily than one which is confined to the thorax[47]. Removal of the cervical thymus from quokka pouch young less than 10 days old causes a persistent lymphocyte depletion in the lymph nodes and spleen and a 30-day delay in the onset of antibody responses to foreign erythrocytes, despite the presence of an intact thoracic thymus. On the other hand, neither thymus is vital to a young quokka: early and complete thymectomy (i.e. from both neck and thorax) only delays or depresses immune responses; it does not eliminate them even if the mother too is thymectomized[6]. Growth, development and survival of thymectomized pouch young are unaffected, and animals which have left the pouch elicit antibodies and reject skin allografts as vigorously as intact controls. Such findings are reminiscent of those reported for thymectomized lambs and amphibians (see above and Chapter 4), and again suggest that an alternative pathway exists for the differentiation of at least some T-cell functions: in the absence of a thymus, immunocompetent T-like cells may be spawned in other tissues, albeit by a sluggish process requiring many weeks for completion.

6.4 ADULT IMMUNE SYSTEMS

6.4.1 The Major Histocompatibility Complex (MHC)

The genes of the eutherian MHC have been the subject of intensive research. Some of their products clearly play a key role in initiating immune responses, not only by presenting to mature T-cells antigenic protein fragments derived from pathogens, but also by selecting those T-cell specificities which will be allowed to mature and leave the thymus in the first place. The selected emigrants in turn control cell-mediated effector mechanisms and the clonal expansion of B-cells. The functional importance of the MHC is illustrated by the 'bare lymphocyte' syndrome in humans, where defective MHC expression is accompanied by reduced numbers of T-cells and plasma cells, reduced antibody production, undeveloped lymphoid organs, increased susceptibility to infection and early death[63,168]. However, a similar condition in mice has less disastrous consequences[168].

The MHC molecules which normally bind and display antigenic peptides at cell surfaces are remarkably diverse, particularly in the antigen-binding region. It has been suggested that this variety will allow an effective response to a greater range of antigens within the individual and species by improving the likelihood of a good 'fit' between MHC and antigen, or by minimizing the risk of molecular mimicry[64–67]. Such assertions have enticed some immunologists away from the laboratory and into the field: animal populations or species with monomorphic or oligomorphic MHC should be the least healthy, buoyant or adaptable. Might there also be species-specific MHC molecules linked to species-specific pathogens, or MHC characteristics linked to particular kinds of habitat or modes of life?

Much more is known about the MHC in the mouse *(Mus)* than in any other animal, reflecting the ready availability of inbred and congenic strains in the laboratory, and abundant well-characterized populations in the wild. The mouse MHC (H-2) was also the first to be discovered.

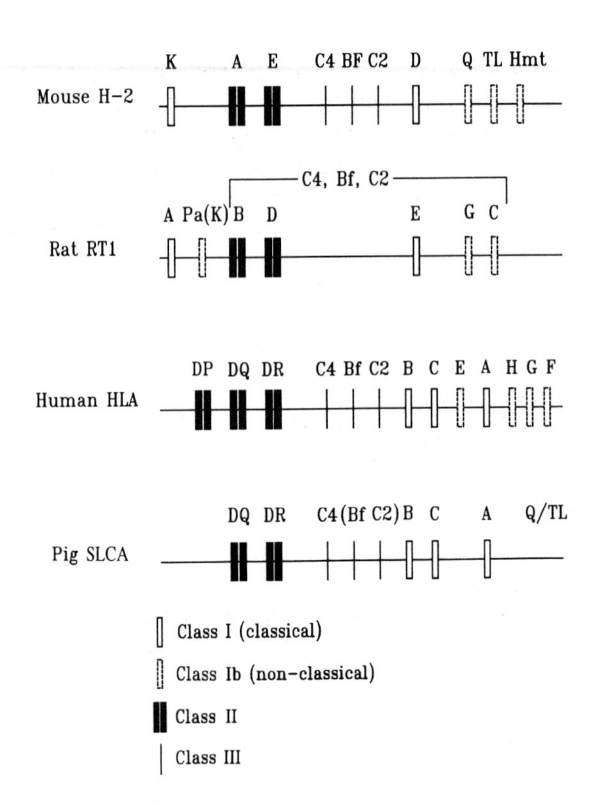

Figure 6.9. Arrangement of the MHC in some eutherian mammals. Based on references 67–71.

The chromosomal organization of its loci is illustrated in Figure 6.9. Note that the classical class I loci K and D are separated by the class II and III regions. The class I and II genes encode proteins involved in antigen presentation to cytotoxic and helper T-cells respectively; class III genes encode a heterogeneous group of proteins, including the complement components C2, C4 and B factor, heat shock proteins and tumour necrosis factors. Whilst each class I molecule has only one MHC-coded (α) polypeptide (attached to an invariant non-MHC-coded β_2-microglobulin), class II molecules have two (α and β). It has been estimated that some of the functional H-2 loci (K, D, Aβ and Eβ) have over 100 alleles occurring at appreciable frequencies, making them the most polymorphic protein-coding loci known in vertebrates[72,73]. Comparisons of the MHC among different species of mouse have indicated that both class I and II polymorphisms have accumulated slowly, in a trans-species fashion, rather than by recurrent mutations: different alleles from one species are often highly divergent at the nucleotide level, but often show close similarities with those in another species[73–75,168]. Mouse-type Aβ alleles (but not Eβ) have also been identified in the rat, indicating their persistence for at least 10 million years[74]. Such findings support the notion that MHC products are functionally important, and that heterozygotes have an improved chance of survival.

Although the organization of the rat *(Rattus norvegicus)* MHC is similar to that of the mouse (Figure 6.9), its degree of class I polymorphism is much lower: analysis of 110 strains of inbred rat and of wild rats in Europe and North America yielded just 14 RT1-A alleles[76]. However, it has been discovered that rats do possess polymorphic 'transporter' genes *(mtp2/Tap2)* which determine the peptides taken up by a class I molecule, thereby modifying its antigen-presenting capability; corresponding transporter genes in mice and humans do not apparently show this level of polymorphism[76,77]. Studies on Syrian (golden) hamsters in the early 1980s suggested a more extreme departure from the mouse model: their class I MHC was reported to be monomorphic[78]. Since this homogeneity did not extend to class II MHC or to non-MHC loci, a 'bottleneck' in the past (involving growth of the population from a very small number of inbreeding individuals) seemed an unlikely explanation. Instead it was suggested that, because wild hamsters live in dry, inhospitable desert and steppe and only briefly encounter other members of the species, they have had less need to develop effective and flexible cytotoxic T-cell responses (e.g. against viruses) than more populous, sociable and adaptable species such as mice and humans[78]. Although this is an attractive idea, more recent studies claim that hamsters do express diverse class I products and have a conventional cytotoxic T-cell repertoire[79,80].

Better examples, perhaps, of successful mammals with monomorphic or oligomorphic MHC are to be found among the mole rats, another group of burrowing rodents. The naked mole rat, *Heterocephalus glaber*, lives in large colonies throughout the arid areas of Northeast Africa. Reproduction in each colony is monopolized by a single female and one or two large males; outsiders are made very unwelcome. As a result there is little or no genetic diversity across a colony or between closely neighbouring ones, but the species nevertheless thrives within its stable and limited habitat[81]. The Balkan mole rat, *Spalax leucodon*, a solitary species, also shows a relatively uniform MHC but a third species, *S. ehrenbergi* (the Mediterranean mole rat), does not: its class I and II are both polymorphic. *S. ehrenbergi*, unlike the others, has managed to spread into diverse climatic zones[82,83]. Relative stability and uniformity of habitat may also explain the limited MHC variation reported in whales, although only a small number of animals from two species were tested[84].

In immunologically demanding and varying environments, mammals with a non-polymorphic MHC do appear to be at risk, though the evidence is again only circumstantial. In 1983, a previously productive and healthy captive population of cheetahs in Oregon was decimated by a coronavirus (causing feline infectious peritonitis) which has only low morbidity in lions and domestic cats; and cheetah cubs in general seem to suffer peculiarly high mortality rates (up to 70%) due in part to disease susceptibility. Skin grafts exchanged between unrelated cheetahs do not show acute rejection; the uniformity in MHC expression indicated by such experiments may reduce the spectrum of viral pathogens that the species is capable of controlling[85]. However, cheetahs show a lack of variability at all other loci examined—the species appears to be the victim of a recent population bottleneck—and homogeneity at non-MHC loci, or a combined effect, could influence their susceptibility to disease. Suggestions that captive breeding of endangered species be designed with the specific goal of maintaining diversity at the MHC ignore this point[86]. Conclusions about the role of the MHC in the health of other genetically uniform animals such as colonial mole rats (see above) are open to similar criticisms.

Early observations on domestic cats indicated

several immunological similarities to cheetahs: they, too, are susceptible to certain viral infections; they appeared to show a uniform MHC; and it was even suggested that they went through a population bottleneck of their own as a result of persecution in the Middle Ages by association with witchcraft and satanism[63]. However, early difficulties in obtaining alloantibodies (from parous cats or by deliberate immunization with peripheral blood lymphocytes) for MHC typing have now been overcome by the use of multiple skin grafts, and serological and molecular analysis reveal hitherto unsuspected variation in the cat's MHC[67,87]. Further efforts in this direction should help our understanding of retroviral feline leukaemia, lymphoma and immune deficiency and enhance the cat's value as a model for human retroviral disease[87].

The large body of data available for the human MHC (second only to the mouse) has provided a useful basis for comparative studies within and beyond the primates. Apes and monkeys are of particular interest because the different species show well-defined degrees of relatedness to humans, although their limited availability is a major obstacle in efforts to assess the diversity of their MHC.

As in mice, products of the classical human MHC class I loci (e.g. HLA-A and B) and class II (e.g. DQα, DRβ) can show high levels of polymorphism, much of which predates speciation. Thus, individual HLA-A and B alleles are more closely related to individual chimp alleles than to other HLA-A or B alleles, and a given human DQα or DRβ allele is more closely related to its chimp or gorilla equivalent than to any other human allelic type[70,74]. It also seems that ecological distinctions between humans and chimps, which include differences in exposure and susceptibility to viral pathogens, have not resulted in clearly distinct MHC sequences[67]. HLA-A and B (but not C) homologues extend to the more distantly related Asian apes, the gibbon and orangutan, indicating that these loci have been maintained as separate entities for at least 30 million years[70].

The cotton-top tamarin, a New World monkey, shows a significant departure from higher primates with respect to class I. Only 11 different class I molecules could be defined from a sample of 79 unrelated animals, and one of these molecules was found on the lymphocytes of every individual examined[88]. Most members of this species are born as non-identical twins which have exchanged haemopoietic cell precursors via placental anastomoses in prenatal life; their more uniform class I might be a device to limit alloreactivity in the chimaeric young. However, they appear to pay an immunological price for this oligomorphism: tamarins succumb easily to pathogens which are 'exotic' to them, for example several human viruses. It was also discovered that the class I tamarin genes expressed are more closely related to human HLA-G (a monomorphic class Ib gene expressed by human trophoblast cells) than to the classical polymorphic HLA-A, B and C genes (see also Section 6.2.2), yet they are presumably involved in the cytotoxic T-cell responses seen in the tamarin: no loci corresponding to HLA-A, B or C have been found. Unlike the tamarin, the owl monkey (a non-chimaeric New World species) expresses polymorphic class I molecules, but these are still related to HLA-G; HLA-A, B and C are again absent. These examples suggest an interchangeability between classical and Ib genes during evolution[70,89].

A phylogenetic analysis of DNA sequences using genes from six different mammals (including rodents, primates and ungulates) indicated that the class Ib genes in one species are more closely related to classical genes from the same species than to Ib genes from a different order or family[90]. It has been argued that Ib proteins include many non-functional products of genes that were once active in antigen presentation but are now inactive and in various states of decay[66], but in the light of observations on human HLA-G and rat Pa/K (Section 6.2.2) this does not apply to all of them; their role may lie more with NK or $\gamma\delta$ T-cells than conventional $\alpha\beta$ T-cells.

The evidence that class I genes vary more widely in their numbers and nature than class II, and that they have been utilized for more diverse and novel purposes (see also Sections 6.2.2 and

6.2.3) suggests that they are more dynamic in evolutionary terms than class II, and they may be further removed from the ancestral condition[66]. The functional class I genes of humans and mouse occupy strikingly different positions within the MHC (Figure 6.9), and they appear to have evolved independently in the two species; the class II and III regions and genes, on the other hand, have retained their identities across the evolutionary divide[66,70]. Moreover, homologues to DQ and DR genes have been described in rabbits, cats, dogs, sheep, cattle, horses and whales, and DP also in rabbits, cats and dogs[67,74,91]. This suggests that the different isotypic class II loci emerged (probably by repeated gene duplication) before the eutherian mammals radiated. The main difference in class II among present-day eutherians is the relative contribution of DP, DQ and DR equivalents: DP is of limited importance in humans and a mere remnant in mice, for example, but has the major class II role in the mole rat *S. ehrenbergi*; DR is missing in mole rats and dominant in humans[92].

The families of class II genes (DA, DB) described in the red-necked wallaby are unlike their eutherian counterparts, suggesting that they evolved from different ancestral genes[93]. However, bearing in mind the patchy representation of class II families in eutherians, this conclusion is tentative. MHC studies on pedigreed grey opossums are also at a preliminary stage: several class I products have been defined[94], but very little is known about class II (mixed lymphocyte reactions between unrelated opossums are reported to be weak)[95].

Given that class I regions of mouse and human are radically different from each other, which provides the better 'prototype' for the remaining eutherians? Current descriptions of the MHC in other Orders are somewhat fragmentary, but the indications are that they more closely resemble HLA than H-2[63,68,96]. This is evident in the pig, for example (Figure 6.9), a species which is emerging as a very useful large-animal model for biomedical research[67–69,97,169]. The distinctive properties of H-2 probably reflect the rodents' ability to evolve rapidly. Alternatively, an early departure from mainstream eutherian evolution may have been responsible, although there is no consensus on this (see Section 6.1 and Figure 6.1). Some major features of H-2 are peculiar to the Muridae[67,70].

The cellular distribution of MHC products is similar in human and mouse but is not identical. Class I occur on virtually all nucleated cells in both species, and on murine but not human erythrocytes; class II products are mainly limited to B-cells, macrophages and some epithelial cells. Class II products seem to occur on larger numbers of T-cells obtained from cats, dogs, horses, pigs and guinea pigs than from mice or humans[68,96,98,99]. These apparent species differences in class II expression could have significant functional implications, although they (and some discrepancies within species) may partly reflect differences in sensitivity of the tests employed, or different levels of exposure of the cells to cytokines.

The argument that pathogens have been the principal force in shaping the MHC and maintaining its polymorphism is a fairly convincing one: the role of class I and II genes in antigen presentation to T-cells is well documented, the antigen-binding site is established as a focus of diversity, and species with a uniform MHC seem unable to respond effectively to novel immunological threats. On the other hand, clear demonstrations of an association between a specific allele (or haplotype) and a life-threatening pathogen have proved elusive[63,168]. Until recently, the only good example was the link between the B^{21} allele and resistance to Marek's disease in chickens[100], although it has now been shown that certain HLA alleles (at HLA-B and DR-DQ loci) have a significantly reduced frequency in West African children suffering from severe forms of malaria compared with healthy controls who had been exposed to the disease[70,101]. Progress in identifying MHC determinants of disease resistance in farm animals has also been slow[96]. These difficulties may sometimes have been due to inadequate sample size or MHC typing methods, or different causes of apparently similar disease, but it may be that many infectious pathogens will present such a complicated array of epitopes—none of which would play an overriding role in disease progression—that hosts of different genotype can launch responses which differ more subtly, or not at all, in their effectiveness. Conversely, a clear association would imply a limited number of epitopes capable of evoking a protective response[70] and, as a corollary, antigenically simplified 'new generation' vaccines would have varying efficacy in recipients of different MHC genotype. The above issues have an important bearing on the prospects for gene transfer as a means of improving disease resistance in livestock.

Alternative, non-immunological explanations for MHC polymorphism have also been proffered. The MHC may, for example, provide a positive stimulus in foetal-maternal interactions (see Section 6.2.2), but this fails to explain why the chicken MHC is also polymorphic[102]. Another suggestion is that MHC diversity is maintained by a preference for heterozygotes in mating: mice can smell the difference in urine from other mice which differ only at MHC loci (class I or II), and they prefer to mate with MHC-disparate partners[70,103]; single MHC gene mutations can alter the odour of the urine[104]. However, it is difficult to see how MHC-controlled mate selection operates in those species with little or no sense of smell, unless other mating cues (e.g. visual) are under its influence. Although MHC-specific body scents have been reported in humans[105], further parallels with the mouse example are certainly difficult to draw: the smells were detected by pregnant or menstruating women, and were described by one as 'extremely aversive'! It is also possible that control of mating cues by the MHC acts as a special device in a restricted number of species but nevertheless serves the same general purpose, namely to maximize disease resistance[70,103].

6.4.2 Immunological Capabilities of Marsupials and Monotremes Compared to Eutherians

Information about the immune system in adult marsupials and monotremes is fragmentary, mostly old and derived from a small number of representatives[183]. Conclusions about the relative levels of immunological sophistication across the Class must therefore remain tentative, although recent developments give some hope for a resurgence of interest in marsupial immunology (Section 6.1). It also remains to be seen whether reported peculiarities in the susceptibility of marsupials to infectious disease have immunological explanations[47,106].

Delayed-type hypersensitivity reactions in quokkas and opossums *(Didelphis virginiana)* are claimed to be less vigorous than in rodents[46,47,106], but those seen in another didelphid, *D. marsupialis*, are just as efficient[107]. According to one report, interleukin-1 (IL-1) produced by a third species of opossum *(Monodelphis domestica)* shows no serological or functional cross-reactivity with the human or mouse product, but a human IL-1α cDNA probe did hybridize with RNA from opossum macrophages[108].

Four distinct immunoglobulin classes resembling eutherian IgM, G, A and E have been reported in quokkas, and IgG1 and 2 subclasses have been identified[109]. Markers for IgM and IgG2 constant region epitopes (identified by anti-quokka Ig) were widely distributed in Australasian species and were also found in the American opossum *D. virginiana*, suggesting a structural conservatism of these marsupial Igs; an anti-quokka IgG1, however, only reacted with closely related diprotodonts[47]. Quokka IgA has biological and physicochemical properties which are close to eutherian IgA, but the identity of the isotype(s) responsible for the immediate-type hypersensitivity reactions observed in quokkas is ambiguous[47]. In contrast to many eutherians, the quokka also apparently lacks histamine in blood leucocytes and peritoneal cells. The presence of a human-like IgG in the serum of a monotreme, *Tachyglossus* (echidna), means that this immunoglobulin class did not evolve to meet the need for placental transfer of antibody[110].

The primary response of echidnas to *Salmonella adelaide* flagella is comparable to rats in its kinetics and total amount of antibody produced. However, secondary responses are erratic and less impressive than rodents or rabbits, and IgG does not completely replace IgM[7]. Similarly, antibody responses of both quokkas and opossums show persistent IgM, only a slow conversion to IgG production, and poorly developed secondary responses[46,106,111]. It has often been concluded from these observations that the eutherian humoral immune system is more sophisticated, but such generalizations are misleading: the much-studied *D. virginiana* has shown more immunological defects than other marsupial species[106], whilst primary and secondary antibody responses in bats are weaker than responses of rabbits and rodents to the same stimuli and IgM production is more persistent[112,113]. It is tempting to speculate that the important role of bats as reservoirs of infection is linked to limitations in their antibody output related to their phylogenetic status and/or their regular and dramatic loss of body temperature when they roost. Nothing is known about killer T-cells in bats, but delayed-type hypersensitivity responses are sometimes defective[114].

6.4.3 Lymphoid Tissues

This section will focus on major structural differences in definitive secondary lymphoid tissues across the Class, and discuss possible functional implications. Developmental aspects, particularly those involving primary lymphoid tissues, have been considered in Section 6.3.

The most remarkable differences, both in architecture and number, are seen in lymph nodes. The echidna possesses chains of simple nodes, which are especially prominent in the cervical, axillary and pelvic regions. Each node usually consists of a single antigen-trapping follicle (less than 2mm diameter) with a germinal centre, suspended by a vascular bundle from the walls of lymphatic vessels; the node thus 'floats' in the lymph (Figure 6.10). In marsupials and eutherians on the other hand, each node has several lymphoid follicles plus distinct medullary and cortical regions through which the lymphatic fluid is obliged to percolate. The latter arrangement

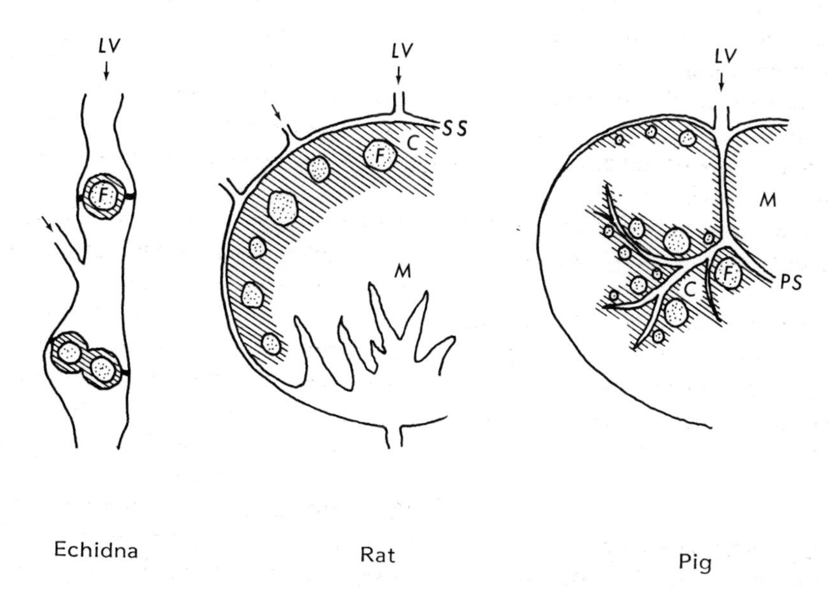

Figure 6.10. Mammalian lymph nodes: schematic diagrams to compare structure in different species (after Diener *et al.*[115] and Binns[119]). In the echidna, separate nodes (equivalent to follicles, F) are suspended in the lumen of lymphatic vessels (LV). In the rat, afferent lymphatic vessels run into a subcapsular sinus (SS) which surrounds and supplies the whole node. In the pig, an afferent lymphatic sends lymph along a deep and branching paratrabecular sinus (PS) (equivalent to subcapsular sinus). C = cortex, M = medulla.

increases the likelihood of secondary encounters between antigen and specific recognizing cells in the higher mammals[7,46,106,115] (see also Section 6.4.2). Lymph nodes of such sophistication are not seen in any non-mammalian vertebrates.

Didelphid opossums possess less than 50 lymph nodes; these occur singly at each drainage site[46]. Bats are similar in this respect, but large eutherians (e.g. human) have many more, often occurring together at a given site[116]. Thus, large volumes of lymph can be filtered more thoroughly.

Ruminants are unusual in possessing not only conventional lymph nodes but also haemal nodes. The latter are found in the pelvic cavity, are histologically similar to lymph nodes (Figure 6.11) but filter blood instead of lymph. Their immunological role is uncertain, but appears to be unlike that of lymph nodes or spleen because their balance of lymphocyte subpopulations is quite distinct: a higher percentage of $\gamma\delta$ T-cells and fewer $CD8^+$ cells occur in the haemal nodes of adult sheep compared with their mesenteric lymph nodes or spleen, and they contain significant numbers of $CD45^-$ cells[117].

The peripheral area of each echidna lymph node is well vascularized by post-capillary venules whose walls of cuboidal epithelium often contain small lymphocytes[115]. Very similar vessels have been reported in the echidna appendix and in the lymph nodes, appendix, tonsils and Peyer's patches of didelphid opossums as well as the corresponding lymphoid organs of diverse eutherians; the post-capillary venules are probably acting as points of entry for blood lymphocytes into these tissues for the Class as a whole[118].

The lymph nodes of pigs show significant departures from the eutherian 'norm'. The follicle-containing 'cortex' of the node is deep-lying, while the 'medulla' is peripheral and lacks sinuses and cords (Figure 6.10). Instead of the conven-

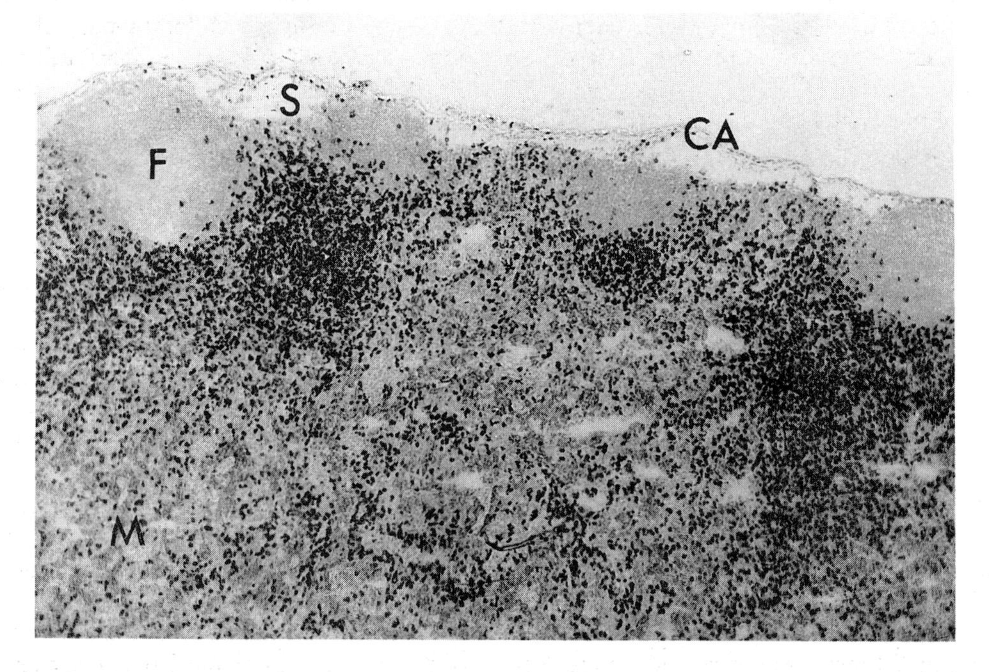

Figure 6.11. Section of a haemal node from a sheep (immunoperoxidase staining of $CD4^+$ lymphocytes counterstained with haematoxylin; ×100). The node consists of a capsule (CA), subcapsular sinus (S), a medulla (M), and a cortex containing follicles (F) and heavily stained deep cortical units. Photo by courtesy of B.H. Thorp, reproduced from reference 117 with kind permission from Elsevier Science Ltd, The Boulevard, Langford Lane, Kidlington OX5 1GB, UK.

tional discharge of immunocompetent cells from the node via the medulla and efferent lymph channels, postcapillary venules in the paracortex provide the major routes of exit as well as entry. Thus, recirculating porcine lymphocytes follow a blood–lymph node–blood route. As in rodent and marsupial nodes, B-cells occupy the follicular cortex while the paracortical zone is T-dependent[119,120].

It has been suggested that a peripherally located medulla is less likely to become clogged or restricted in an active lymph node and therefore less likely to impede the flow of lymph than the conventional arrangement. The inverted node may also allow more specific (receptor-mediated) control of emigrating cells since they have to pass back into postcapillary venules. However, malignant and pathogen-bearing cells that manage to leave would not be subject to any local filtering or surveillance conducted by other lymph nodes in a 'downstream' position; more distant protective sites within the blood vascular system (e.g. spleen, bone marrow and lungs) may therefore assume greater significance in pigs than other mammals[120].

The homing behaviour of a lymphocyte depends on the cell's origin, function, degree of maturity, experience of antigen and state of activation as well as the identity of the target lymphoid tissue[121,170]. There is increasing interest in the nature and specificity of interactions between receptors on the lymphocyte surface and complementary ligands (addressins) on high endothelial surfaces which facilitate adhesion upon arrival in a tissue; factors which control the subsequent redistribution of a cell to specific zones within that tissue; and molecular factors governing length of stay. It is a pity that large mammals such as pigs and sheep do not match rodents in the availability of standardized reagents for recognizing lymphocyte subsets or binding factors (although efforts are being made in this direction[169], because their sheer size and robustness make them excellent models for studying lymphocyte movements and meeting places: cell traffic can be monitored in individual arteries, veins and lymphatics; functioning lymphoid tissues which have been subjected to minimal physiological interference can be examined and compared; regional responses to various antigens can be readily defined; and embryos as well as adults can be manipulated[50,120].

Whilst lymphoid architecture and lymphocyte routes do show differences among species (e.g. in lymph nodes), there appears to be considerable conservation of migration determinants. It has been reported, for example, that the homing receptors on bovine, mouse and human lymphocytes have close molecular and functional similarities[122,123]. Human lymphocytes can home to mouse lymphoid tissues; bovine and chicken lymphocytes stick to rodent postcapillary venules; sheep and pig lymphocytes can move into and out of each other's lymph nodes along the same routes as those taken by syngeneic cells[120]. Thus the signposts are the same, even when the routes are different. However, lymphocytes are less able to enter and concentrate in a tissue if they are more foreign (e.g. horse or rat versus sheep cells in a pig) or from a different part of the body. Does this imply a selective distribution of some signposts or an inability of unfamiliar travellers to read them all?

In recent years the skin (Figure 6.7) has come to be regarded as an active component of the mammalian immune system: antigens which manage to penetrate into its deeper layers are taken up by local antigen-presenting cells for subsequent triggering of T-cells[72,124,171]. In addition, cytokine-induced endothelial addressins have been identified which bind skin-seeking T-cell subsets at sites of cutaneous inflammation[125,178]. Significant variations in cutaneous immune function are to be expected, since there are such marked regional and species differences in the amount of hair present, thickness of epidermis, presence of sweat glands and susceptibility to skin disorders.

Unlike the skin, the mucosal membranes of the body present a poor mechanical barrier to antigen entry, but they are defended by an extensive network of mucosa-associated lymphoid tissues (MALT). The best-studied of these occur in the gut (GALT) as Peyer's patches, where they play

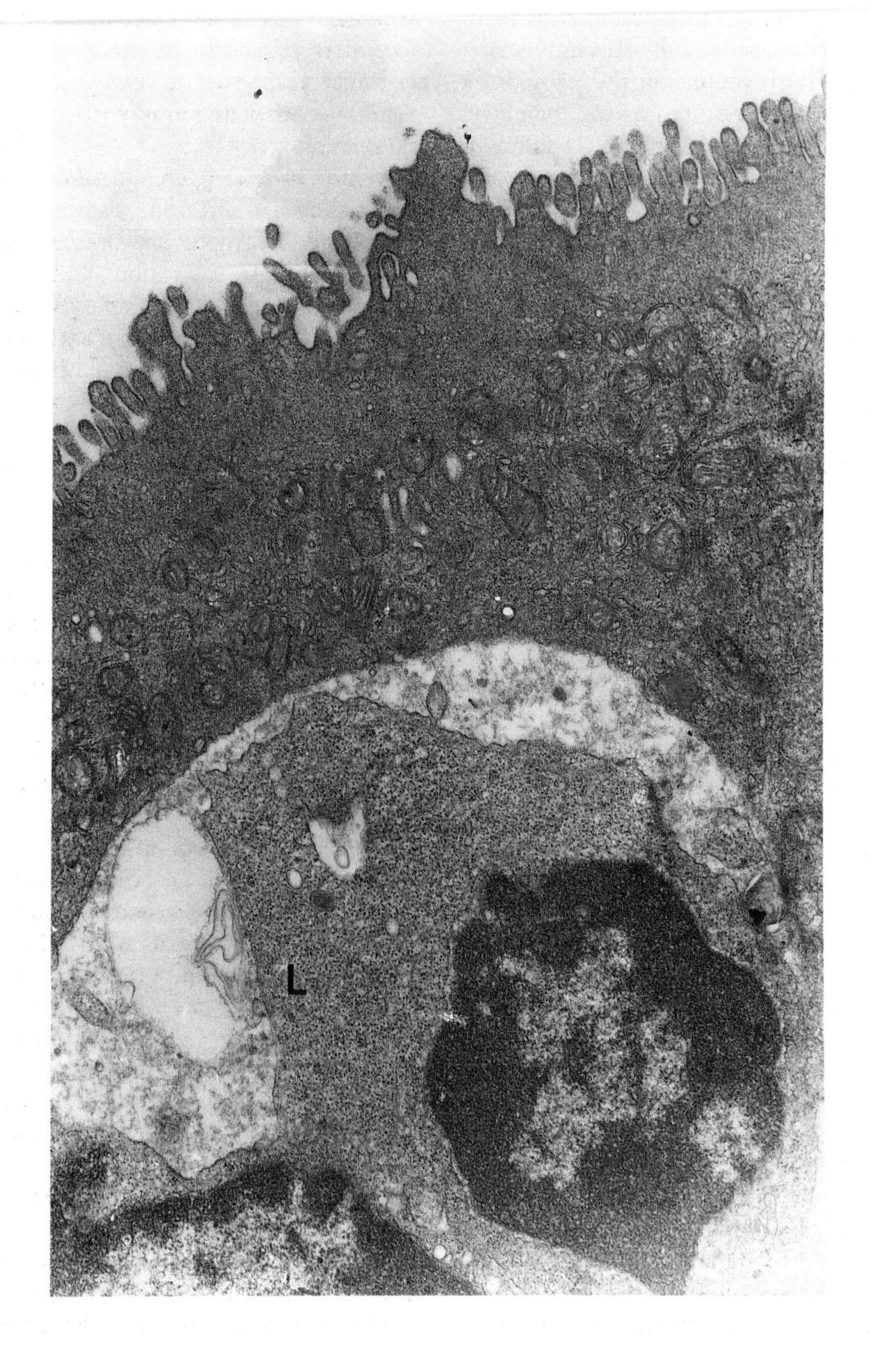

Figure 6.12. Electronmicrograph of an opossum (*Didelphis albiventris*) M-cell presenting short microvilli (×11 000). The M-cell encloses a lymphocyte (L), and the space between the two is occupied by material of low electron density, comparable to the content of endocytic vesicles; this suggests antigen 'feeding'. Photo by courtesy of H.B. Coutinho, reproduced from reference 127 by permission of Memorias do Instítuto Oswaldo Cruz.

an important role in secretory IgA production (for discussion of their role as bursa equivalents see Section 6.3). Their typical domed structure has been described in the adult gut of diverse eutherians and in marsupial mice (*Antechinus* spp.), although the size, number and location of the patches and the balance of T- and B-cells present differ among species[55,120,126]. The vasculature of *Antechinus* Peyer's patches is not as prominent as in rodents, nor could high endothelial venules be observed in their interfollicular regions. However, high venules are not an invariant feature of eutherian GALT[49]. The Peyer's patches described in opossums *(D. albiventris)* lack a conspicuous dome, but specialized epithelial cells are present. As in eutherians, these M-cells enfold lymphocytes and appear to 'feed' them with luminal antigen[55,127] (Figures 6.12 and 6.13). However, the apparent differentiation of plasma cells *in situ* (Figures 6.13 and 6.14) is not seen in rats (H.B. Coutinho, personal communication).

Well-developed Peyer's patches have also been noted along the intestine of echidnas, though their epithelial ultrastructure has not been examined. Very similar lymphoid nodules have been seen in the cloacal region, prompting speculation that echidna GALT acts as a bursa equivalent in early life then switches to a secondary lymphoid function in adulthood. However, developmental and functional studies are needed to substantiate these claims[7].

The nasopharyngeal lymphoid tissues (NALT), like Peyer's patches, are part of the mucosal immune system. In humans they comprise the adenoids and tonsils; similar structures have been described in monkeys, horses and cattle[128]. In pigs there are, in addition to tonsils, large conical lymphoid papillae on the dorsal side of the tongue which again seem to be involved in sampling the local antigenic environment and attracting subpopulations of lymphocytes. Rodents lack equivalents of palatine and lingual tonsils, but well-organized lymphoid accumulations (adenoid equivalents?) have been identified at the entrance of the nasopharyngeal duct[128]. Their lymphocyte content and the binding characteristics of their high endothelial venules are quite different from Peyer's patches. Animal models of NALT should help us to evaluate more closely the immunological consequences of tonsillar removal; identification of immunological changes caused by infections or air pollutants which damage nasopharyngeal linings could also be facilitated.

Judging by the extent of follicular development and lymphocyte infiltration, NALT has a more important role in defending respiratory surfaces than similar structures in the bronchial region (BALT). In some mammals (e.g. humans), BALT is difficult to find at all except in certain cases of immunological malfunction; in cattle its occurrence and morphology are very variable, while in rats and rabbits it can at least be detected reliably and regularly[129]. The reasons for these wide species differences are unclear: it seems unlikely that different patterns of antigenic assault on the respiratory linings are solely responsible. Moreover, the bronchial defences are not simply duplicating those nearer the entrance of the airways, at least in rats: BALT shows significant numbers of IgA-bearing B-cells, whereas NALT is more concerned with cellular immunity[128].

The spleen is an unusual secondary lymphoid tissue in that it is not specialized to monitor local entry of antigen across body surfaces, but acts as a systemic filter. High endothelial vessels are missing in this organ, and the homing receptors described above play no part in its uptake of blood lymphocytes[121,130]: the fact that it processes a very large number of recirculating lymphocytes each day may make chance encounters between appropriate antigen-recognizing cells sufficiently frequent, and selective devices unnecessary. On the other hand, different routes have been charted for lymphocyte subpopulations migrating through the splenic tissues, and a guidance role has been suggested for reticular cells[130].

The rat provides the most favoured model for the human spleen. Its white pulp shows comparable subcompartments (albeit with different configurations), and housed within them are lymphoid cells with similar surface markers. Greater differences are seen in the white pulp of

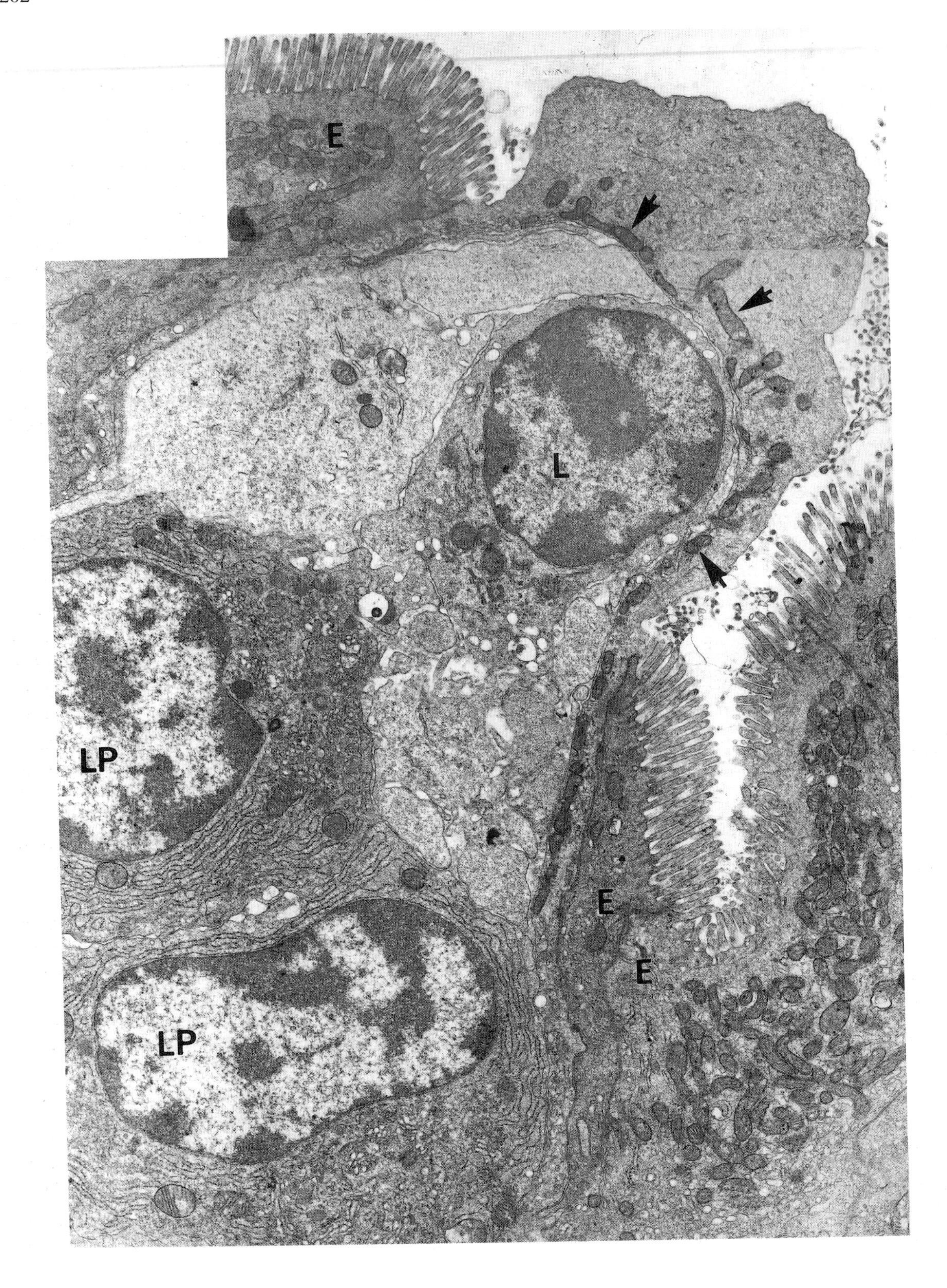
E
L
LP
E
E
LP

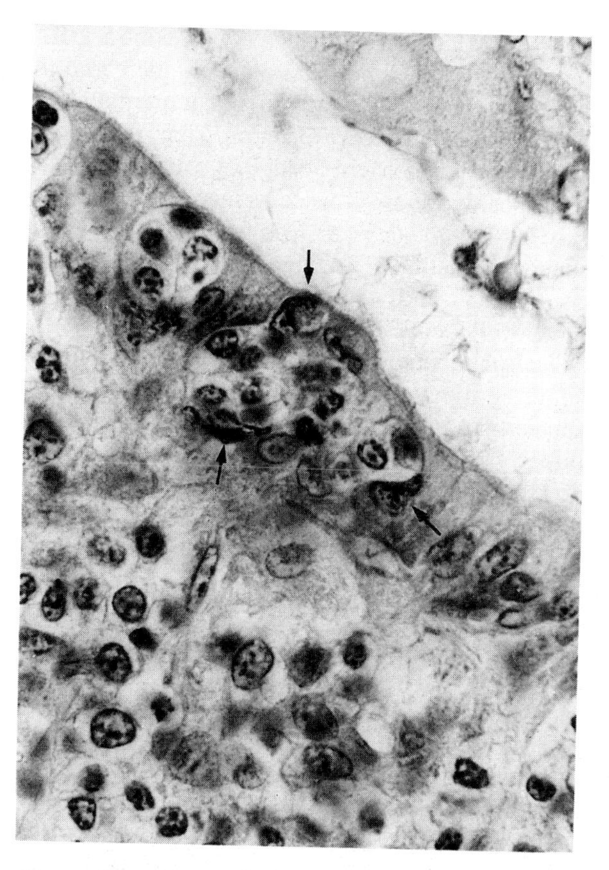

Figure 6.14. Opossum M-cells containing IgA-positive plasma cells (arrowed) (×800). Immunoperoxidase with haematoxylin counterstain. Reproduced by courtesy of H.B. Coutinho.

the mouse[130]. Nevertheless, primate, rodent and rabbit spleens are all rich in lymphoid tissue, their red pulps are similar in structure and they only have a minor role in red cell storage. In contrast, horse and artiodactyl spleens have prominent trabeculae and enlarged red pulps containing a lot of smooth muscle. These 'storage' spleens are designed to transfer large numbers of red cells into the circulation on demand, and they appear to be a late evolutionary development[131].

Monotreme and marsupial spleens have not been studied in such detail, but a gross structure (i.e. triradiate or T-shaped) which is reminiscent of reptilian spleens has been noted[7,46].

The thymus gland continues to play an important role in the differentiation and processing of lymphocytes in adult mammals (but see also Section 6.5). Its structure is highly conserved: in monotremes and marsupials as well as eutherians it comprises a series of lobules, each with a cortical and medullary zone[7,49,59] (Figure 6.8C). In pigs, significant numbers of lymphocytes are known to recirculate back into thymic tissue, entering via the high endothelium of blood vessels near the cortico-medullary junction[120]. The immunological significance of this migration is unknown.

6.5 THE IMMUNE SYSTEM OF AGEING MAMMALS

For reasons of cost, convenience and control, the laboratory mouse has been and is likely to remain the mammalian model most favoured for immunological studies of ageing (see Section 6.1). Different mouse strains are interesting because they show different patterns of disease as they age, and different causes and timing of death. Other ageing rodents have also been studied, notably rats: the latter offer the practical advantage of size (e.g. for repetitive blood sampling or provision of tissue), though they are obviously more costly to maintain than mice and there are fewer probes and standards available for them. Occasional reports do exist for other species, e.g. rabbits and dogs[132], but detailed comparisons of the ageing process among most mammals are hampered by a paucity of 'longitudinal' information: in the absence of regular and detailed tests throughout the life of each species and strain under scrutiny, peaks of response and

Figure 6.13 (opposite). Electronmicrograph of another opossum M-cell (lacking microvilli) (×11 000). The cells within it include a lymphocyte (L) and two others (LP) nearer the basal region which are differentiating plasma cells. Elongated mitochondria (arrowed) are seen clustered in the apical and lateral regions of the M-cell. Photo by courtesy of H.B. Coutinho, reproduced from reference 127 by permission of Memorias do Instítuto Oswaldo Cruz.

rates of decline will remain matters of controversy and guesswork.

In recent years, significant advances have been made in assessing the immunocompetence of humans with age. However, interpretation is complicated by the fact that elderly subjects have sometimes been diseased or medicated, they represent a selected group, and it is difficult to find appropriate controls. It is a moot point whether information derived from such outbred individuals should be regarded as a problem or an asset.

The most visible target of ageing within the immune system is the T-cell compartment. A significant age-related decline has been reported for delayed-type hypersensitivity reactions, cytotoxic T-cell function, delivery of help, and responses to T-cell mitogens and alloantigens[133–135,179]. The decline can be due to decreased numbers of cells, subtle shifts in the balance of T-subpopulations, and functional deterioration, the relative importance of which differs with species, strain and tissue[133,136–139]. Peripheral T-cells from the widely studied 'systemic' immune system (e.g. spleen) of the ageing mouse show earlier and more severe dysfunctions than T-cells from the gut (GALT)[137,140], whilst cells from a given population can also deteriorate at different rates. The latter produces a 'mosaic' comprised of fully functioning cells mixed with an increasing proportion of cells which fail to respond to activating stimuli (e.g. by proliferating or by generating effector clones)[138]. The absence of simple, coordinated changes applies also to suppressor cells.

Intensive studies on mouse and human IL-2 indicate an age-related impairment in the capacity of cells both to receive and transmit immunological signals, although detailed comparative studies on other interleukins and the T-cell receptor complex are still awaited. A decrease in IL-2 production has been reported for aged mouse splenocytes and human peripheral blood lymphocytes, together with a reduced expression of high- and low-affinity IL-2 receptors. While the 'mosaic' model of cell deterioration seems to hold for IL-2 production and reception in mice[138], some workers report that in humans all the cells gradually lose their ability to synthesize IL-2 and to display IL-2 receptor sites[133,141]. In contrast, IL-2 output is said to be maintained or increased in old rat splenocytes, with no net impairment in high-affinity IL-2 receptor expression: fewer old rat cells carried these receptors, but those that were positive showed twice as many on their surfaces[141]. Although such results certainly suggest species variations, findings within species lack consistency, and several methodological pitfalls have been identified[141].

The phenomenon of thymic involution was recognized many years ago. In all of the eutherian species studied the gland begins to shrink at puberty, lymphoid and epithelial tissues are progressively replaced by fat, and levels of thymic hormones are reduced[136,142,143]. It is widely believed that there is a causal relationship between these events and the peripheral T-cell dysfunctions which follow them. Early workers charted a dramatic shrinkage of the human thymus from early adulthood, but a more recent view is that it often remains sizeable well into old age and is still capable of T-cell support, albeit at a reduced level: the gland's demise is greater in diseased and stressed individuals than in the healthy aged[9,143]. The mouse thymus also becomes a less hospitable environment for differentiating T-cells with age; development and involution of the gland are slower in long-lived than short-lived strains[136,137].

The ontogeny of the thymus gland in opossums *(Monodelphis domestica)* is similar to eutherians. Adult opossums reach reproductive age at about 6 months, beyond which point the thymus begins to involute and lymphoid tissue is progressively replaced by fat (Figure 6.8C and 6.8D). As in humans, the speed of this replacement differs considerably in different individuals: a 5-month-old opossum may show an atrophic thymus, while at the other extreme a 34-month-old animal may not[59]. A similar sequence of thymic changes is seen in marsupial mice (*Antechinus* spp.), but here the process is greatly accelerated in relation to the animals' state of maturity: the gland reaches its maximum size shortly before weaning; involution beings only 2 weeks after the youngster has left the pouch; and thymic lymphoid tissue has become completely replaced by fat 2 months before puberty is reached[144]. A point of particular interest is that this remarkably rapid degeneration of the gland does not seem to compromise the well-being of adult female *Antechinus*: they continue with an atrophic thymus for at least half of their lifespan. Males, on the other hand, die much earlier (and synchronously) due to high levels of stress-inducing glucocorticoids after the breeding season has finished, but it is not known whether the early loss of the thymus is a predisposing factor[6]. It has been suggested that a survey of other marsupial species might clarify the relationship between thymic involution and longevity[6]. In quokkas, neonatal thymectomy causes premature deaths: most experimental animals die before they are 3 years old, compared with the normal lifespan of 10 years. However, unlike some laboratory rodents their early death is not associated with a wasting disease[145].

Immunological decline with age is not limited to the T-cell compartment. Defects in humoral immunity due partly to regulatory changes and partly to intrinsic B-cell defects have been widely reported in ageing rodents and humans. Old B-cells show an altered expression and mobility of surface receptors and a reduced ability to respond to T-independent antigens and mitogens[133,137,142]. The total numbers of B-cells do not change appreciably with age but, judging by mouse studies, the number capable of responding diminishes[138]. Most groups report increased IgA and IgG output and a decline in IgM; in humans, heightened IgG levels are due primarily to increases in IgG1 and IgG3, while in mice IgG1 and IgG2b subclasses are preferentially increased[141]. Studies on mice also suggest that there are age-related shifts in the usage by B-cells of variable region genes, in favour of autoantibody production and at the expense of humoral immunity[138,141]. B cell subsets are of interest here: $CD5^+$ cells, whose repertoire is skewed towards autoantibody production, retain or increase their activity in old mice; $CD5^-$ cells on the other hand, which produce the majority of antibodies directed against foreign antigens, lose much of theirs[181].

The evidence for age-related changes in leucocytes other than B- and T-cells is relatively limited, and inconclusive. IL-1 secretion by monocytes from elderly people and rats is not decreased, but peritoneal macrophages from aged mice do show diminished IL-1, IL-6 and TNF production and a reduced ability to heal wounds[133,137,141,146]. Studies on humans and rats have provided evidence both for and against an age-dependent decline in natural killer (NK) activity, while neutrophil function in aged humans is reported to be maintained[147,148,179].

It is clear from the above description that the onset, rate and extent of immunological decline vary with cell type, the individual and the species; there is no evidence for a synchronized or ubiquitous trigger. It is also worth noting that while the loss of immunocompetence is accompanied by an increased susceptibility to infections and malignancies as an individual grows old, immunological failures seem to be manifestations rather than fundamental causes of the ageing process. An inability to control pathogens may cause death, but germ-free animals (rodents) die as early as their conventionally reared counterparts, sometimes earlier[149]. Similarly, autoimmune pathology cannot be considered a primary cause of normal ageing, even though accelerated ageing is a feature of diseases such as systemic lupus erythematosus (SLE): the increased output of autoantibodies found in ageing individuals is not associated with an increase in clinical autoimmune disease[142,150,180]. Susceptibility to tumours in old age may depend partly on defective anti-tumour immune mechanisms, but the extent and significance of such changes need further investigation; other physiological changes (e.g. of a hormonal or nutritional nature) are likely to be important contributory factors, as well as age-related DNA modifications responsible for tumorigenesis[133]. Even in the absence of physiological deterioration, the likelihood of natural exposure to carcinogens and completion of tumour induction periods would still be greatest in old animals.

In view of the multicentric nature of age-related decline, simple immunological remedies

Figure 6.15. The effects of dietary restriction. The rat on the left has been fed *ad libitum*; its littermate on the right has been kept to half the normal body weight by restricting the number of food pellets it consumes each day. The latter is likely to live 40% longer and show an inhibition or delay of age-associated diseases. Photo by courtesy of R.A. Moore.

involving, for example, administration of a thymic hormone or IL-2 will not be very effective. More radical, multisystem intervention by means of dietary restriction[135,151,152] has, on the other hand, proved highly effective in prolonging the life and health of laboratory rats (Figure 6.15) and mice—but can it work for humans?!

6.6 CONCLUDING REMARKS: THE USE OF MAMMALIAN MODELS

In summarizing the dominant position of the laboratory mouse in immunological research, an analogy can be drawn with the car industry. The most popular model is not necessarily the best in its class, though it is certainly well advertised, relatively cheap to buy and run, boasts a comprehensive supply and service network, and has a comfortable and reliable air about it. Many customers choose it through habit, because it has sold well in the past, or because they have been seduced by 'high-tech' versions, but they would be wrong in assuming that it is without quirks or that no other vehicle could better serve a specialist need. Some potentially exciting alternatives, however, are hindered by lack of resources and involve a greater gamble.

It has been suggested that both new and existing models are luxuries we would be better off without. In an article asking 'Is man the odd mouse out?' a tumour immunologist remarked some years ago that he seldom heard others 'decrying the fact that their technical skills and intellectual abilities are wasted on rodents and really should be used to alleviate human misery'. He concluded that 'immunology can be and should be a human science'[153]. This ignores the fact that some immunologists wish to alleviate misery in non-human species; that direct immunological manipulation and testing of humans (or for that matter valuable domestic animals) are more circumscribed; and that *in vitro* techniques can give misleading results. Most immunologists would maintain that basic studies on murine haemopoiesis, autoimmunity and immunodeficiency, for example, *have* provided important insights into human disease[47,154,155]. On the other hand some mechanisms, such as regulation of IgG subclasses, do show a significant evolutionary divergence between mice and humans: knowledge of murine interleukins is unlikely to help IgG-deficient children[156]. Similarly, the large volume of immunological data collected on pregnancy loss in mice cannot be applied with confidence to the treatment of recurrently aborting women[157].

Rodent models of human microbial, parasitic and malignant diseases have been widely sought in order to identify effective means of immunological control. Ideally, the pattern of tissue invasion and the nature of the response engendered after inoculation into the animal will closely resemble the natural condition seen in humans, but this goal is often only partially realized. Increasingly, mice with severe combined immunodeficiency (SCID) are being used as hosts for human pathogens that would normally fail to 'take'[155]. These mice can also be provided with human cells (the immunological equivalent of the kit-car!). Thus normal human lymphocytes or stem cells, HIV-infected human cells or malignant human cells can be assessed and manipulated freely *in vivo*[154,155,158,172]. Unfortunately, the foreign components of the SCID-Hu model do not blend perfectly: NK cells of the mouse host can respond to human xenografted cells, and donated human lymphocytes react to mouse xenoantigens, albeit weakly. It is also reported that virtually all SCID mice are 'leaky', i.e. they generate a few of their own B- and T-cells[158,159]. Further exotic immunological models have recently arrived through developments in transgenic technology, whereby foreign genes, including those of man, can be introduced into animal germ lines. Transgenic mice have been offered as models for a variety of human diseases, for example AIDS and AIDS-related disorders, HLA-associated disease, cancers, and abnormalities of cytokine production; also as a means of generating better quality tissue typing reagents[160,161].

Although the technical sophistication achieved in mouse models is particularly advanced, we cannot simply assume that inbred or mutant mice are 'representative'. Development of a broader spectrum of models would enable a more balanced view of the immune system to emerge, and allow us to distinguish features of central importance from detailed variants which are the end products of adaptive radiation—the main routes from the culs de sac. The student of human or veterinary immunology would therefore be more likely to benefit from the information provided.

Imbalances in current research effort are evident not only in the species being used for study but also in the kind of immunological questions being asked. Much is known about the cells and molecules of the immune system, relatively little about the way they behave or are controlled *in situ*; and we know much more about immunological components derived from the blood and spleen than from mucosa-associated lymphoid tissues, even though infections of concern in the field are commonly those of the respiratory and intestinal tracts. Clearly, physiological and regional analyses of the immune system are most easily conducted on large animal models such as sheep and pigs rather than small rodents[50,119,120], but the large animal work would benefit from an improved provision of reagents. Hopes that observation of blood leucocytes might provide an accurate 'window' on immunological events occurring elsewhere in the body have not been fulfilled[162].

While some animal models have emerged for reasons of chance, cost, convenience and/or technical wizardry, the use of others (notably non-human primates) has been justified mainly on phylogenetic grounds. Our attempts to understand and control AIDS, for example, have been assisted by detailed study of HIV and related pathogens in chimps and monkeys, as well as in mice, rabbits, cats and ungulates[163,173]. The use of primates for such work demands particular care and sensitivity, but also illustrates the important general point that biomedical demands cannot be divorced from questions of animal welfare or public opinion: cramped or barren conditions of captivity, and the plunder of rare or dwindling wild populations, are neither in the animals' nor the scientists' interest. Thus, a reliance on buoyant, self-sustaining species, together with a willingness to enlarge and enrich the environment of captive animals, should be important aspects of comparative work in the future.

Acknowledgements

I am indebted to Bob Moore for help with line diagrams, and to various other colleagues who generously provided photos (see figure captions).

6.7 REFERENCES

1. Medawar, P.B. (1953). Some immunological and endocrinological problems raised by the evolution of viviparity in vertebrates. *Symp. Soc. Exp. Biol.* **7**, 320–338.
2. Lillegraven, J.A., Thompson, S.D., McNab, B.K. and Patton, J.L. (1987). The origin of eutherian mammals. *Biol. J. Linn. Soc.* **32**, 281–336.
3. Tyndale-Biscoe, C.H. and Renfree, M.B. (1987). *Reproductive Physiology of Marsupials.* University Press, Cambridge.

4. Colbert, E.H. and Morales, M. (1991). *Evolution of the Vertebrates*, 4th edn. Wiley, New York.
5. Tizard, I.R. (1992). *Veterinary Immunology*, 4th edn. W.B. Saunders, Philadelphia.
6. Tyndale-Biscoe, C.H. and Janssens, P.A. (eds). (1988). *The Developing Marsupial: Models for Biomedical Research*. Springer-Verlag, Berlin.
7. Griffiths, M. (1978). *The Biology of the Monotremes*. Academic Press, New York.
8. Romer, A.S. and Parsons, T.S. (1986). *The Vertebrate Body*, 6th edn. Saunders College Publ., Philadelphia.
9. Crighton, D.B. (ed.) (1984). *Immunological Aspects of Reproduction in Mammals*. Butterworths, London.
10. Kurpisz, M. (ed.) (1990). Immunological and molecular mechanisms in reproduction. *Arch. Immunol. Ther. Exp.* **38**(1–2).
11. Talwar, G.P. (ed.) (1990). Reproduction. *Curr. Opinion in Immunol.* **2**, 723 ff.
12. James, K. and Hargreave, T.B. (1984). Immunosuppression by seminal plasma and its possible clinical significance. *Immunol. Today* **5**, 357–363.
13. Mukherjee, D.C., Agrawal, A.K., Manjunath, R. and Mukherjee, A.B. (1983). Suppression of epididymal sperm antigenicity in the rabbit by uteroglobin and transglutaminase *in vitro*. *Science* **219**, 989–991.
14. Hunter, R.H.F. (1980). *Physiology and Technology of Reproduction in Female Domestic Animals*. Academic Press, London.
15. Jacoby, D.R., Olding, L.B. and Oldstone, M.B.A. (1984). Immunologic regulation of fetal-maternal balance. *Adv. Immunol.* **35**, 157–208.
16. Rodger, J.C. and Drake, B.L. (1987). The enigma of the fetal graft. *Amer. Sci.* **75**, 51–58.
17. Chaouat, G. (ed.) (1987). Reproductive immunology: materno-fetal relationship; fundamental investigations and future applied prospects. *Colloque INSERM* **154**, Paris.
18. Chaouat, G. and Mowbray, J. (eds.) (1991). Cellular and molecular biology of the materno-fetal relationship. *Colloque INSERM* **212**, Paris.
19. Wegmann, T.G. (ed.) (1991). Reproduction. *Curr. Opinion in Immunol.* **3**, 759 ff.
20. Donaldson, W.L., Zhang, C.H., Oriel, J.G. and Antczak, D.F. (1990). Invasive equine trophoblast expresses conventional class I major histocompatibility antigens. *Development* **110**, 63–71.
21. Johnson, P.M. (1989). Immunological intercourse at the feto-maternal interface? *Immunol. Today* **10**, 215–218.
22. King, A. and Loke, Y.W. (1991). On the nature and function of human uterine granular lymphocytes. *Immunol. Today* **12**, 432–435.
23. Stewart, I.J. (1991). Granulated metrial gland cells: pregnancy-specific leukocytes? *J. Leukocyte Biol.* **50**, 198–207.
24. Giordano, C. (1990). Immunobiology of normal and diabetic pregnancy. *Immunol. Today* **11**, 301–303.
25. Faulk, W.P. (1981). Trophoblast survival. *Transplantation* **32**, 1–4.
26. Fehilly, C.B., Willadsen, S.M. and Tucker, E.M. (1984). Interspecific chimaeras between sheep and goat. *Nature* **307**, 634–636.
27. Summers, P.M., Shephard, A.M., Hodges, J.K. *et al.* (1987). Successful transfer of the embryos of Przewalski's horses (*Equus przewalskii*) and Grant's zebra (*E. burchelli*) to domestic mares (*E. caballus*). *J. Reprod. Fertil.* **80**, 13–20.
28. Clark, D.A., Slapsys, R.M., Croy, B.A. and Rossant, J. (1983). Suppressor cell activity in uterine decidua correlates with success or failure of murine pregnancies. *J. Immunol.* **131**, 540–542.
29. Kamel, S. and Wood, G.W. (1990). Immunoregulatory activity of cells from lymph nodes draining the uterus of allopregnant mice. *J. Reprod. Immunol.* **17**, 239–252.
30. Kamel, S. and Wood, G.W. (1991). Failure of *in vitro* expanded hyperimmune cytotoxic T lymphocytes to affect survival of mouse embryos *in vivo*. *J. Reprod. Immunol.* **19**, 69–84.
31. Blackburn, D.G., Taylor, J.M. and Padykula, H.A. (1988). Trophoblast concept as applied to therian mammals. *J. Morphol.* **196**, 127–136.
32. Dawson, T.J. (1983). *Monotremes and Marsupials: the Other Mammals. Studies in Biology 150*. Arnold, London.
33. Van Oorschot, R.A.H. and Cooper, D.W. (1988). Lack of evidence for complement-dependent cytotoxic antibodies to fetal paternally derived antigens in the marsupial *Macropus eugenii* (tammar wallaby). *Amer. J. Reprod. Immunol. Microbiol.* **17**, 145–148.
34. Samples, N.K., Vandeberg, J.L. and Stone, W.H. (1986). Passively acquired immunity in the newborn of a marsupial (*Monodelphis domestica*). *Amer. J. Reprod. Immunol. Microbiol.* **11**, 94–97.
35. Yadav, M., Stanley, N.F. and Waring, H. (1972). The microbial flora of the gut of the pouch-young and the pouch of a marsupial, *Setonix brachyurus*. *J. Gen. Microbiol.* **70**, 437–442.
36. Outteridge, P.M. and Lee, C.S. (1988). The defence mechanisms of the mammary gland of domestic ruminants. *Progr. Vet. Microbiol. Immunol.* **4**, 165–196.
37. Hanson, L.A. (1982). The mammary gland as an immunological organ. *Immunol. Today* **3**, 168–171.

38. Rodewald, R. (1980). Immunoglobulin transmission in mammalian young and the involvement of coated vesicles. In *Coated Vesicles*, Ockleford, C.D. and Whyte, A. (eds). University Press, Cambridge, pp. 69–101.
39. Newby, T.J., Stokes, C.R. and Bourne F.J. (1982). Immunological activities of milk. *Vet. Immunol. Immunopathol.* **3**, 67–94.
40. Head, J.R. (1977). Immunobiology of lactation. *Seminars in Perinatol.* **1**, 195–210.
41. Parham, P. (1989). MHC meets mother's milk. *Nature* **337**, 118–119.
42. Gill, T.J. and Wegmann, T.G. (1987). *Immunoregulation and Fetal Survival.* Oxford University Press, New York.
43. Deane, E.M., Cooper, D.W. and Renfree, M.B. (1990). Immunoglobulin G levels in fetal and newborn tammar wallabies (*Macropus eugenii*). *Reprod. Fertil. Dev.* **2**, 369–375.
44. Cockson, A. and McNeice, R. (1980). Survival in the pouch: the role of macrophages and maternal milk cells. *Comp. Biochem. Physiol.* **66A**, 221–225.
45. Sheldrake, R.F. and Husband, A.J. (1985). Intestinal uptake of intact maternal lymphocytes by neonatal rats and lambs. *Res. Vet. Sci.* **39**, 10–15.
46. Hunsaker, D. (ed.) (1977). *The Biology of Marsupials.* Academic Press, New York, pp. 349–386.
47. Gershwin, M.E. and Cooper, E.L. (eds). (1978). *Animal Models of Comparative and Developmental Aspects of Immunity and Disease.* Pergamon Press, New York.
48. Solomon, J.B. (1971). *Foetal and Neonatal Immunology.* North-Holland, Amsterdam.
49. Morrison, W.I. (ed.) (1986). *The Ruminant Immune System in Health and Disease.* University Press, Cambridge.
50. Morris, B. and Miyasaka, M. (eds). (1985). *Immunology of the Sheep.* Editiones Roche, Basel.
51. Fahey, K.J. and Morris, B. (1978). Humoral immune responses in foetal sheep. *Immunology* **35**, 651–661.
52. Sawyer, M., Moe, J. and Osburn, B.I. (1978). Ontogeny of immunity and leukocytes in the ovine foetus and evaluation of immunoglobulins related to congenital infection. *Amer. J. Vet. Res.* **39**, 643–648.
53. Aleksandersen, M., Nicander, L. and Landswerk, T. (1991). Ontogeny, distribution and structure of aggregated lymphoid follicles in the large intestine of sheep. *Dev. Comp. Immunol.* **15**, 413–422.
54. Reynolds, J.D. and Morris, B. (1983). The evolution and involution of Peyer's patches in foetal and postnatal sheep. *Eur. J. Immunol.* **13**, 627–635.
55. Landswerk, T., Halleraker, M., Aleksandersen, M. *et al.* (1991). The intestinal habitat for organized lymphoid tissues in ruminants: comparative aspects of structure, function and development. *Vet. Immunol. Immunopathol.* **28**, 1–16.
56. Reynolds, J.D., Kennedy, L., Peppard, J. and Pabst, R. (1991). Ileal Peyer's patch emigrants are predominantly B cells and travel to all lymphoid tissues in sheep. *Eur. J. Immunol.* **21**, 283–289.
57. Hein, W.R. and Mackay, C.R. (1991). Prominence of $\gamma\delta$T cells in the ruminant immune system. *Immunol. Today* **12**, 30–34.
58. Press, C., McClure, S. and Landswerk, T. (1991). Computer-assisted morphometric analysis of absorptive and follicle-associated epithelia of Peyer's patches in sheep fetuses and lambs indicates the presence of distinct T cell and B cell components. *Immunology* **72**, 386–392.
59. Hubbard, G.B., Saphire, D.G., Hackleman, S.M. *et al.* (1991). Ontogeny of the thymus gland of a marsupial (*Monodelphis domestica*). *Lab. Anim. Sci.* **41**, 227–233.
60. Fadem, B.H., Hill, H.Z., Huselton, C.A. and Hill, G.J. (1988). Transplantation, growth and regression of mouse melanoma xenografts in neonatal marsupials. *Cancer Invest.* **6**, 403–408.
61. Stear, M.J. (1980). Lymphocyte antigens in sheep. PhD thesis, University of Edinburgh.
62. Hines, H.C. and Newman, M.J. (1981). Production of foetally stimulated lymphocytotoxic antibodies by multiparous cows. *Anim. Blood Groups and Biochem. Genetics* **12**, 201–206.
63. Klein, J. (1986). *Natural History of the Major Histocompatibility Complex.* Wiley, New York.
64. Doherty, P.C. and Zinkernagel, R.M. (1975). Enhanced immunological surveillance in mice heterozygous at the H-2 gene complex. *Nature* **256**, 50–52.
65. Nagy, Z.A., Lehmann, P.V., Falcioni, F. *et al.* (1989). Why peptides? Their possible role in the evolution of MHC-restricted T cell recognition. *Immunol. Today* **10**, 132–138.
66. Lawlor, D.A., Zemmour, J., Ennis, P.D. and Parham, P. (1990). Evolution of class I MHC genes and proteins: from natural selection to thymic selection. *Ann. Rev. Immunol.* **8**, 23–63.
67. Srivastava, R., Ram, B.P. and Tyle, P. (eds) (1991). *Immunogenetics of the MHC.* VCH, New York.
68. Warner, C.M., Rothschild, M.F. and Lamont,

S.J. (eds). (1988). *The Molecular Biology of the Major Histocompatibility Complex of Domestic Animal Species*. Iowa State University Press, Ames.
69. Xu, Y.X., Rothschild, M.F. and Warner, C.M. (1992). Mapping of the SLA complex of miniature swine. *Mammalian Genome* **2**, 2–10.
70. Klein, J. and Klein, D. (1991). *Molecular Evolution of the Major Histocompatibility Complex*. NATO advanced research workshop on MHC evolution, Key Biscayne, Florida, in cooperation with Springer-Verlag.
71. Trowsdale, J., Ragoussis, J. and Campbell, R.D. (1991). Map of the human MHC. *Immunol. Today* **12**, 443–446.
72. Klein, J. (1990). *Immunology*. Blackwell, Oxford.
73. Figueroa, F., Gunther, E. and Klein, J. (1988). MHC polymorphism predating speciation. *Nature* **335**, 265–267.
74. Möller, G. (ed.) (1990). Phylogeny of the major histocompatibility complex. *Immunol. Revs.* **113**, 5–241.
75. McConnell, T.J., Talbot, W.S., McIndoe, R.A. and Wakeland, E.K. (1988). The origin of MHC class II gene polymorphism within the genus *Mus*. *Nature* **332**, 651–654.
76. Parham, P. (1992). Flying the first class flag. *Nature* **357**, 193–194.
77. Powis, S.J., Deverson, E.V., Coadwell, W.J., *et al.* (1992). Effect of polymorphism of an MHC-linked transporter on the peptides assembled in a class I molecule. *Nature* **357**, 211–215.
78. Streilein, J.W. and Duncan, W.R. (1983). On the anomalous nature of the major histocompatibility complex in Syrian hamsters. *Transplant. Proc.* **15**, 1540–1545.
79. Watkins, D.I., Chen, Z.W., Hughes, A.L. *et al.* (1990). Syrian hamsters express diverse MHC class I gene products. *J. Immunol.* **145**, 3483–3490.
80. Vandenbogaerde, J. and Howard, J.C. (1991). Xenogeneic responses *in vitro* in the Syrian hamster *Mesocricetus auratus*. 1. Evidence for a normal T cell repertoire. *Int. Immunol.* **3**, 49–56.
81. Faulkes, C.G., Abbott, D.H. and Mellor, A.L. (1990). Investigation of genetic diversity in wild colonies of naked mole-rats (*Heterocephalus glaber*) by DNA fingerprinting. *J. Zool. Lond.* **221**, 87–97.
82. Nizetic, D., Figueroa, F., Nevo, E. and Klein, J. (1985). Major histocompatibility complex of the mole-rat. 2. Restriction fragment polymorphism. *Immunogenetics* **22**, 55–67.
83. Nizetic, D., Stevanovic, M., Soldatovic, B. *et al.* (1988). Limited polymorphism of both classes of MHC genes in four different species of the Balkan mole rat. *Immunogenetics* **28**, 91–98.
84. Trowsdale, J., Groves, V. and Arnason, A. (1989). Limited MHC polymorphism in whales. *Immunogenetics* **29**, 19–24.
85. O'Brien, S.J., Roelke, M.E., Marker, L. *et al.* (1985). Genetic basis for species vulnerability in the cheetah. *Science* **227**, 1428–1434.
86. Miller, P.S. and Hedrick, P.W. (1991). MHC polymorphism and the design of captive breeding programs—simple solutions are not the answer. *Conservation Biol.* **5**, 556–558.
87. Winkler, C., Schultz, A., Cevario, S. and O'Brien, S. (1989). Genetic characterization of FLA, the cat major histocompatibility complex. *Proc. Nat. Acad. Sci. US.* **86**, 943–947.
88. Watkins, D.I., Hodi, F.S. and Letvin, N.L. (1988). A primate species with limited major histocompatibility complex class I polymorphism. *Proc. Nat. Acad. Sci. US.* **85**, 7714–7718.
89. Watkins, D.I., Chen, Z.W., Hughes, A.C. *et al.* (1990). Evolution of the MHC class I genes of a New World primate from ancestral homologues of human non-classical genes. *Nature* **346**, 60–63.
90. Hughes, A.L. and Nei, M. (1989). Evolution of the major histocompatibility complex—independent origin of non-classical class I genes in different groups of mammals. *Molec. Biol. Evol.* **6**, 559–579.
91. Puri, N.K. and Brandon, M.R. (1987). Sheep MHC class II molecules. 2. Identification and characterization of four distinct subsets of sheep MHC class II molecules. *Immunology* **62**, 575–580.
92. Nizetic, D., Figueroa, F., Dembic, Z. *et al.* (1987). Major histocompatibility gene organization in the mole rat *Spalax ehrenbergi*: evidence for transfer of function between class II genes. *Proc. Nat. Acad. Sci. US* **84**, 5828–5832.
93. Schneider, S., Vincek, V., Tichy, H. *et al.* (1991). MHC class II genes of a marsupial, the red-necked wallaby (*Macropus rufogriseus*). Identification of new gene families. *Molec. Biol. Evol.* **8**, 753–766.
94. Stone, W.H., Samples, N.K., Kraig, E. and Vandeberg, J.L. (1987). Definition of class I antigens of the MHC in a marsupial *(Monodelphis domestica). Arch. Zootecnica* **36**, 219–236.
95. Infante, A.J., Samples, N.K., Croix, D.A. *et al.* (1991). Cellular immune response of a marsupial, *Monodelphis domestica. Dev. Comp. Immunol.* **15**, 189–199.
96. Stear, M.J., Mallard, B.A., Newman, M.J. and Wilkie, B.N. (1989). The current status of the major histocompatibility system: definition in

cattle, goats, horses, pigs and sheep. *Int. J. Anim. Sci.* **4**, 32–44.
97. Tumbleson, M.E. (ed.) (1986). *Swine in Biomedical Research* (3 vols.). Plenum, New York.
98. Doxiadis, I., Krumbacher, K., Neefjes, J.J. *et al.* (1989). Biochemical evidence that the DLA-B locus codes for a class II determinant expressed on all canine peripheral blood lymphocytes. *Exp. Clin. Immunogenetics* **6**, 219–224.
99. Rideout, B.A., Moore, P.F. and Pedersen, N.C. (1990). Distribution of MHC class II antigens in feline tissues and peripheral blood. *Tissue Antigens* **36**, 221–227.
100. Briles, W.E., Stone, H.A. and Cole, R.K. (1977). Marek's disease: effects of B histocompatibility alloalleles in resistant and susceptible chickens. *Science* **195**, 193–195.
101. Hill, A.V.S., Allsopp, C.E.M., Kwiatowski, D. *et al.* (1991). Common West African HLA antigens are associated with protection from severe malaria. *Nature* **352**, 595–600.
102. Klein, J., Kasahara, M., Gutknecht, J. and Figueroa, F. (1990). Origin and function of MHC polymorphism. *Chem. Immunol.* **49**, 35–50.
103. Potts, W.K., Manning, C.J. and Wakeland, E.K. (1991). Mating patterns in seminatural populations of mice influenced by MHC genotype. *Nature* **352**, 619–621.
104. Yamazaki, K., Beauchamp, G.J., Bard, J. and Boyse, E.A. (1990). Single MHC gene mutations alter urine odour constitution in mice. In *Chemical Signals in Vertebrates 5*, Macdonald, D.W. *et al.* (eds). University Press, Oxford. pp. 255–259.
105. Ferstl, R., Eggert, F., Pause, B. *et al.* (1991). Immune-system signalling to the other's brain: MHC specific body scents in humans. In *Neuronal Control of Bodily Function* Vol. 6, *Peripheral Signalling of the Brain*, Frederickson, R.C.A. (ed.). Hogrefe and Huber, Toronto, pp. 497–501.
106. Marchalonis, J.J. (ed.) (1976). *Comparative Immunology.* Blackwell, Oxford.
107. Franco, A.M.R., Alencar, A. and Deane, M.P. (1987). DTH reaction in opossum *(Didelphis)* inoculated with *Leishmania. Mem. Inst. Oswaldo Cruz* **82** (Suppl. 1), 152.
108. Brozek, C.M. and Ley, R.D. (1991). Production of interleukin-1 in a South American opossum *(Monodelphis domestica). Dev. Comp. Immunol.* **15**, 401–412.
109. Stanley, N.F. (1983). Immunological studies on the quokka. *J. Roy. Soc. W. Australia* **66**, 28–31.
110. Atwell, J.L. and Marchalonis, J.J. (1977). Immunolgobulin gamma chains of a monotreme mammal, the echidna *(Tachyglossus aculeatus)*: amino acid composition and partial amino acid sequence. *J. Immunogenetics* **4**, 73–80.
111. Croix, D.A., Samples, N.K., Vandeberg, J.L. and Stone, W.H. (1989). Immune response of a marsupial *(Monodelphis domestica)* to sheep red blood cells. *Dev. Comp. Immunol.* **13**, 73–78.
112. Hatton, B.A., Allen, R. and Sulkin, S.E. (1968). Immune response in Chiroptera to bacteriophage ϕX174. *J. Immunol.* **101**, 141–150.
113. Chakraborty, A.K. and Chakravarty, A.K. (1984). Antibody-mediated immune response in the bat, *Pteropus giganteus. Dev. Comp. Immunol.* **8**, 415–423.
114. Chakraborty, A.K. and Chakravarty, A.K. (1983). Dichotomy of lymphocyte population and cell-mediated immune responses in a fruit bat, *Pteropus giganteus. J. Indian Inst. Sci.* **64**, 157–168.
115. Diener, E., Ealey, E.H.M. and Legge, J.S. (1967). Phylogenetic studies on the immune response. 3. Autoradiographic studies on the lymphoid system of the Australian echidna *Tachyglossus aculeatus. Immunology* **13**, 339–347.
116. Kampmeier, O.F. (1969). *Evolution and Comparative Morphology of the Lymphatic System.* C.C. Thomas, Springfield, Illinois.
117. Thorp, B.H., Seneque, S., Staute, K. and Kimpton, W.G. (1991). Characterization and distribution of lymphocyte subsets in sheep hemal nodes. *Dev. Comp. Immunol.* **15**, 393–400.
118. Miller, J.J. (1969). Studies on the phylogeny and ontogeny of the specialized lymphatic tissue venules. *Lab. Invest.* **21**, 484–490.
119. Binns, R.M. (1980). Pig lymphocytes—behaviour, distribution and classification. *Monogr. Allergy* **16**, 19–37.
120. Husband, A.J. (ed.) (1988). *Migration and Homing of Lymphoid Cells, Vol. 2.* CRC/Wolfe, London.
121. Yednock, T.A. and Rosen, S.D. (1989). Lymphocyte homing. *Adv. Immunol.* **44**, 313–378.
122. Bosworth, B.T., St. John, T., Gallatin, W.M. and Harp, J.A. (1991). Sequence of the bovine CD44 cDNA: comparison with human and mouse sequences. *Molec. Immunol.* **28**, 1131–1136.
123. Bosworth, B.T. and Harp, J.A. (1992). Evidence for conservation of peripheral lymphocyte homing receptors between the bovine, murine and human species. *Vet. Immunol. Immunopathol.* **33**, 79–88.
124. Edelson, R.L. and Fink, J.M. (1985). The

immunologic function of skin. *Sci. Amer.* **252**, 34–41.
125. Mackay, C.R. (1991). Skin-seeking memory T cells. *Nature* **349**, 737–738.
126. Poskitt, D.C., Duffey, K., Barnett, J. *et al.* (1984). The gut-associated lymphoid system of two species of Australian marsupial mice, *Antechinus swainsonii* and *A. stuartii*. *Aust. J. Exp. Biol. Med. Sci.* **62**, 81–88.
127. Coutinho, V.B., Coutinho, H.B., Bobalinho, T.I. *et al.* (1990). Histological and ultrastructural studies of the marsupial *Didelphis albiventris* Peyer's patches. *Mem. Inst. Oswaldo Cruz* **85**, 435–443.
128. Kuper, C.F., Koornstra, P.J., Hameleers, D.M.H. *et al.* (1992). The role of nasopharyngeal lymphoid tissue. *Immunol. Today* **13**, 219–224.
129. Pabst, R. (1992). Is BALT a major component of the human lung immune system? *Immunol. Today* **13**, 119–121.
130. Claasen, E. (1991). Histological organization of the spleen: implications for immune functions in different species. *Res. Immunol.* **142**, 315–372.
131. McCuskey, R.S., Weiss, L. *et al.* (1985). New trends in spleen research. *Experientia* **41**, 143–248.
132. Kay, M.M.B. and Baker, L.S. (1979). Cell changes associated with declining immune function. In *Physiology and Cell Biology of Aging (Aging, Vol. 8)*, Cherkin, A. *et al.* (eds). Raven, New York.
133. Albarede, J.L. and Vellas, P. (eds). (1991). *Facts and Research in Gerontology. L'Année Gerontologique*, Serdi, Paris.
134. De Weck, A.L. (ed.) (1984). *Lymphoid Cell Functions in Ageing*. Proc. EURAGE Symp., Rijswijk.
135. Johnson, B.C., Good, R.A. *et al.* (1990). Cellular and molecular bases for the influence of nutrition on ageing and longevity. *Proc. Soc. Exp. Biol. Med.* **193**, 1–38.
136. Makinodan, T. and Kay, M.M.B. (1980). Age influence on the immune system. *Adv. Immunol.* **29**, 287–330.
137. Thoman, M.L. and Weigle, W.O. (1989). The cellular and subcellular bases of immunosenescence. *Adv. Immunol.* **46**, 221–262.
138. Miller, R.A. (1989). The cell biology of aging: immunological models. *J. Gerontol.* **44**, B4–8.
139. Dubiski, S., Ponnappan, U. and Cinader, B. (1989). Strain polymorphism in progression of aging: changes in CD4, CD8 bearing subpopulations. *Immunol. Letters* **23**, 1–8.
140. Koyama, K., Hosokawa, T. and Aoike, A. (1990). Aging effect on the immune functions of murine gut-associated lymphoid tissues. *Dev. Comp. Immunol.* **14**, 465–473.
141. Rothstein, M. (ed.) (1990). *Review of Biological Research in Aging 4*. Wiley, New York.
142. Weksler, M.E. and Siskind, G.W. (1984). The cellular basis of immune senescence. *Monogr. Dev. Biol.* **17**, 110–121.
143. Kendall, M.D. (ed.) (1981). *The Thymus Gland*. Academic Press, London.
144. Poskitt, D.C., Barnett, J., Duffey, K. *et al.* (1984). Involution of the thymus in marsupial mice. *Dev. Comp. Immunol.* **8**, 483–488.
145. Ashman, R.B., Waring, H., and Stanley, N.F. (1973). Adult mortality after neonatal thymectomy in a marsupial *Setonix brachyurus* (quokka). *Proc. Soc. Exp. Biol. Med.* **144**, 819–821.
146. Effros, R.B., Svoboda, K. and Walford, R.L. (1991). Influence of age and caloric restriction on macrophage IL-6 and TNF production. *Lymphokine and Cytokine Res.* **10**, 347–351.
147. Solomon, G.F., Fiatarone, M.A., Benton, D. *et al.* (1988). Psychoimmunologic and endorphin function in the aged. *Ann. N.Y. Acad. Sci.* **521**, 43–58.
148. Riley, M.L., Turner, R.J., Evans, P.M. and Merry, B.J. (1989). Failure of dietary restriction to influence natural killer activity in old rats. *Mechs. Ageing Dev.* **50**, 81–93.
149. Bellamy, D. (1981). Ageing: with particular reference to the use of the house mouse as a mammalian model. In *Biology of the House Mouse*, Berry, R.J. (ed.), *Zoo. Soc. Lond. Symp.* **47**, 267–300. Academic Press, London.
150. Rose, M.R. (1991). *Evolutionary Biology of Aging*. University Press, Oxford.
151. Weindruch, R. and Walford, R.L. (1988). *The Retardation of Aging and Disease by Dietary Restriction*. C.C. Thomas, Springfield, Illinois.
152. Holehan, A.M. and Merry, B.J. (1986). The experimental manipulation of ageing by diet. *Biol. Revs.* **61**, 329–368.
153. Habeshaw, J.A. (1980). Tumour immunology: is man the odd mouse out? *Immunol. Today* **1**(2), iv–v.
154. Albright, J.F. and Oppenheim, J.J. (1991). Contributions of basic immunology to human health. *FASEB J.* **5**, 265–270.
155. Greiner, D.L., Rajan, T.V. and Shultz, L.D. (1992). Animal models for immunodeficiency diseases. *Immunol. Today* **13**, 116–117.
156. Callard, R.E. and Turner, M.W. (1990). Cytokines and Ig switching: evolutionary divergence between mice and humans. *Immunol. Today* **11**, 200–203.
157. Toder, V. *et al.* (1990). Commentaries on immunology of spontaneous abortion. *Res. Immunol.* **141**, 259–261.

158. Mosier, D.E. (1991). Adoptive transfer of human lymphoid cells to severely immunodeficient mice: models for normal human immune function, autoimmunity, lymphomagenesis and AIDS. *Adv. Immunol.* **50**, 303–325.
159. Bosma, M.J. and Carroll, A.M. (1991). The SCID mouse mutant: definition, characterization and potential uses. *Ann. Rev. Immunol.* **9**, 323–350.
160. Merlino, G.T. (1991). Transgenic animals in biomedical research. *FASEB J.* **5**, 2996–3001.
161. Arnold, B., Goodnow, C., Hengartner, H. and Hämmerling, G. (1990). The coming of transgenic mice: tolerance and immune reactivity. *Immunol. Today* **11**, 69–72.
162. Westermann, J. and Pabst, R. (1990). Lymphocyte subsets in the blood: a diagnostic window on the immune system? *Immunol. Today* **11**, 406–410.
163. Letwin, N.L. (1990). Animal models for AIDS. *Immunol. Today* **11**, 322–326.
164. Allen, W.R. (1975). The influence of fetal genotype upon endometrial cup development and PMSG and progestagen production in equids. *J. Reprod. Fert.* Suppl. 23, 405–413.
165. Allen, W.R. (1982). Immunological aspects of the endometrial cup reaction and the effects of xenogeneic pregnancy in horses and donkeys. *J. Reprod. Fert.* Suppl. 31, 57–94.
166. Kydd, J., Boyle, M.S., Allen, W.R. *et al.* (1985). Transfer of exotic equine embryos to domestic horses and donkey. *Equine Vet. J.* Suppl. 3, 80–84.
167. Chaouat, G. (ed.) (1993). *Immunology of Pregnancy*. CRC Press, London.
168. Klein, J., Satta, Y. and O'hUigin, C. (1993). The molecular descent of the major histocompatibility complex. *Ann. Rev. Immunol.* **11**, 269–295.
169. Lunney, J.K. (1993). Characterization of swine leucocyte differentiation antigens. *Immunol. Today* **14**, 147–148.
170. Picker, L.J. and Butcher, E.C. (1992). Physiological and molecular mechanisms of lymphocyte homing. *Ann. Rev. Immunol.* **10**, 561–591.
171. Bos, J.D. and Kapsenberg, M.L. (1993). The skin immune system: progress in cutaneous biology. *Immunol. Today* **14**, 75–78.
172. Aldrovandi, G.M., Feuer, G., Gao, L. *et al.* (1993). The SCID-hu mouse as a model for HIV-1 infection. *Nature* **363**, 732–736.
173. Kindt, T.J., Hirsch, V.M., Johnson, P.R. and Sawasdikosol, S. (1992). Animal models for acquired immunodeficiency syndrome. *Adv. Immunol.* **52**, 425–474.
174. Skibinski, G., Kelly, R.W., Harkiss, D. and James, K. (1992). Immunosuppression by human seminal plasma–extracellular organelles (prostasomes) modulate activity of phagocytic cells. *Amer. J. Reprod. Immunol.* **28,** 97–103.
175. Stewart, I.J. and Clarke, J.R. (1993). Uterine morphology and the distribution of granulated metrial cells in the virgin and pregnant short-tailed field vole, *Microtus agrestis*. *J. Anat.* **182,** 75–85.
176. Wegmann, T.G., Lin, H., Guilbert, L. and Mosmann, T.R. (1993). Bidirectional cytokine interactions in the maternal–fetal relationship—is successful pregnancy a TH2 phenomenon? *Immunol. Today* **14,** 353–356.
177. Hampson, J., McLaughlin, P.J. and Johnson, P.M. (1993). Low affinity receptors for TNF-α, IFN γ and GM-CSF are expressed on human syncytiotrophoblast. *Immunology* **79,** 485–490.
178. Walcheck, B., Watts, G. and Jutila, M.A. (1993). Bovine $\gamma\delta$ T cells bind E-selectin via a novel glycoprotein receptor—first characterization of a lymphocyte E-selectin interaction in an animal model. *J. Exp. Med.* **178,** 853–863.
179. Rosenberg, I.H. (ed.) (1992). Symposium on nutrition and aging. *Nutr. Revs.* **50,** 347–489.
180. Zeitz, H.J. (ed.) (1993). Immunological problems in the aged. *Immunol. Allergy Clinics N. America* **13,** 517–712.
181. Hu, A., Ehleiter, D., Ben-Yehuda, A. *et al.* (1993). Effect of age on the expressed B-cell repertoire: role of B-cell subsets. *Int. Immunol.* **5,** 1035–1039.
182. Allen, W.R., Kydd, J.H., Boyle, M.S. and Antczak, D.F. (1987). Extraspecific donkey-in-horse pregnancy as a model of early fetal death. *J. Reprod. Fert.* **35** (suppl.), 197–209.
183. Jurd, R.D. (1994). "Not proper mammals": immunity in monotremes and marsupials. *Comp. Immunol. Microbiol. Infect. Dis.* **17,** in press.

INDEX

Note: page references in *italics* refer to Figures; those in **bold** refer to Tables

Index compiled by Annette Musker